THE
GREAT WAR
AND
MODERN MEMORY

By the same author

Theory of Prosody in Eighteenth-Century England
Poetic Meter and Poetic Form
The Rhetorical World of Augustan Humanism:
Ethics and Imagery from Swift to Burke
Samuel Johnson and the Life of Writing

Editor
English Augustan Poetry
The Ordeal of Alfred M. Hale

Co-Editor
Eighteenth-Century English Literature

THE
GREAT WAR
AND
MODERN
MEMORY

Paul Fussell

New York and London
OXFORD UNIVERSITY PRESS

First published, 1975
This reprint, 1989, of the original 1975 edition

ISBN 0-19-501918-0

Grateful acknowledgment is made to the following for permission to quote material in copyright:
Alison & Busby, Ltd., for excerpts from Vernon Scannell, *Selected Poems*, copyright © 1971 by Vernon Scannell. Mrs. George Bambridge, A. P. Watt & Son, and Doubleday & Co., Inc., for "The Beginner" from *Rudyard Kipling's Verse: Definitive Edition*, copyright 1891, 1927, by Rudyard Kipling. Carcanet Press, Ltd., and Defour Editions (USA) for "Trench Poets" from Edgell Rickword, *Collected Poems and Translations*, copyright © Edgell Rickword 1975. City Lights Books for excerpt from Allen Ginsberg, *Howl and Other Poems*, copyright 1956, 1959, by Allen Ginsberg. Collins-Knowlton-Wing, Inc., for excerpts from Robert Graves, *Collected Poems, 1955*, copyright © 1955 by Robert Graves; *Fairies and Fusiliers*, copyright 1918 by Alfred A. Knopf, Inc.; and *Good-bye to All That*, copyright © 1929, 1957, by Robert Graves. EMI Ltd. and The Parlophone Co. for transcription from

Printed in the United States of America

To the Memory of
Technical Sergeant Edward Keith Hudson, ASN 36548772
Co. F, 410th Infantry
Killed beside me in France
March 15, 1945

Preface

This book is about the British experience on the Western Front from 1914 to 1918 and some of the literary means by which it has been remembered, conventionalized, and mythologized. It is also about the literary dimensions of the trench experience itself. Indeed, if the book had a subtitle, it would be something like "An Inquiry into the Curious Literariness of Real Life." I have focused on places and situations where literary tradition and real life notably transect, and in doing so I have tried to understand something of the simultaneous and reciprocal process by which life feeds materials to literature while literature returns the favor by conferring forms upon life. And I have been concerned with something more: the way the dynamics and iconography of the Great War have proved crucial political, rhetorical, and artistic determinants on subsequent life. At the same time the war was relying on inherited myth, it was generating new myth, and that myth is part of the fiber of our own lives.

In suggesting the forms of that myth, I have tried to supply contexts, both actual and literary, for writers who have most effectively memorialized the Great War as a historical experience with conspicuous imaginative and artistic meaning. These writers I take to be the classic memoirists Siegfried Sassoon, Robert Graves, and Edmund Blunden. I have dealt also with poets of very high literary consciousness like David Jones, Isaac Rosenberg, and of course Wilfred Owen. And to see what the ordinary man has to say about it all, I have compared the scores of amateur memoirs lodged in the collections of the Imperial War Museum.

Correctly or not, the current idea of "the Great War" derives primarily from images of the trenches in France and Belgium. I have thus stayed there with the British infantry, largely disregarding events in Mesopotamia, Turkey, Africa, and Ireland, and largely ignoring air and naval warfare. By thus narrowing my view, I have hoped to sharpen it to probe into the origins of what some future "medievalist" may call The Matter of Flanders and Picardy.

My work on this book has been eased by the kindness of three institutions. The Research Council of Rutgers University again has my thanks, especially its amiable and responsive Associate Director, C. F. Main, who has also read the proofs. I am indebted to the National Endowment for the Humanities for a Senior Fellowship during the academic year 1973–74. And I owe a great deal to the Imperial War Museum and its courteous staff, especially D. G. Lance, Keeper of Libraries and Archives, and Roderick Suddaby, Chief of the Documents Section. Although occasionally I have dissented from some of their findings, I could never have written without Bernard Bergonzi's *Heroes' Twilight*, John H. Johnston's *English Poetry of the First World War*, and Arthur E. Lane's *An Adequate Response*. An inquirer into this subject is lucky to have such predecessors. He is lucky too to have as a general guide the *History of the First World War* by the admirable B. H. Liddell Hart, prince of modern military critics. Throughout I have drawn on the criticism of Northrop Frye. I am grateful for his work. For answering queries and for donating ideas and criticism, I want to thank Paul Bertram, Christopher Bond, Catharine Carver, Maurice Charney, Joseph Frank, Robert Hollander, Alfred L. Kellogg, John McCormick, Peter McCormick, Lawrence Millman, Julian Moynahan, George A. Panichas, Richard Quaintance, James Raimes, George Robinson, and Mary E. Slayton. I remember with delight Leo Cooper and Tom Hartman, generous companions during a tour of the Somme. As usual, I am indebted to Richard Poirier, who contributed some materials stemming from his own authoritative work on Norman Mailer. I am grateful to the following authors and heirs for permission to quote unpublished material as well as for the documents and memorabilia they have kindly sent me: Mrs. David Angus, G. Bricknall, William Collins, Howard H. Cooper, M. J. H. Drummond, Marjorie H. Gilbertson, the Rev. S. Horsley, Colonel R. Macleod, R. R. B. Mecredy, D. M. Mitchell, W. R. Price, Peter Reeve, M. E. Ryder, and Clive Watts.

As always, it is to Betty that I owe the most. She has sustained me generously with interest, encouragement, criticism, and much more.

P. F.

Rutgers University
January 1975

Contents

Illustrations

THE
GREAT WAR
AND
MODERN MEMORY

I
A Satire of Circumstance

THOMAS HARDY, CLAIRVOYANT

By mid-December, 1914, British troops had been fighting on the Continent for over five months. Casualties had been shocking, positions had settled into self-destructive stalemate, and sensitive people now perceived that the war, far from promising to be "over by Christmas," was going to extend itself to hitherto unimagined reaches of suffering and irony. On December 19, 1914, Lytton Strachey published a piece in the *New Statesman* focusing on "the tragedies of whole lives and the long fatalities of human relationships." His language was dark. He spoke of events *remorseless, terrible, gruesome*. He noted that "the desolation is complete" and recalled a phrase of Gibbon's appropriate to the kind of irony he was contemplating: "the abridgment of hope." "If there is joy . . . . , it is joy that is long since dead; and if there are smiles, they are sardonical." [1]

But actually Strachey was not writing about the war at all. In his 2000 words he doesn't mention it. Instead, he is reviewing Thomas Hardy's most recent volume of poems, *Satires of Circumstance*, published in November, 1914, but containing—with the exception of the patriotic and unironic "Men Who March Away," hastily added as a "Postscript"—only poems written before the war. Many emanate from Hardy's personal experience as far back as 1870.

As if by uncanny foresight, Hardy's volume offers a medium for perceiving the events of the war just beginning. It does so by establishing a terrible irony as the appropriate interpretative means. Although in these poems the killer is tuberculosis rather than the machine gun, their ambience of mortal irony is one with which, in the next four years, the British will become wholly familiar. The materials of the poems are largely funerary: they are full of graves, headstones, "clay cadavers," coffins, skeletons, and rot. The favorite rhetorical situation is the speaking of the dead. Voices from the grave—like that of the speaker in John McCrae's "In Flanders Fields"—sadly or sardonically recall, admonish, and regret. They utter brief shapely ironic memoirs. And irony of situation is the substance of even those poems, like "The Convergence of the Twain" (first published in May, 1912), which involve no voice from the grave. A typical poem in the collection is "Channel Firing," written five months before the war. Here the occupants of a seaside cemetery confuse offshore naval gunnery practice with the thunders of the Day of Judgment, only to be reassured by God that they are mistaken: "The world is as it used to be"—namely, brutal and stupid. The actual Judgment Day, He tells them, will be considerably warmer and more punitive.

A characteristic irony of situation—a wry enactment of Gibbon's "abridgment of hope"—is the one in "Ah, Are You Digging on My Grave?" first published in December, 1913. Aware of a scratching sound above, the voice from the grave asks repeatedly who it is who digs at her grave. Is it her lover? No, a voice answers; he was married yesterday and is busy. Is it one of her kinfolk planting memorial flowers? No, they know that planting flowers does no good. Is it then perhaps her "enemy" (a word which public events will soon weight uniquely) "prodding sly" in an easy revenge? No, her enemy, she is told, thinks her no longer worth hating "And cares not where you lie." Finally "giving up," the speaker learns the identity of the digger from the digger himself: he is her "little dog." This news moves her to utter a stanza rich with "prewar" complacency:

> "Ah, yes! *You* dig upon my grave . . .
>   Why flashed it not on me
> That one true heart was left behind!
> What feeling do we ever find
> To equal among human kind
>   A dog's fidelity?"

But the dog deprives his mistress of even this comforting connection with the world she's left behind:

> "Mistress, I dug upon your grave
>   To bury a bone, in case

> I should be hungry near this spot
> When passing on my daily trot.
> I am sorry, but I quite forgot
> It was your resting-place."

If that points back to the eighteenth century, a poem like "Your Last Drive" (written in December, 1912) reaches toward Robert Frost. Here the irony of situation arises from a collision between innocence and awareness. The speaker recalls how his friend recently admired the night view of the lighted village from a place on the approaching road:

> . . . you told of the charm of that haloed view.

The road from which the friend admired the bright town of the living happened to run past the local cemetery,

> Where eight days later you were to lie.

Even the narrator, skilled in irony, could not have foretold how rudely soon the admirer of lights would remove to the dark town of the dead, to be now

> . . . past love, praise, indifference, blame.

The contrast between before and after here will remind us of the relation between, say, the golden summer of 1914 and the appalling December of that year, although an even more compelling paradigm of that contrast is a poem Hardy wrote in 1913, "After a Journey." The speaker seeks the idea of his dead beloved by revisiting the dramatic seacoast sites of their affair. He imagines her spirit saying,

> Summer gave us sweets, but autumn wrought division.

We know that regardless of literal fact or the special needs of the times, "summer" in poems belongs conventionally to gladness and "autumn" to melancholy. But what will happen a year later will compel this traditional figure toward the joltingly literal. Hardy's very private experience will be appropriated then as a very public one.

A more obvious rendering of the irony of benign ignorance is "In the Cemetery," one of the fifteen brief "Satires of Circumstance" which give the volume its title. Now the speaker is the cemetery caretaker, who explains to a bystander how preposterous is the quarreling of a group of mothers over whose child lies in which grave. Actually, says the caretaker, when a main drain had to be laid across the cemetery,

> ". . . we moved the lot some nights ago,
> And packed them away in the general foss
> With hundreds more. But their folks don't know."

Single grave, mass grave, main drain—it's all one:

> ". . . as well cry over a new-laid drain
> As anything else, to ease your pain!"

The idea of mass graves seems to pertain especially to the twentieth century. There are 2500 British war cemeteries in France and Belgium. The sophisticated observer of the rows of headstones will do well to suspect that very often the bodies below are buried in mass graves, with the headstones disposed in rows to convey the illusion that each soldier has his individual place.[2] As Hardy prophesies,

> ". . . all their children were laid therein
> At different times, like sprats in a tin."

The one ultimate Satire of Circumstance, as Hardy knows, is mortality itself, "the wonder and the wormwood of the whole." Death's material attendants and conditions, however noisome, fascinate universally, even as Edmund Blunden recognizes, recalling with gentle irony and sympathy a small satire of circumstance fifteen years after Hardy. He is remembering a wrecked French civilian churchyard close to the front line and considering its odd appeal to the "morbid curiosity" of the troops, who crowded to gape at the open vaults and graves:

> Greenish water stood in some of these pits; bones and skulls and decayed cerements there attracted frequent soldiers past the "No Loitering" noticeboard.

Under the circumstances, an odd attraction indeed.

> Why should these mortalities lure those who ought to be trying to forget mortality, ever threatening them? Nearly corpses ourselves, by the mere fact of standing near Richebourg Church, how should we find the strange and the remote in these corpses? [3]

I am not really arguing that Hardy, the master of situational irony, "wrote" the Great War, although if wars were written the author of *Time's Laughing-Stocks* and *The Dynasts* could certainly have written this one. From his imagination was available more or less ready-made—and, certainly, well *a priori*—a vision, an action, and a tone superbly suitable for rendering an event constituting an immense and unprecedented Satire of Circumstance. A traditional "tragic satire" (like Johnson's "The Vanity of Human Wishes") is an accumulation of numerous small constituent satires. Likewise the great tragic satire which was the war will be seen to consist of its own smaller constituent satires, or ironic actions. Thus the literary Blunden regards a battlefield, thoroughly torn up and littered with German equipment, as "this satire in iron brown and field grey." [4]

Glancing back thirty-one years later, Siegfried Sassoon recalled that during the war Hardy had been his "main admiration among living writers," and he acknowledged the debt of his satirical poems about the war to the prewar ironies of *Satires of Circumstance*. [5] Fit to take its place as a sixteenth Satire of Circumstance entirely consonant with Hardy's is a poem by Sassoon enacting this plot: "Brother officer giving white-haired mother fictitious account of her cold-footed son's death at the front." [6] The poem is "The Hero":

> "Jack fell as he'd have wished," the Mother said,
> And folded up the letter that she'd read.
> "The Colonel writes so nicely." Something broke
> In the tired voice that quavered to a choke.
> She half looked up. "We mothers are so proud
> Of our dead soldiers." Then her face was bowed.
>
> Quietly the Brother Officer went out.
> He'd told the poor old dear some gallant lies
> That she would nourish all her days, no doubt.
> For while he coughed and mumbled, her weak eyes
> Had shone with gentle triumph, brimmed with joy,
> Because he'd been so brave, her glorious boy.
>
> He thought how "Jack," cold-footed, useless swine,
> Had panicked down the trench that night the mine
> Went up at Wicked Corner; how he'd tried
> To get sent home, and how, at last, he died,
> Blown to small bits. And no one seemed to care
> Except that lonely woman with white hair.

Two nights before participating in the attack on the Somme—perhaps the most egregious ironic action of the whole war—Sassoon found himself "huddled up in a little dog-kennel of a dug-out, reading *Tess of the D'Urbervilles*." [7] Clearly, there are some intersections of literature with life that we have taken too little notice of.

THE WAR AS IRONIC ACTION

Every war is ironic because every war is worse than expected. Every war constitutes an irony of situation because its means are so melodramatically disproportionate to its presumed ends. In the Great War eight

million people were destroyed because two persons, the Archduke
Francis Ferdinand and his Consort, had been shot. The Second World
War offers even more preposterous ironies. Ostensibly begun to guarantee the sovereignty of Poland, that war managed to bring about Poland's
bondage and humiliation. Air bombardment, which was supposed to
shorten the war, prolonged it by inviting those who were its targets to
cast themselves in the role of victim-heroes and thus stiffen their resolve.

But the Great War was more ironic than any before or since. It was a
hideous embarrassment to the prevailing Meliorist myth which had dominated the public consciousness for a century. It reversed the Idea of
Progress. The day after the British entered the war Henry James wrote a
friend:

> The plunge of civilization into this abyss of blood and darkness . . . is a
> thing that so gives away the whole long age during which we have supposed
> the world to be, with whatever abatement, gradually bettering, that to have
> to take it all now for what the treacherous years were all the while really
> making for and *meaning* is too tragic for any words.[8]

James's essential point was rendered in rowdier terms by a much smaller
writer, Philip Gibbs, as he remembered the popularity during the war of
what today would be called Black Humor. "The more revolting it was,"
he says, "the more . . . [people] shouted with laughter":

> It was . . . the laughter of mortals at the trick which had been played on
> them by an ironical fate. They had been taught to believe that the whole ob-
> ject of life was to reach out to beauty and love, and that mankind, in its
> progress to perfection, had killed the beast instinct, cruelty, blood-lust, the
> primitive, savage law of survival by tooth and claw and club and ax. All po-
> etry, all art, all religion had preached this gospel and this promise.
>   Now that ideal was broken like a china vase dashed to the ground. The
> contrast between That and This was devastating. . . . The war-time humor
> of the soul roared with mirth at the sight of all that dignity and elegance
> despoiled.[9]

The British fought the war for four years and three months. Its potential of ironic meaning, considered not now in relation to the complacen-
cies of the past but in itself alone, emerges when we consider its events
chronologically. The five last months of 1914, starting August 4, when
the British declared war on the Central Powers, began with free maneuver in Belgium and Northern France and ended with both sides locked
into the infamous trench system. Before this stalemate, the British
engaged in one major retreat and fought two large battles, although
battles is perhaps not the best word, having been visited upon these

events by subsequent historiography in the interest of neatness and the assumption of something like a rational causality. To call these things *battles* is to imply an understandable continuity with earlier British history and to imply that the war makes sense in a traditional way. As Esmé Wingfield-Stratford points out, "A vast literature has been produced in the attempt to bring [the Great War] into line with other wars by highlighting its so-called battles by such impressive names as Loos, Verdun, the Somme, and Passchendaele. . . ." [10] This is to try to suggest that these events parallel Blenheim and Waterloo not only in glory but in structure and meaning.

The major retreat was the Retreat from Mons on August 24, necessitated when Sir John French's four divisions—the whole of the British force engaged—found themselves outflanked. In early September this retreat merged into the first of the "battles," known as the Marne, where the British and the French gradually stopped the German advance on Paris, although at a cost of half a million casualties on each side. Prevented from going through to Paris, the Germans sought an opening further north, and each side now began trying to turn its enemy's western flank with the object of winning the war rapidly and economically; it was still thought by some that this was a compassable object. The ensuing maneuvers during late October and early November are variously misnamed "The First Battle of Ypres" and "The Race to the Sea"—that is, to the Belgian seaports. The journalistic formula "The Race to the ———" was ready to hand, familiar through its use in 1909 to describe Peary's "Race to the (North) Pole" against Cook. Rehabilitated and applied to these new events, the phrase had the advantage of a familiar sportsmanlike, Explorer Club overtone, suggesting that what was happening was not too far distant from playing games, running races, and competing in a thoroughly decent way.

By the middle of November these exertions had all but wiped out the original British army. At the beginning of the war, a volunteer had to stand five feet eight to get into the army. By October 11 the need for men was such that the standard was lowered to five feet five. And on November 5, after the thirty thousand casualties of October, one had to be only five feet three to get in.[11] The permanent trenchline had been dug running from Nieuport, on the Belgian coast, all the way to the Swiss border, with the notorious Ypres Salient built in. The perceptive could already see what the war was going to be like. As early as October, 1914, Captain G. B. Pollard wrote home, using gingerly a novel word whose implications would turn more and more ghastly as time went on: "It's absolutely certainly a war of 'attrition,' as somebody said here the other day, and we have got to stick it longer than the other side and go on producing men, money, and material until they cry quits, and that's all about it, as far as I can see." [12] Lord Kitchener was one who agreed with Captain Pollard. Near the end of October he issued his call for 300,000 volunteers. Most of them would be expended on the Somme in 1916. The first Christmas of the war saw an absolute deadlock in the trenches. Both British and German soldiers observed an informal, *ad hoc* Christmas Day truce, meeting in No Man's Land to exchange cigarets and to take snapshots. Outraged, the Staff forbad this ever to happen again.

The new year, 1915, brought the repeated failure of British attempts to break through the German line and to unleash the cavalry in pursuit. They failed, first, because of insufficient artillery preparation—for years no one had any idea how much artillery fire would be needed to destroy the German barbed wire and to reach the solid German deep dugouts; second, because of insufficient reserves for exploiting a suddenly apparent weakness; and third, because the British attacked on excessively narrow frontages, enabling every part of the ground gained to be brought under retaliatory artillery fire.

However, the first failed attack of 1915 was not British but German. The area selected was near Ypres, and the fracas has been named the Second Battle of Ypres, or simply Second Ypres. On April 22, after discharging chlorine gas from cylinders, the Germans attacked and advanced three miles. But then they faltered for lack of reserves. Gas had first been used by the Germans on October 27, 1914, when they fired a prototype of modern tear gas from artillery near Ypres. The German use of gas—soon to be imitated by the British—was thought an atrocity by the ignorant, who did not know that, as Liddell Hart points out, gas is "the least inhumane of modern weapons." Its bad press was the result of its novelty: "It was novel and therefore labelled an atrocity by a world which condones abuses but detests innovations." [13] In the late-April attack at Ypres the British were virtually unprotected against gas—the "box respirator" was to come later—and even though the line was substantially held, the cost was 60,000 British casualties.

A few weeks later it was the British turn. On March 10 the first of the aborted British offensives was mounted at Neuve Chapelle. The attack was only 2000 yards wide, and, although it was successful at first, it died for lack of reserves and because the narrow frontage invited too much retributive German artillery. Again the British tried, on May 15 at Festubert, and with similar results: initial success turned to disaster. Going through the line was beginning to look impossible. It was thus essential to entertain hopes of going around it, even if going around took one as far away as Gallipoli, 2200 miles southeast of the Western Front, where the troops had begun landing on April 25.

Imagining themselves instructed by these occasions of abridged hope at Neuve Chapelle and Festubert, the British mounted a larger attack near Loos on September 15. Six divisions went forward at once, and this time the attack was preceded not only by the customary artillery barrage but by the discharge of what Robert Graves tells us was euphemized as "the accessory"—cylinders of chlorine gas.[14] Most of it blew back into the British trenches, and the attack was another failure which even the *Official History* later stigmatized as a "useless slaughter of infantry." [15] The proceedings at Loos were called off eleven days after they had started, but not before 60,000 more British casualties had been added to the total.

Now volunteers were no longer sufficient to fill the ranks. In October Lord Derby's "scheme"—a genteel form of conscription—was promulgated, and at the beginning of 1916, with the passing of the Military Service Act, England began to train her first conscript army, an event which could be said to mark the beginning of the modern world. Clearly the conscripts were needed, for things were going badly everywhere. The assault at Gallipoli was proving as unsuccessful as the assaults elsewhere, and at the end of 1915 the forces there were withdrawn with nothing gained.

The need for a stiffening of home-front morale at the beginning of 1916 can be gauged by the Poet Laureate's issuing in January an anthology of uplifting literary passages of a neo-Platonic tendency titled *The Spirit of Man*. Such was the military situation, Robert Bridges implied in his Introduction, that "we can turn to seek comfort only in the quiet confidence of our souls." We will thus "look instinctively to the seers and poets of mankind, whose sayings are the oracles and prophecies of loveliness and lovingkindness." The news from Belgium and France, not to mention Turkey, was making it more and more necessary to insist, as Bridges does, that "man is a spiritual being, and the proper work of his mind is to interpret the world according to his higher nature. . . ." Such an outlook is now indispensable, for we are confronted with "a grief that is intolerable constantly to face, nay impossible to face without that trust in God which makes all things possible." [16]

The comforts purveyed by *The Spirit of Man* were badly needed, for 1915 had been one of the most depressing years in British history. It had been a year not only of ironic mistakes but of a grossly unimaginative underestimation of the enemy and of the profound difficulties of siege

warfare. Poor Sir John French had to be sent home, to be replaced by
Sir Douglas Haig as commander of British forces. One doesn't want to
be too hard on Haig, who doubtless did all he could and who has been
well calumniated already. But it must be said that it now appears that
one thing the war was testing was the usefulness of the earnest Scottish
character in a situation demanding the military equivalent of wit and invention. Haig had none. He was stubborn, self-righteous, inflexible, in-
tolerant—especially of the French—and quite humorless. And he was
provincial: at his French headquarters he insisted on attending a Church
of Scotland service every Sunday. Bullheaded as he was, he was the perfect commander for an enterprise committed to endless abortive assaulting. Indeed, one powerful legacy of Haig's performance is the conviction
among the imaginative and intelligent today of the unredeemable defec-
tiveness of all civil and military leaders. Haig could be said to have established the paradigm. His want of imagination and innocence of artistic culture have seemed to provide a model for Great Men ever since.

To Haig the lesson of 1915 was clear and plain. A successful attack
leading to a breakthrough would have to be infinitely larger and wider
and stronger and better planned than had been imagined. With this kind
of attack in view, Haig and his staff spent the first six months of 1916
preparing an immense penetration of the German line on the Somme
which he was confident would end the war. The number of men destined for the attack, equal to twenty-six World War II infantry divisions,
constituted a seven-to-one superiority over the Germans. While the planning was underway, France was engaged at Verdun. Its defense bled her
so badly that henceforth the main offensive effort on the Western Front
had to be British. There were not enough French left, and those remaining were so broken in spirit that the mutinies of May, 1917, given the
stingy French leave and recreation policy, might have been predicted.
The ironic structure of events was becoming conventional, even Har-
dyesque: if the pattern of things in 1915 had been a number of small optimistic hopes ending in small ironic catastrophes, the pattern in 1916
was that of one vast optimistic hope leading to one vast ironic catastrophe. The Somme affair, destined to be known among the troops as the
Great Fuck-Up, was the largest engagement fought since the beginnings
of civilization.

By the end of June, 1916, Haig's planning was finished and the attack
on the Somme was ready. Sensing that this time the German defensive
wire must be cut and the German front-line positions obliterated, Haig
bombarded the enemy trenches for a full week, firing a million and a half
shells from 1537 guns. At 7:30 on the morning of July 1 the artillery

shifted to more distant targets and the attacking waves of eleven British divisions climbed out of their trenches on a thirteen-mile front and began walking forward. And by 7:31 the mere six German divisions facing them had carried their machine guns upstairs from the deep dugouts where during the bombardment they had harbored safely—and even comfortably—and were hosing the attackers walking toward them in orderly rows or puzzling before the still uncut wire. Out of the 110,000 who attacked, 60,000 were killed or wounded on this one day, the record so far. Over 20,000 lay dead between the lines, and it was days before the wounded in No Man's Land stopped crying out.

The disaster had many causes. Lack of imagination was one: no one imagined that the Germans could have contrived such deep dugouts to hide in while the artillery pulverized the ground overhead, just as no one imagined that the German machine gunners could get up the stairs and mount their guns so fast once the bombardment moved away at precisely 7:30. Another cause was traceable to the class system and the assumptions it sanctioned. The regulars of the British staff entertained an implicit contempt for the rapidly trained new men of "Kitchener's Army," largely recruited among workingmen from the Midlands. The planners assumed that these troops—burdened for the assault with 66 pounds of equipment—were too simple and animal to cross the space between the opposing trenches in any way except in full daylight and aligned in rows or "waves." It was felt that the troops would become confused by more subtle tactics like rushing from cover to cover, or assault-firing, or following close upon a continuous creeping barrage.

A final cause of the disaster was the total lack of surprise. There was a hopeless absence of cleverness about the whole thing, entirely characteristic of its author. The attackers could have feinted: they could have lifted the bombardment for two minutes at dawn—the expected hour for an attack—and then immediately resumed it, which might have caught the seduced German machine gunners unprotected up at their open firing positions. But one suspects that if such a feint was ever considered, it was rejected as unsporting. Whatever the main cause of failure, the attack on the Somme was the end of illusions about breaking the line and sending the cavalry through to end the war. Contemplating the new awareness brought to both sides by the first day of July, 1916, Blunden wrote eighteen years later: "By the end of the day both sides had seen, in a sad scrawl of broken earth and murdered men, the answer to the question. No road. No thoroughfare. Neither race had won, nor could win, the War. The War had won, and would go on winning." [17]

Regardless of this perception, the British attempt on the Somme continued mechanically until stopped in November by freezing mud. A month earlier the British had unveiled an innovation, the tank, on the road between Albert and Bapaume, to the total surprise and demoralization of the enemy. But only thirty-two had been used, and this was not enough for a significant breakthrough. A terrible gloom overcame everyone at the end of 1916. It was the bottom, even worse than the end of 1915. "We are going to lose this war," Lloyd George was heard to say.[18] And the dynamics of hope abridged continued to dominate 1917 with two exceptions, the actions at Messines in June and at Cambrai in November.

On January 1, 1917, Haig was elevated to the rank of Field Marshal, and on March 17, Bapaume—one of the main first-day objectives of the Somme jump-off nine months before—was finally captured. The Germans had proclaimed their intention of practicing unrestricted submarine warfare in the Atlantic on February 1, and by April 6 this had brought a declaration of war from the United States. Henceforth the more subtle Allied strategists knew that winning the war would be only a matter of time, but they also knew that, since the United States was not ready, the time would not be short.

Meanwhile, something had to be done on the line. On April 9 the British again tried the old tactic of head-on assault, this time near Arras in an area embracing the infamous Vimy Ridge, which for years had dominated the southern part of the Ypres Salient. The attack, pressed for five days, gained 7000 yards at a cost of 160,000 killed and wounded. The same old thing. But on June 7 there was something new, something finally exploiting the tactic of surprise. Near Messines, south of Ypres, British miners had been tunneling for a year under the German front lines, and by early June they had dug twenty-one horizontal mineshafts stuffed with a million pounds of high explosive a hundred feet below crucial points in the German defense system. At 3:10 in the morning these mines were set off all at once. Nineteen of them went up, and the shock wave jolted Lloyd George in Downing Street 130 miles away. Two failed to explode. One of these went off in July, 1955, injuring no one but forcibly reminding citizens of the nearby rebuilt town of Ploegsteert of the appalling persistence of the Great War. The other, somewhere deep underground near Ploegsteert Wood, has not gone off yet.

The attack at Messines following these explosions had been brilliantly planned by General Sir Herbert Plumer, who emerges as a sort of intellectual's hero of the British Great War. In sad contrast to Haig, he was unmilitary in appearance, being stout, chinless, white-haired, and pot-

Plumer, the King, Haig. (Imperial War Museum)

bellied. But he had imagination. His mines totally surprised the Germans, ten thousand of whom were permanently entombed immediately. Seven thousand panicked and were taken prisoner. Nine British divisions and seventy-two tanks attacked straightway on a ten-mile front. At the relatively low cost of 16,000 casualties they occupied Vimy Ridge.

If Messines showed what imagination and surprise could do, the attack toward Passchendaele, on the northern side of the Ypres Salient, indicated once more the old folly of reiterated abortive assaulting. Sometimes dignified as the Third Battle of Ypres, this assault, beginning on July 31, was aimed, it was said, at the German submarine bases on the Belgian coast. This time the artillery was relied on to prepare the ground for the attack, and with a vengeance: over ten days four million shells were fired. The result was highly ironic, even in this war where irony was a staple. The bombardment churned up the ground; rain fell and turned the dirt to mud. In the mud the British assaulted until the attack finally attenuated three and a half months later. Price: 370,000 British dead and wounded and sick and frozen to death. Thousands literally drowned in the mud. It was a reprise of the Somme, but worse. Twenty years later Wyndham Lewis looked back on Passchendaele as an all-but-inevitable collision between two "contrasted but as it were complementary types of *idée fixe*": the German fondness for war, on the one hand, and British muddle-headed "doggedness," on the other. These, he says, "found their most perfect expression on the battlefield, or battle-bog, of Passchendaele." Onomatopoeic speculations bring him finally to a point where again we glimpse Hardy as the presiding spirit:

> The very name [Passchendaele], with its suggestion of *splashiness* and of *passion* at once, was subtly appropriate. This nonsense could not have come to its full flower at any other place but at *Passchendaele*. It was pre-ordained. The moment I saw the name on the trench-map, intuitively I knew what was going to happen.[19]

Ever since the first use of tanks in the autumn of 1916 it had been clear that, given sufficient numbers, here was a way of overcoming the gross superiority provided an entrenched enemy by the machine gun. But not until the attack near Cambrai on November 20 were tanks used in sufficient quantity. Now 381 of them coughed and crawled forward on a six-mile front, and this time with impressive success. But as usual, there were insufficient reserves to exploit the breakthrough.

The next major event was a shocking reversal. During the last half of 1917 the Germans had been quietly shifting their eastern forces to the Western Front. Their armistice with the Bolsheviks gave them the opportunity of increasing their western forces by 30 per cent. At 4:30 on the morning of March 21, 1918, they struck in the Somme area, and on a forty-mile front. It was a stunning victory. The British lost 150,000 men almost immediately, 90,000 as prisoners; and total British casualties rose to 300,000 within the next six days. The Germans plunged forty miles into the British rear.

The impact of this crisis on home-front morale can be inferred from London newspaper reaction. The following is typical:

> WHAT CAN I DO?
> How the Civilian May Help in This Crisis.
>
> Be cheerful.
>   . . .
> Write encouragingly to friends at the front.
>   . . .
> Don't repeat foolish gossip.
> Don't listen to idle rumors.
> Don't think you know better than Haig.[20]

Haig, back-pedaling, felt sufficiently threatened to issue on April 12 his famous "Backs to the Wall" Order of the Day. This registered the insecurity of the British position in some very rigid and unencouraging terms: "Every position must be held to the last man. There must be no retirement. With our backs to the wall and believing in the justice of our cause, each one must fight on to the end." In its dogged prohibition of maneuver or indeed of any tactics, this can stand as the model for Hitler's later orders for the ultimate defense of positions like El Alamein and Stalingrad. There are conventions and styles in Orders of the Day just as for any literary documents.

Hardy would have been pleased to know that of this famous order one corporal noted: "We never received it. We to whom it was addressed, the infantry of the front line, were too scattered, too busy trying to survive, to be called into any formation to listen to orders of the day."[21]

During May and June the Germans advanced to great effect near the rivers Lys and Marne. But unwittingly they were engaged in demonstrating the most ironic point of all, namely, that successful attack ruins troops. In this way it is just like defeat. This is a way of reiterating Blunden's point that it is the war that wins. The spectacular German advance finally stopped largely for this reason: the attackers, deprived of the sight of "consumer goods" by years of efficient Allied blockade, slowed down and finally halted to loot, get drunk, sleep it off, and peer about. The champagne cellars of the Marne proved especially tempting. The German Rudolf Binding records what happened when the attack reached Albert:

> Today the advance of our infantry suddenly stopped near Albert. Nobody could understand why. Our airmen had reported no enemy between Albert and Amiens. . . . I jumped into a car with orders to find out what was causing the stoppage in front. . . . As soon as I got near [Albert] I began to see curious sights. Strange figures, which looked very little like soldiers, and certainly showed no sign of advancing, were making their way back out of the town. There were men driving cows before them . . . ; others who carried a hen under one arm and a box of notepaper under the other. Men carrying a bottle of wine under their arm and another one open in their hand. Men who had torn a silk drawing-room curtain from off its rods and were dragging it to the rear. . . . More men with writing-paper and colored note-books. . . . Men dressed up in comic disguise. Men with top-hats on their heads. Men staggering. Men who could hardly walk.[22]

By midsummer it was apparent that the German army had destroyed itself by attacking successfully. On August 8, designated by Ludendorff "The Black Day of the German Army," the Allies counterattacked and broke through. In the German rear they found that maneuver was now possible for the first time since the autumn of 1914. From here to the end their advance was rapid as the German forces fell apart.

The German collapse was assisted by American attacks in September at the St. Mihiel Salient and between the River Meuse and the Argonne Forest. Simultaneously the British were advancing near St. Quentin-Cambrai and the Belgians near Ghent. Despite exhaustion and depletion on all sides—half the British infantry were now younger than nineteen—the end was inevitable. On November 9, 1918, the Kaiser having fled, Germany declared herself a republic and two days later signed the Armistice in the Forest of Compiègne. The war had cost the Central Powers three and a half million men. It had cost the Allies over five million.

NEVER SUCH INNOCENCE AGAIN

Irony is the attendant of hope, and the fuel of hope is innocence. One reason the Great War was more ironic than any other is that its beginning was more innocent. "Never such innocence again," observes Philip Larkin, who has found himself curiously drawn to regard with a wondering tenderness not the merely victimized creatures of the nearby Second World War but the innocents of the remote Great War, those sweet, generous people who pressed forward and all but solicited their own destruction. In "MCMXIV," written in the early sixties, Larkin contemplates a photograph of the patient and sincere lined up in early August outside a recruiting station:

> Those long uneven lines
> Standing as patiently
> As if they were stretched outside
> The Oval or Villa Park,
> The crowns of hats, the sun
> On moustached archaic faces
> Grinning as if it were all
> An August Bank Holiday lark. . . .

The shops are shut, and astonishingly, the Defense of the Realm Act not yet having been thought of,

>   . . . the pubs
>   Wide open all day. . . .

The class system is intact and purring smoothly:

> The differently-dressed servants
> With tiny rooms in huge houses,
> The dust behind limousines. . . .

"Never such innocence," he concludes:

> Never before or since,
> As changed itself to past
> Without a word—the men
> Leaving the gardens tidy,
> The thousands of marriages
> Lasting a little while longer:
> Never such innocence again.

Far now from such innocence, instructed in cynicism and draft-dodging by the virtually continuous war since 1936, how can we forbear condescending to the eager lines at the recruiting stations or smiling at news like this, from the *Times* of August 9, 1914:

> At an inquest on the body of Arthur Sydney Evelyn Annesley, aged 49, formerly a captain in the Rifle Brigade, who committed suicide by flinging himself under a heavy van at Pimlico, the Coroner stated that worry caused by the feeling that he was not going to be accepted for service led him to take his life.

The Central London Recruiting Depot, August 1914. (Culver Pictures)

But our smiles are not appropriate, for that was a different world. The certainties were intact. Britain had not known a major war for a century, and on the Continent, as A. J. P. Taylor points out, "there had been no war between the Great Powers since 1871. No man in the prime of life knew what war was like. All imagined that it would be an affair of great marches and great battles, quickly decided." [23]

Furthermore, the Great War was perhaps the last to be conceived as taking place within a seamless, purposeful "history" involving a coherent stream of time running from past through present to future. The shrewd recruiting poster depicting a worried father of the future being asked by his children, "Daddy, what did *you* do in the Great War?" assumes a future whose moral and social pressures are identical with those of the past. Today, when each day's experience seems notably *ad hoc*, no such appeal would shame the most stupid to the recruiting office. But the Great War took place in what was, compared with ours, a static world, where the values appeared stable and where the meanings of abstractions seemed permanent and reliable. Everyone knew what Glory was, and what Honor meant. It was not until eleven years after the war that Hemingway could declare in *A Farewell to Arms* that "abstract words such as glory, honor, courage, or hallow were obscene beside the concrete names of villages, the numbers of roads, the names of rivers, the numbers of regiments and the dates." [24] In the summer of 1914 no one would have understood what on earth he was talking about.

Certainly the author of a personal communication in the *Times* two days before the declaration of war would not have understood:

PAULINE—Alas, it cannot be. But I will dash into the great venture with all that pride and spirit an ancient race has given me. . . .

The language is that which two generations of readers had been accustomed to associate with the quiet action of personal control and Christian self-abnegation ("sacrifice"), as well as with more violent actions of aggression and defense. The tutors in this special diction had been the boys' books of George Alfred Henty; the male-romances of Rider Haggard; the poems of Robert Bridges; and especially the Arthurian poems of Tennyson and the pseudo-medieval romances of William Morris. We can set out this "raised," essentially feudal language in a table of equivalents:

A friend is a   *comrade*
Friendship is   *comradeship*, or *fellowship*

A horse is a	*steed,* or *charger*
The enemy is	*the foe,* or *the host*
Danger is	*peril*
To conquer is to	*vanquish*
To attack is to	*assail*
To be earnestly brave is to be	*gallant*
To be cheerfully brave is to be	*plucky*
To be stolidly brave is to be	*staunch*
Bravery considered after the fact is	*valor*
The dead on the battlefield are	*the fallen*
To be nobly enthusiastic is to be	*ardent*
To be unpretentiously enthusiastic is to be	*keen*
The front is	*the field*
Obedient soldiers are	*the brave*
Warfare is	*strife*
Actions are	*deeds*
To die is to	*perish*
To show cowardice is to	*swerve*
The draft-notice is	*the summons*
To enlist is to	*join the colors*
Cowardice results in	*dishonor*
Not to complain is to be	*manly*
To move quickly is to be	*swift*
Nothing is	*naught*
Nothing but is	*naught, save*
To win is to	*conquer*
One's chest is one's	*breast*
Sleep is	*slumber*
The objective of an attack is	*the goal*
A soldier is a	*warrior*
One's death is one's	*fate*
The sky is	*the heavens*
Things that glow or shine are	*radiant*
The army as a whole is	*the legion*
What is contemptible is	*base*
The legs and arms of young men are	*limbs*
Dead bodies constitute	*ashes,* or *dust*
The blood of young men is	*"the red/Sweet wine of youth"*—R. Brooke.

This system of "high" diction was not the least of the ultimate casualties of the war. But its staying power was astonishing. As late as 1918 it was still possible for some men who had actually fought to sustain the old rhetoric. Thus Sgt. Reginald Grant writes the Dedication of his book *S.O.S. Stand To* (1918):

> In humble, reverent spirit I dedicate these pages to the memory of the lads who served with me in the "Sacrifice Battery," and who gave their lives that those behind might live, and, also, in brotherly affection and esteem to my brothers, Gordon and Billy, who are still fighting the good fight and keeping the faith.

Another index of the prevailing innocence is a curious prophylaxis of language. One could use with security words which a few years later, after the war, would constitute obvious *double entendres*. One could say *intercourse*, or *erection*, or *ejaculation* without any risk of evoking a smile or a leer. Henry James's innocent employment of the word *tool* is as well known as Browning's artless misapprehensions about the word *twat*. Even the official order transmitted from British headquarters to the armies at 6:50 on the morning of November 11, 1918, warned that "there will be no intercourse of any description with the enemy." Imagine daring to promulgate that at the end of the Second War! In 1901 the girl who was to become Christopher Isherwood's mother and whose fiancé was going to be killed in the war could write in her diary with no self-consciousness: "Was bending over a book when the whole erection [a toque hat she had been trimming] caught fire in the candles and was ruined. So vexed!" She was an extraordinarily shy, genteel, proper girl, and neither she nor her fiancé read anything funny or anything not entirely innocent and chaste into the language of a telegram he once sent her after a long separation: "THINKING OF YOU HARD." [25] In this world "he ejaculated breathlessly" was a tag in utterly innocent dialogue rather than a moment in pornographic description.

Indeed, the literary scene is hard to imagine. There was no *Waste Land*, with its rats' alleys, dull canals, and dead men who have lost their bones: it would take four years of trench warfare to bring these to consciousness. There was no *Ulysses*, no *Mauberley*, no *Cantos*, no Kafka, no Proust, no Waugh, no Auden, no Huxley, no Cummings, no *Women in Love* or *Lady Chatterley's Lover*. There was no "Valley of Ashes" in *The Great Gatsby*. One read Hardy and Kipling and Conrad and frequented worlds of traditional moral action delineated in traditional moral language.

Although some memories of the benign last summer before the war can be discounted as standard romantic retrospection turned even rosier by egregious contrast with what followed, all agree that the prewar summer was the most idyllic for many years. It was warm and sunny, eminently pastoral. One lolled outside on a folding canvas chaise, or swam, or walked in the countryside. One read outdoors, went on picnics, had tea served from a white wicker table under the trees. You could leave your books on the table all night without fear of rain. Siegfried Sassoon was busy fox hunting and playing serious county cricket. Robert Graves went climbing in the Welsh mountains. Edmund Blunden took country walks near Oxford, read Classics and English, and refined his pastoral diction. Wilfred Owen was teaching English to the boys of a French family living near Bordeaux. David Jones was studying illustration at Camberwell Art School. And for those like Strachey who preferred the pleasures of the West End, there were splendid evening parties, as well as a superb season for concerts, theater, and the Russian ballet.

For the modern imagination that last summer has assumed the status of a permanent symbol for anything innocently but irrecoverably lost. Transferred meanings of "our summer of 1914" retain the irony of the original, for the change from felicity to despair, pastoral to anti-pastoral, is melodramatically unexpected. Elegizing the "Old South" in America, which could be said to have disappeared around 1950, David Lowe writes in 1973:

> We never thought that any of this would change; we never thought of change at all. But we were the last generation of the Old South; that spring in the early fifties was our summer of 1914. . . . Like those other generations who were given to witness the guillotining of a world, we never expected it. And like that of our counterparts, our world seemed most beautiful just before it disappeared.[26]

Out of the world of summer, 1914, marched a unique generation. It believed in Progress and Art and in no way doubted the benignity even of technology. The word *machine* was not yet invariably coupled with the word *gun*.

It was not that "war" was entirely unexpected during June and July of 1914. But the irony was that trouble was expected in Ulster rather than in Flanders. It was expected to be domestic and embarrassing rather than savage and incomprehensible. Of the diary his mother kept during 1914, Christopher Isherwood notes that it has "the morbid fascination of a document which records, without the dishonesty of hindsight, the day-by-day approach to a catastrophe by an utterly unsuspecting victim. Meanwhile, as so often happens, this victim expects and fears a different catastrophe—civil war in Ulster—which isn't going to take place." Kathleen Isherwood writes in her diary on July 13: "The papers look fearfully serious. . . . Sir Edward Carson says 'if it be not peace with honor it must be war with honor.' " The rhetoric seems identical with that of the early stages of the war itself. I have omitted only one sentence in the middle: "Ulster is an armed camp." [27] Alec Waugh remembers the farewell address of his school headmaster: "There were no clouds on my horizon during those long July evenings, and when the Chief in his farewell speech spoke of the bad news in the morning papers, I thought he was referring to the threat of civil war in Ireland." [28] Even in a situation so potent with theatrical possibilities as the actual war was to become, for ironic melodrama it would be hard to improve on the Cabinet meeting of July 24, with the map of Ireland spread out on the big table. "The fate of nations," says John Terraine, "appeared to hang upon parish boundaries in the counties of Fermanagh and Tyrone." [29] To them, enter Sir Edward Grey ashen-faced, in his hand the Austro-Hungarian ultimatum to Servia: *coup de théâtre*.

In nothing, however, is the initial British innocence so conspicuous as in the universal commitment to the sporting spirit. Before the war, says Osbert Sitwell,

> we were still in the trough of peace that had lasted a hundred years between two great conflicts. In it, such wars as arose were not general, but only a brief armed version of the Olympic Games. You won a round; the enemy won the next. There was no more talk of extermination, or of Fights to a Finish, than would occur in a boxing match. [30]

It is this conception of war as strenuous but entertaining that permeates Rupert Brooke's letters home during the autumn and winter of 1914–15. "It's all great fun," he finds. [31] The classic equation between war and sport—cricket, in this case—had been established by Sir Henry Newbolt in his poem "Vitaï Lampada," a public-school favorite since 1898:

> There's a breathless hush in the Close tonight—
> Ten to make and the match to win—
> A bumping pitch and a blinding light,
> An hour to play and the last man in.
> And it's not for the sake of a ribboned coat,
> Or the selfish hope of a season's fame,
> But his Captain's hand on his shoulder smote—
> "Play up! play up! and play the game!"

In later life, the former cricket brave exhorts his colonial troops beset by natives:

> The sand of the desert is sodden red—
> Red with the wreck of a square that broke;

> The Gatling's jammed and the Colonel dead,
>   And the regiment blind with dust and smoke;
> The river of death has brimmed his banks,
>   And England's far, and Honor a name;
> But the voice of a schoolboy rallies the ranks:
>   "Play up! play up! and play the game!"

The author of these lines was a lifetime friend of Douglas Haig. They had first met when they were students together at Clifton College, whose cricket field provides the scene of Newbolt's first stanza. Much later Newbolt wrote, "When I looked into Douglas Haig I saw what is really great—perfect acceptance, which means perfect faith." This version of Haig brings him close to the absolute ideal of what Patrick Howarth has termed *homo newboltiensis*, or "Newbolt Man": honorable, stoic, brave, loyal, courteous—and unaesthetic, unironic, unintellectual and devoid of wit. To Newbolt, the wartime sufferings of such as Wilfred Owen were tiny—and whiny—compared with Haig's: "Owen and the rest of the broken men," he says, "rail at the Old Men who sent the young to die: they have suffered cruelly, but in the nerves and not the heart—they haven't the experience or the imagination to know the extreme human agony. . . ." [32] Only Newbolt Man, skilled in games, can know that.

Cricket is fine for implanting the right spirit, but football is even better. Indeed, the English young man's fondness for it was held to be a distinct sign of his natural superiority over his German counterpart. That was Lord Northcliffe's conclusion in a quasi-official and very popular work of propaganda, *Lord Northcliffe's War Book:*

> Our soldiers are individual. They embark on little individual enterprises. The German . . . is not so clever at these devices. He was never taught them before the war, and his whole training from childhood upwards has been to obey, and to obey in numbers.

The reason is simple:

> He has not played individual games. Football, which develops individuality, has only been introduced into Germany in comparatively recent times. [33]

The English tank crews, Lord Northcliffe finds, "are young daredevils who, fully knowing that they will be a special mark for every kind of Prussian weapon, enter upon their task in a sporting spirit with the same cheery enthusiasm as they would show for football." [34] One thing notable about Prussians is that they have an inadequate concept of playing the game. Thus Reginald Grant on the first German use of chlorine gas:

"It was a new device in warfare and thoroughly illustrative of the Prussian idea of playing the game." [35]

One way of showing the sporting spirit was to kick a football toward the enemy lines while attacking. This feat was first performed by the 1st Battalion of the 18th London Regiment at Loos in 1915. It soon achieved the status of a conventional act of bravado and was ultimately exported far beyond the Western Front. Arthur ("Bosky") Borton, who took part in an attack on the Turkish lines near Beersheeba in November, 1917, proudly reported home: "One of the men had a football. How it came there goodness knows. Anyway we kicked off and rushed the first [Turkish] guns, dribbling the ball with us." [36] But the most famous football episode was Captain W. P. Nevill's achievement at the Somme attack. Captain Nevill, a company commander in the 8th East Surreys, bought four footballs, one for each platoon, during his last London leave before the attack. He offered a prize to the platoon which, at the jump-off, first kicked its football up to the German front line. Although J. R. Ackerley remembered Nevill as "the battalion buffoon," [37] he may have been shrewder than he looked: his little sporting contest did have the effect of persuading his men that the attack was going to be, as the staff had been insisting, a walkover. A survivor observing from a short distance away recalls zero hour:

> As the gun-fire died away I saw an infantryman climb onto the parapet into No Man's Land, beckoning others to follow. [Doubtless Captain Nevill or one of his platoon commanders.] As he did so he kicked off a football. A good kick. The ball rose and travelled well towards the German line. That seemed to be the signal to advance. [38]

Captain Nevill was killed instantly. Two of the footballs are preserved today in English museums.

That Captain Nevill's sporting feat was felt to derive from the literary inspiration of Newbolt's poem about the cricket-boy hero seems apparent from the poem by one "Touchstone" written to celebrate it. This appears on the border of an undated field concert program preserved in the Imperial War Museum:

THE GAME

A Company of the East Surrey Regiment is reported to have dribbled four footballs—the gift of their Captain, who fell in the fight—for a mile and a quarter into the enemy trenches.

> On through the hail of slaughter,
> Where gallant comrades fall,

> Where blood is poured like water,
>   They drive the trickling ball.
> The fear of death before them
>   Is but an empty name.
> True to the land that bore them—
>   The SURREYS play the game.

And so on for two more stanzas. If anyone at the time thought Captain Nevill's act preposterous, no one said so. The nearest thing to such an attitude is a reference in the humorous trench newspaper *The Wipers Times* (Sept. 8, 1917), but even here the target of satire is not so much the act of Captain Nevill as the rhetoric of William Beach Thomas, who served as the *Daily Mail's* notoriously fatuous war correspondent. As the famous correspondent "Teech Bomas," he is made to say of Nevill's attack: "On they came kicking footballs, and so completely puzzled the Potsdammers. With one last kick they were amongst them with the bayonet, and although the Berliners battled bravely for a while, they kameraded with the best."

Modern mass wars require in their early stages a definitive work of popular literature demonstrating how much wholesome fun is to be had at the training camp. The Great War's classic in this genre is *The First Hundred Thousand*, written in 1915—originally in parts for *Blackwood's Magazine*—by "Ian Hay," i.e. Ian Hay Beith. It is really very good, nicely written and thoroughly likable. It gives a cheerful half-fictionalized account of a unit of Kitchener's Army, emphasizing the comedies of training and the brave, resourceful way the boys are playing the game and encountering the absurdities of army life with spirit and humor. ("Are we downhearted? NO!") The appeal of the book is to readers already appreciative of Kipling's fantasy of school high-jinks, *Stalky & Co.* (1899). Hay finally mentions trench casualties, but in such a way as to make them seem no more serious than skinned knees. The Second World War classic in this genre—at least in America—is Marion Hargrove's *See Here, Private Hargrove*, published in May, 1942. It performed the same function as Hay's book: it reassured the folks at home and at the same time persuaded the troops themselves that they were undergoing really quite an amusing experience. Interestingly, Hargrove's book appeared at about the same time after the start of its war as Hay's did after its. Little had happened yet to sour the jokes.

The innocent army depicted by Hay actually did resemble closely the real army being trained in 1914. It was nothing if not sincere, animated by the values of doing one's very best and getting on smartly. C. E. Montague remembers that real, constitutional lazy fellows would buy little cram-books of drill out of their pay and sweat them up at night so as to get on the faster. Men warned for a guard next day would agree among themselves to get up an hour before the pre-dawn Revéillé to practice among themselves . . . in the hope of approaching the far-off, longed-for ideal of smartness, the passport to France.[39]

It was an army whose state of preparation for what faced it can be estimated from the amount of attention the officers' *Field Service Pocket Book* (1914) devoted to topics like "Care of Transport Camels" and "Slinging Camels On To a Ship." As Douglas Haig used to say, two machine guns were ample for any battalion. And he thought the power of bullets to stop horses had been greatly exaggerated. People were so innocent that they were embarrassed to pronounce the new stylish foreign word *camouflage*. They had known so little of *debris* that they still put an acute accent over the *e*.

IRONY AND MEMORY

The innocent army fully attained the knowledge of good and evil at the Somme on July 1, 1916. That moment, one of the most interesting in the whole long history of human disillusion, can stand as the type of all the ironic actions of the war. What could remain of confidence in Divine assistance once it was known what Haig wrote his wife just before the attack: "I feel that every step in my plan has been taken with the Divine help"? "The wire has never been so well cut," he confided to his diary, "nor the artillery preparation so thorough." [40] His hopes were those of every man. Private E. C. Stanley recalls: "I was very pleased when I heard that my battalion would be in the attack. I thought this would be the last battle of the war and I didn't want to miss it. I remember writing to my mother, telling her I would be home for the August Bank Holiday." [41] Even the weather cooperated to intensify the irony, just as during the summer of 1914. "On the first of July," Sassoon says, "the weather, after an early morning mist, was of the kind commonly called heavenly." [42] Thirteen years after that day Henry Williamson recalled it vividly:

I see men arising and walking forward; and I go forward with them, in a glassy delirium wherein some seem to pause, with bowed heads, and sink carefully to their knees, and roll slowly over, and lie still. Others roll and roll, and scream and grip my legs in uttermost fear, and I have to struggle to break away, while the dust and earth on my tunic changes from grey to red.

And I go on with aching feet, up and down across ground like a huge ruined honeycomb, and my wave melts away, and the second wave comes up, and also melts away, and then the third wave merges into the ruins of the first and second, and after a while the fourth blunders into the remnants of the others, and we begin to run forward to catch up with the barrage, gasping and sweating, in bunches, anyhow, every bit of the months of drill and rehearsal forgotten, for who could have imagined that the "Big Push" was going to be this? [43]

What assists Williamson's recall is precisely the ironic pattern which subsequent vision has laid over the events. In reading memoirs of the war, one notices the same phenomenon over and over. By applying to the past a paradigm of ironic action, a rememberer is enabled to locate, draw forth, and finally shape into significance an event or a moment which otherwise would merge without meaning into the general undifferentiated stream.

This mechanism of irony-assisted recall is well illustrated by the writing of Private Alfred M. Hale. He was a genteel, delicate, monumentally incompetent middle-aged batman, known somewhat patronizingly as "our Mr. Hale" in the Royal Flying Corps installations where he served. Four years after the war, he composed a 658-page memoir of his agonies and humiliations, dwelling on his palpable unfitness for any kind of military life and on the constant ironic gap between what was expected of him and what he could perform. At one camp it was his job to heat water for the officers' ablutions. At the same time, he was strictly forbidden to gather fuel for heating water, since the only source of fuel was the lumber of numerous derelict barracks in the camp. Frustrated almost to madness by this conflict of obligations, by the abuse now from one set of officers for the insufficiently heated water, now from another for his tearing up and incinerating the barracks piece by piece, Hale confesses to an anxiety fully as agonizing as that faced by troops in an assault. "Heating water," he remembers, "was a sort of punishment for every sin I have ever committed, I should say." Writing his aggrieved memoir, he knows that he is dwelling excessively on his water-heating problems, incontinently returning to them again and again. He tries to break away and resume his narrative: "I said I was going to turn to other matters." But it is exactly the irony of his former situation that keeps calling him back: "In truth it is the irony of things, as they were in those days, that has forced me back on my tracks, as it has a habit of doing, whenever writing of what I then went through." [44]

Another private, Gunner Charles Bricknall, recalling the war many years later, likewise behaves as if his understanding of the irony attending events is what enables him to recall them. He was in an artillery battery being relieved by a new unit fresh from England:

There was a long road leading to the front line which the Germans occasionally shelled, and the shells used to drop plonk in the middle of it. This new unit assembled right by the wood ready to go into action in the night.

What rises to the surface of Bricknall's memory is the hopes and illusions of the newcomers:

They was all spick and span, buttons polished and all the rest of it.

He tries to help:

We spoke to a few of the chaps before going up and told them about the Germans shelling the road, but of course they was not in charge, so up they went and the result was they all got blown up.

Contemplating this ironic issue, Bricknall is moved to an almost Dickensian reiterative rhetoric:

Ho, what a disaster! We had to go shooting lame horses, putting the dead to the side of the road, what a disaster, which could have been avoided if only the officers had gone into action the hard way [i.e., overland, avoiding the road]. That was something I shall never forget.[45]

It is the *if only* rather than the slaughter that helps Bricknall "never forget" this. A slaughter by itself is too commonplace for notice. When it makes an ironic point it becomes memorable.

Bricknall was a simple man from Walsall, Staffordshire, who died in 1968 at the age of 76. He was, his son tells me, "a man; a real man; a real soldier from Walsall." [46] Sir Geoffrey Keynes, on the other hand, John Maynard's brother, was a highly sophisticated scholar, surgeon, author, editor, book collector, and bibliographer, with honorary doctorates from Oxford, Cambridge, Edinburgh, Sheffield, Birmingham, and Reading.[47] In 1968 he recalled an incident of January 26, 1916. A German shell landed near a British artillery battery and killed five officers, including the major commanding, who were standing in a group. "I attended as best I could to each of them," he remembers, "but all were terribly mutilated and were dead or dying." He then wonders why he remembers so clearly this relatively minor event: "Far greater tragedies were happening elsewhere all the time. The long, drawn-out horrors of Passchendaele were to take place not far away." It is, he concludes, the small ironic detail of the major's dead dog that enables him to "see these things as clearly today as if they had just happened": "The pattern of war is shaped in the individual mind by small individual experiences, and I can see these things as clearly today as if they had just happened, down to the body of the major's terrier bitch . . . lying near her master." [48]

In gathering material for his book *The First Day on the Somme* in 1970, Martin Middlebrook took pains to interview as many of the survivors as he could find. They too use the pattern of irony to achieve their "strongest recollections." Thus Private E. T. Radband: "My strongest recollection: all those grand-looking cavalrymen, ready mounted to follow the breakthrough. What a hope!" And Corporal J. H. Tansley: "One's revulsion to the ghastly horrors of war was submerged in the belief that this war was to end all wars and Utopia would arise. What an illusion!" [49]

"There are some contrasts war produces," says Hugh Quigley, "which art would esteem hackneyed or inherently false." [50] And, we can add, which the art of memory organizes into little ironic vignettes, satires of circumstance more shocking, even, than Hardy's. Here is one from Blunden's *Undertones of War*:

> A young and cheerful lance-corporal of ours was making some tea [in the trench] as I passed one warm afternoon. Wishing him a good tea, I went along three fire-bays; one shell dropped without warning behind me; I saw its smoke faint out, and I thought all was as lucky as it should be. Soon a cry from that place recalled me; the shell had burst all wrong. Its butting impression was black and stinking in the parados where three minutes ago the lance-corporal's mess-tin was bubbling over a little flame. For him, how could the gobbets of blackening flesh, the earth-wall sotted with blood, with flesh, the eye under the duckboard, the pulpy bone be the only answer?

And irony engenders worse irony:

> At this moment, while we looked with dreadful fixity at so isolated a horror, the lance-corporal's brother came round the traverse.[51]

Another example, again of an ironic family tragedy. Here the narrator is Max Plowman, author (under the pseudonym "Mark VII") of the memoir *A Subaltern on the Somme* (1928). The commanding officer of the front-line company in which Plowman is serving has received "a piteous appeal," a letter from "two or three influential people in a Northern town, setting forth the case of a mother nearly demented because she has had two of her three sons killed in the trenches since July 1 [1916], and is in mortal fear of what may happen to the sole surviving member of the family, a boy in our company named Stream." The authors of this letter ask if anything can be done. The company commander "is helpless at the moment, but he has shown the letter to the colonel, who promises to see what can be done next time we are out." The reader will be able to construct the rest of the episode himself. A few days later "Sergeant Brown . . . comes to the mouth of the dug-out to report that a big shell dropped right in the trench, killing one man, though who it was he doesn't yet know: the body was blown to pieces. No one else was hurt." [52]

The irony which memory associates with the events, little as well as great, of the First World War has become an inseparable element of the general vision of war in our time. Sergeant Croft's ironic patrol in Mailer's *The Naked and the Dead* (1948) is one emblem of that vision. The unspeakable agonies endured by the patrol in order to win—as it imagines—the whole campaign take place while the battle is being easily won elsewhere. The patrol's contribution ("sacrifice," it would have been called thirty years earlier) has not been needed at all. As Polack puts it: "We broke our ass for nothin'." [53]

There is continuity too in a favorite ironic scene which the Great War contributes to the Second. A terribly injured man is "comforted" by a friend unaware of the real ghastliness of the friend's wounds. The classic Great War scene of this kind is a real "scene": it is Scene 3, Act III, of R. C. Sherriff's play of 1928, *Journey's End*, which had the amazing run of 594 performances at the Savoy Theater. The dying young Second Lieutenant James Raleigh (played by the twenty-eight-year-old Maurice Evans) is carried down into the orderly-room dugout to be ministered to by his old public-school football idol, Captain Dennis Stanhope:

RALEIGH. Something—hit me in the back—knocked me clean over—sort of— winded me—I'm all right now. (*He tries to rise*)
STANHOPE. Steady, old boy. Just lie there quietly for a bit.
RALEIGH. I'll be better if I get up and walk about. It happened once before— I got kicked in just the same place at Rugger; it—it soon wore off. It—it just numbs you for a bit.

. . .

STANHOPE. I'm going to have you taken away.
RALEIGH. Away? Where?
STANHOPE. Down to the dressing-station—then hospital—then home. (*He smiles*) You've got a Blighty one, Jimmy.

. . .

(*There is quiet in the dug-out for a time. Stanhope sits with one hand on Raleigh's arm, and Raleigh lies very still. Presently he speaks again—hardly above a whisper*)
Dennis—
STANHOPE. Yes, old boy?
RALEIGH. Could we have a light? It's—it's so frightfully dark and cold.
STANHOPE. (*rising*) Sure! I'll bring a candle and get another blanket.
(*Stanhope goes out R, and Raleigh is alone, very still and quiet. . . . A tiny sound comes from where Raleigh is lying—something between a sob and a moan; his L hand drops to the floor. Stanhope comes back with a blanket. He takes a candle from the table and carries it to Raleigh's bed. He puts it on the box beside Raleigh and speaks cheerfully*)
Is that better, Jimmy? (*Raleigh makes no sign*) Jimmy— [54]

The most conspicuous modern beneficiary of this memorable scene is Joseph Heller. Alfred Kazin has accurately distinguished the heart of *Catch-22* from the distracting vaudeville surrounding it: "The impressive emotion in *Catch-22*," he says, "is not 'black humor,' the 'totally absurd' . . . but horror. Whenever the book veers back to its primal scene, a bombardier's evisceration in a plane being smashed by flak, a scene given us directly and piteously, we recognize what makes *Catch-22* disturbing." What makes it disturbing, Kazin decides, is the book's implying, by its Absurd farce, that in the last third of the twentieth century, after the heaping of violence upon violence, it is no longer possible to " 'describe war' in traditional literary ways." [55] But what is notable about Heller's "primal scene" is that it does "describe war" in exactly a traditional literary way. It replays Sherriff's scene and retains all its Great War irony.

Heller's unforgettable scene projects a terrible dynamics of horror, terrified tenderness, and irony. Yossarian has gone to the tail of the plane to help the wounded gunner, the "kid" Snowden: "Snowden was lying on his back on the floor with his legs stretched out, still burdened cumbersomely by his flak suit, his flak helmet, his parachute harness and his Mae West. . . . The wound Yossarian saw was in the outside of Snowden's thigh." It was "as large and deep as a football, it seemed." Yossarian masters his panic and revulsion and sets to work with a tourniquet. "He worked with simulated skill and composure, feeling Snowden's lack-luster gaze resting upon him." Cutting away Snowden's trouser-leg, Yossarian is pleased to discover that the wound "was not nearly as large as a football, but as long and wide as his hand. . . . A long sigh of relief escaped slowly through Yossarian's mouth when he saw that Snowden was not in danger of dying. The blood was already coagulating inside the wound, and it was simply a matter of bandaging him up and keeping him calm until the plane landed."

Cheered by these hopes, Yossarian goes to work "with renewed confidence and optimism." He competently sprinkles sulfanilimide into the wound as he has been taught and binds it up, making "the whole thing fast with a tidy square knot. It was a good bandage, he knew, and he sat back on his heels with pride . . . and grinned at Snowden with spontaneous friendliness." It is time for ironic reversal to begin:

"I'm cold," Snowden moaned. "I'm cold."

"You're going to be all right, kid," Yossarian assured him, patting his arm comfortingly. "Everything's under control."

Snowden shook his head feebly. "I'm cold," he repeated, with eyes as dull and blind as stone. "I'm cold."

"There, there," said Yossarian. . . . "There, there. . . ."

And soon everything proves to be not under control at all:

> Snowden kept shaking his head and pointed at last, with just the barest
> movement of his chin, down toward his armpit. . . . Yossarian ripped open
> the snaps of Snowden's flak suit and heard himself scream wildly as Snow-
> den's insides slithered down to the floor in a soggy pile and just kept drip-
> ping out.

Yossarian "wondered how in the world to begin to save him."

> "I'm cold," Snowden whimpered. "I'm cold."
> "There, there," Yossarian mumbled mechanically in a voice too low to be
> heard. "There, there."

And the scene ends with Yossarian covering the still whimpering Snowden with the nearest thing he can find to a shroud:

> "I'm cold," Snowden said. "I'm cold."
> "There, there," said Yossarian. "There, there." He pulled the rip cord of
> Snowden's parachute and covered his body with the white nylon sheets.
> "I'm cold."
> "There, there." [56]

This "primal scene" works because it is undeniably horrible, but its
irony, its dynamics of hope abridged, is what makes it haunt the memory. It embodies the contemporary equivalent of the experience offered
by the first day on the Somme, and like that archetypal original, it can
stand as a virtual allegory of political and social cognition in our time. I
am saying that there seems to be one dominating form of modern understanding; that it is essentially ironic; and that it originates largely in the
application of mind and memory to the events of the Great War.

II
The Troglodyte World

THE TRENCH SCENE

The idea of "the trenches" has been assimilated so successfully by metaphor and myth ("Georgian complacency died in the trenches") that it is not easy now to recover a feeling for the actualities. *Entrenched*, in an expression like *entrenched power*, has been a dead metaphor so long that we must bestir ourselves to recover its literal sense. It is time to take a tour.

From the winter of 1914 until the spring of 1918 the trench system was fixed, moving here and there a few hundred yards, moving on great occasions as much as a few miles. London stationers purveying maps felt secure in stocking "sheets of 'The Western Front' with a thick wavy black line drawn from North to South alongside which was printed 'British Line.'" [1] If one could have gotten high enough to look down at the whole line at once, one would have seen a series of multiple parallel excavations running for 400 miles down through Belgium and France, roughly in the shape of an *S* flattened at the sides and tipped to the left. From the North Sea coast of Belgium the line wandered southward, bulging out to contain Ypres, then dropping down to protect Béthune, Arras, and Albert. It continued south in front of Montidier, Compiègne, Soissons, Reims, Verdun, St. Mihiel, and Nancy, and finally attached its southernmost end to the Swiss border at Beurnevisin, in Alsace. The top forty miles—the part north of Ypres—was held by the Belgians; the next ninety miles, down to the river Ancre, were British; the French held the rest, to the south.

Henri Barbusse estimates that the French front alone contained about 6250 miles of trenches.[2] Since the French occupied a little more than half the line, the total length of the numerous trenches occupied by the British must come to about 6000 miles. We thus find over 12,000 miles of trenches on the Allied side alone. When we add the trenches of the Central Powers, we arrive at a figure of about 25,000 miles, equal to a trench sufficient to circle the earth. Theoretically it would have been possible to walk from Belgium to Switzerland entirely below ground, but although the lines were "continuous," they were not entirely seamless: occasionally mere shell holes or fortified strong-points would serve as a connecting link. Not a few survivors have performed the heady imaginative exercise of envisioning the whole line at once. Stanley Casson is one who, imagining the whole line from his position on the ground, implicitly submits the whole preposterous conception to the criterion of the "normally" rational and intelligible. As he remembers, looking back from 1935,

> Our trenches stood on a faint slope, just overlooking German ground, with a vista of vague plainland below. Away to right and left stretched the great lines of defense as far as eye and imagination could stretch them. I used to wonder how long it would take for me to walk from the beaches of the North Sea to that curious end of all fighting against the Swiss boundary; to try to guess what each end looked like; to imagine what would happen if I passed a verbal message, in the manner of the parlor game, along to the next man on my right to be delivered to the end man of all up against the Alps. Would anything intelligible at all emerge? [3]

Another imagination has contemplated a similar absurd transmission of sound all the way from north to south. Alexander Aitken remembers the Germans opposite him celebrating some happy public event in early June, 1916, presumably either the (ambiguous) German success at the naval battle of Jutland (May 31–June 1) or the drowning of Lord Kitchener, lost on June 5 when the cruiser *Hampshire* struck a mine and sank off the Orkney Islands. Aitken writes, "There had been a morning in early June when a tremendous tin-canning and beating of shell-gongs had begun in the north and run south down their lines to end, without doubt, at Belfort and Mulhausen on the Swiss frontier." [4] Impossible to believe, really, but in this mad setting, somehow plausible.

The British part of the line was normally populated by about 800 battalions of 1000 men each. They were concentrated in the two main sectors of the British effort: the Ypres Salient in Flanders and the Somme area in Picardy. Memory has given these two sectors the appearance of

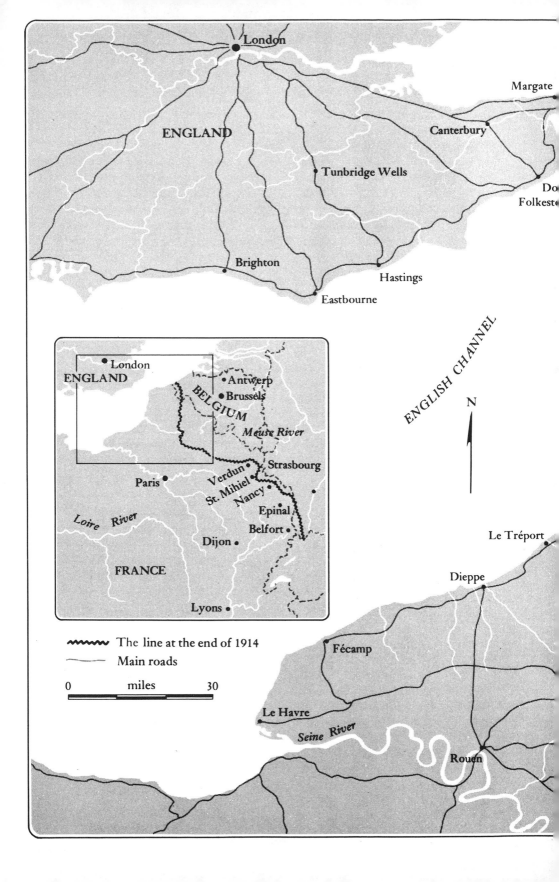

ENGLAND

London

Margate

Canterbury

Tunbridge Wells

Do
Folkesto

Brighton

Hastings

Eastbourne

ENGLISH CHANNEL

N

London

ENGLAND

Antwerp

Brussels

BELGIUM

Meuse River

Strasbourg

Paris

Verdun

St. Mihiel

Nancy

Epinal

Belfort

Dijon

FRANCE

Lyons

Le Tréport

Dieppe

Fécamp

The line at the end of 1914

Main roads

0   miles   30

Le Havre

Seine River

Rouen

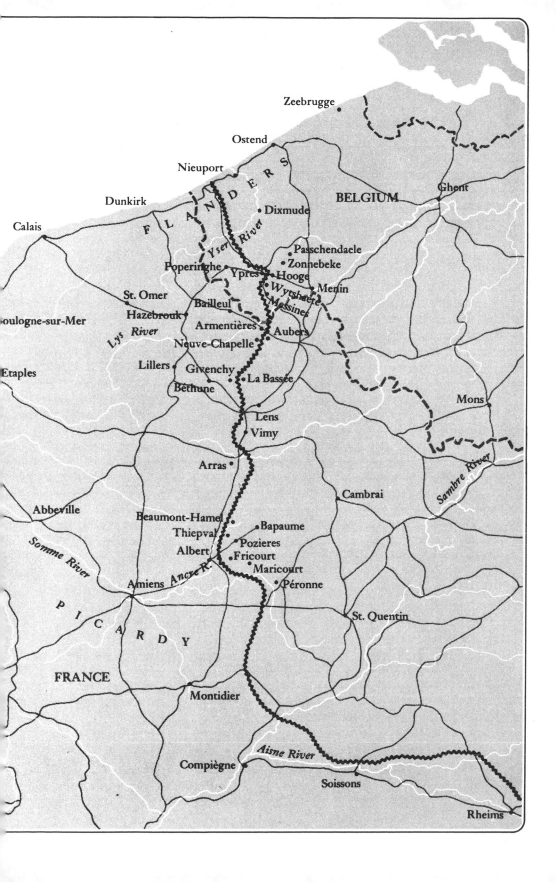

two distinguishable worlds. The Salient, at its largest point about nine miles wide and projecting some four miles into the German line, was notable for its terrors of concentrated, accurate artillery fire. Every part of it could be covered from three sides, and at night one saw oneself almost surrounded by the circle of white and colored Very lights sent up by the Germans to illuminate the ground in front of their trenches or to signal to the artillery behind them. The "rear area" at Ypres was the battered city itself, where the troops harbored in cellars or in the old fortifications built by Vauban in the seventeenth century. It was eminently available to the German guns, and by the end of the war Ypres was flattened to the ground, its name a byword for a city totally destroyed. Another war later, in 1940, Colin Perry—who was not born until four years after the Great War—could look at the ruins of London and speak of "the Ypres effect of Holborn." [5] If the character of the Ypres sector was concentration and enclosure, inducing claustrophobia even above ground, the Somme was known—at least until July 1, 1916—for its greater amplitude and security. German fire came generally from only one direction; and troops at rest could move further back. But then there was the Somme mud; although the argument about whether the mud wasn't really worse at Ypres was never settled.

Each of these two sectors had its symbolic piece of ruined public architecture. At Ypres it was the famous Cloth Hall, once a masterpiece of medieval Flemish civic building. Its gradual destruction by artillery and its pathetic final dissolution were witnessed by hundreds of thousands, who never forgot this eloquent emblem of what happens when war collides with art. In the Somme the memorable ruined work of architecture, connoting this time the collision of the war with religion and the old pieties, was the battered Basilica in the town of Albert, or "Bert," as the troops called it. The grand if rather vulgar red and white brick edifice had been built a few years before the war, the result of a local ecclesiastic's enthusiasm. Together with his townsmen he hoped that Albert might become another Lourdes. Before the war 80,000 used to come on pilgrimages to Albert every year. The object of veneration inside the church was a statue of the Virgin, said to have been found in the Middle Ages by a local shepherd. But the statue of the Virgin never forgotten by the hordes of soldiers who passed through Albert was the colossal gilded one on top of the battered tall tower of the Basilica. This figure, called Notre Dame des Brebières, originally held the infant Christ in outstretched arms above her; but now the whole statue was bent down below the horizontal, giving the effect of a mother about to throw her child—in disgust? in sacrifice?—into the debris-littered street below. To

Colonel Sir Maurice Hankey, Secretary of the War Committee, it was "a most pathetic sight." [6] Some said that the statue had been bent down by French engineers to prevent the Germans from using it to aim at.[7] But most—John Masefield among them—preferred to think it a victim of German artillery.[8] Its obvious symbolic potential (which I will deal with later) impressed itself even on men who found they could refer to it only facetiously, as "The Lady of the Limp."

The two main British sectors duplicated each other also in their almost symbolic road systems. Each had a staging town behind: for Ypres it was Poperinghe (to the men, "Pop"); for the Somme, Amiens. From these towns troops proceeded with augmenting but usually well-concealed terror up a sinister road to the town of operations, either Ypres itself or Albert. And running into the enemy lines out of Ypres and Albert were the most sinister roads of all, one leading to Menin, the other to Bapaume, both in enemy territory. These roads defined the direction of ultimate attack and the hoped-for breakout. They were the goals of the bizarre inverse quest on which the soldiers were ironically embarked.

But most of the time they were not questing. They were sitting or lying or squatting in place below the level of the ground. "When all is said and done," Sassoon notes, "the war was mainly a matter of holes and ditches." [9] And in these holes and ditches extending for ninety miles, continually, even in the quietest times, some 7000 British men and officers were killed and wounded daily, just as a matter of course. "Wastage," the Staff called it.

There were normally three lines of trenches. The front-line trench was anywhere from fifty yards or so to a mile from its enemy counterpart. Several hundred yards behind it was the support trench line. And several hundred yards behind that was the reserve line. There were three kinds of trenches: firing trenches, like these; communication trenches, running roughly perpendicular to the line and connecting the three lines; and "saps," shallower ditches thrust out into No Man's Land, providing access to forward observation posts, listening posts, grenade-throwing posts, and machine gun positions. The end of a sap was usually not manned all the time: night was the favorite time for going out. Coming up from the rear, one reached the trenches by following a communication trench sometimes a mile or more long. It often began in a town and gradually deepened. By the time pedestrians reached the reserve line, they were well below ground level.

A firing trench was supposed to be six to eight feet deep and four or five feet wide. On the enemy side a parapet of earth or sandbags rose about two or three feet above the ground. A corresponding "parados" a foot or so high was often found on top of the friendly side. Into the sides of trenches were dug one- or two-man holes ("funk-holes"), and there were deeper dugouts, reached by dirt stairs, for use as command posts and officers' quarters. On the enemy side of a trench was a fire-step two feet high on which the defenders were supposed to stand, firing and throwing grenades, when repelling attack. A well-built trench did not run straight for any distance: that would have been to invite enfilade fire. Every few yards a good trench zig-zagged. It had frequent traverses designed to contain damage within a limited space. Moving along a trench thus involved a great deal of weaving and turning. The floor of a proper trench was covered with wooden duckboards, beneath which were sumps a few feet deep designed to collect water. The walls, perpetually crumbling, were supported by sandbags, corrugated iron, or bundles of sticks or rushes. Except at night and in half-light, there was of course no looking over the top except through periscopes, which could be purchased in the "Trench Requisites" section of the main London department stores. The few snipers on duty during the day observed No Man's Land through loopholes cut in sheets of armor plate.

The entanglements of barbed wire had to be positioned far enough out in front of the trench to keep the enemy from sneaking up to grenade-throwing distance. Interestingly, the two novelties that contributed most to the personal menace of the war could be said to be American inventions. Barbed wire had first appeared on the American frontier in the late nineteenth century for use in restraining animals. And the machine gun was the brainchild of Hiram Stevens Maxim (1840–1916), an American who, disillusioned with native patent law, established his Maxim Gun Company in England and began manufacturing his guns in 1889. He was finally knighted for his efforts. At first the British regard for barbed wire was on a par with Sir Douglas Haig's understanding of the machine gun. In the autumn of 1914, the first wire Private Frank Richards saw emplaced before the British positions was a single strand of agricultural wire found in the vicinity.[10] Only later did the manufactured article begin to arrive from England in sufficient quantity to create the thickets of mock-organic rusty brown that helped give a look of eternal autumn to the front.

The whole British line was numbered by sections, neatly, from right to left. A section, normally occupied by a company, was roughly 300 yards wide. One might be occupying front-line trench section 51; or support trench S 51, behind it; or reserve trench SS 51, behind both. But a less formal way of identifying sections of trench was by place or street names with a distinctly London flavor. *Piccadilly* was a favorite; popular also were *Regent Street* and *Strand;* junctions were *Hyde Park Corner* and *Marble Arch.* Greater wit—and deeper homesickness—sometimes surfaced in the naming of the German trenches opposite. Sassoon remembers "Durley" 's account of the attack at Delville Wood in September, 1916: "Our objective was Pint Trench, taking Bitter and Beer and clearing Ale and Vat, and also Pilsen Lane." [11] Directional and traffic control signs were everywhere in the trenches, giving the whole system the air of a parody modern city, although one literally "underground."

The trenches I have described are more or less ideal, although not so ideal as the famous exhibition trenches dug in Kensington Gardens for the edification of the home front. These were clean, dry, and well furnished, with straight sides and sandbags neatly aligned. R. E. Vernède writes his wife from the real trenches that a friend of his has just returned from viewing the set of ideal ones. He "found he had never seen anything at all like it before." [12] And Wilfred Owen calls the Kensington Gardens trenches "the laughing stock of the army." [13] Explaining military routines to civilian readers, Ian Hay labors to give the impression that the real trenches are identical to the exhibition ones and that they are properly described in the language of normal domesticity a bit archly deployed:

> The firing-trench is our place of business—our office in the city, so to speak.
> The supporting trench is our suburban residence, whither the weary toiler
> may betake himself periodically (or, more correctly, in relays) for purposes
> of refreshment and repose.[14]

The reality was different. The British trenches were wet, cold, smelly, and thoroughly squalid. Compared with the precise and thorough German works, they were decidedly amateur, reflecting a complacency about the British genius for improvisation. Since defense offered little opportunity for the display of pluck or swank, it was by implication derogated in the officers' *Field Service Pocket Book.* One reason the British trench system was so haphazard and ramshackle was that it had originally taken form in accord with the official injunction: "The choice of a [defensive] position and its preparation must be made with a view to economizing the power expended on defense in order that the power of offense may be increased." And it was considered really useless to build solid fortifications anyway: "An occasional shell may strike and penetrate the parapet, but in the case of shrapnel the damage to the parapet will be trifling, while in the case of a shell filled with high explosive, the effect will be no worse on a thin parapet than on a thick one. It is, therefore, useless to spend time and labor on making a thick parapet simply to keep

The King inspecting model trenches at the Australian school near Sailly, August 1916. (Imperial War Museum)

out shell." [15] The repeatedly revived hopes for a general breakout and pursuit were another reason why the British trenches were so shabby. A typical soldier's view is George Coppard's:

> The whole conduct of our trench warfare seemed to be based on the concept that we, the British, were not stopping in the trenches for long, but were tarrying awhile on the way to Berlin and that very soon we would be chasing Jerry across country. The result, in the long term, meant that we lived a mean and impoverished sort of existence in lousy scratch holes. [16]

In contrast, the German trenches, as the British discovered during the attack on the Somme, were deep, clean, elaborate, and sometimes even comfortable. As Coppard found on the Somme, "Some of the [German] dugouts were thirty feet deep, with as many as sixteen bunk-beds, as well as door bells, water tanks with taps, and cupboards and mirrors." [17] They also had boarded walls, floors, and ceilings; finished wooden staircases; electric light; real kitchens; and wallpaper and overstuffed furniture, the whole protected by steel outer doors. Foreign to the British style was a German dugout of the sort recalled by Ernst Jünger:

A trench on the Somme. (Imperial War Museum)

At Monchy . . . I was master of an underground dwelling approached by
forty steps hewn in the solid chalk, so that even the heaviest shells at this
depth made no more than a pleasant rumble when we sat there over an interminable game of cards. In one wall I had a bed hewn out. . . . . . At its
head hung an electric light so that I could read in comfort till I was sleepy.
. . . The whole was shut off from the outer world by a dark-red curtain
with rod and rings. . . .[18]

As these examples suggest, there were "national styles" in trenches as in
other things. The French trenches were nasty, cynical, efficient, and
temporary. Kipling remembered the smell of delicious cooking emanating from some in Alsace.[19] The English were amateur, vague, *ad hoc*, and
temporary. The German were efficient, clean, pedantic, and permanent.
Their occupants proposed to stay where they were.

   Normally the British troops rotated trench duty. After a week of
"rest" behind the lines, a unit would move up—at night—to relieve a
unit in the front-line trench. After three days to a week or more in that
position, the unit would move back for a similar length of time to the
support trench, and finally back to the reserve. Then it was time for a

week of rest again. In the three lines of trenches the main business of the soldier was to exercise self-control while being shelled. As the poet Louis Simpson has accurately remembered:

> Being shelled is the main work of an infantry soldier, which no one talks about. Everyone has his own way of going about it. In general, it means lying face down and contracting your body into as small a space as possible. In novels [*The Naked and the Dead* is an example] you read about soldiers, at such moments, fouling themselves. The opposite is true. As all your parts are contracting, you are more likely to be constipated.[20]

Simpson is recalling the Second War, but he might be recalling the First. While being shelled, the soldier either harbored in a dugout and hoped for something other than a direct hit or made himself as small as possible in a funk-hole. An unlucky sentry or two was supposed to be out in the open trench in all but the worst bombardments, watching through a periscope or loophole for signs of an attack. When only light shelling was in progress, people moved about the trenches freely, and we can get an idea of what life there was like if we posit a typical twenty-four hours in a front-line trench.

The day began about an hour before first light, which often meant at about 4:30. This was the moment for the invariable ritual of morning stand-to (short for the archaic formal command for repelling attack, "Stand to Arms"). Since dawn was the favorite time for launching attacks, at the order to stand-to everyone, officers, men, forward artillery observers, visitors, mounted the fire-step, weapon ready, and peered toward the German line. When it was almost full light and clear that the Germans were not going to attack that morning, everyone "stood down" and began preparing breakfast in small groups. The rations of tea, bread, and bacon, brought up in sandbags during the night, were broken out. The bacon was fried in mess-tin lids over small, and if possible smoke-less, fires. If the men were lucky enough to be in a division whose commanding general permitted the issue of the dark and strong government rum, it was doled out from a jar with the traditional iron spoon, each man receiving about two tablespoonsful. Some put it into their tea, but most swallowed it straight. It was a precious thing, and serving it out was almost like a religious ceremonial, as David Jones recalls in *In Parenthesis*, where a corporal is performing the rite:

> O have a care—don't spill the precious
> O don't jog his hand—ministering;
> do take care.
> O please—give the poor bugger elbow room.

Larger quantities might be issued to stimulate troops for an assault, and one soldier remembers what the air smelled like during a British attack: "Pervading the air was the smell of rum and blood." [21] In 1922 one medical officer deposed before a parliamentary committee investigating the phenomenon of "shell shock": "Had it not been for the rum ration I do not think we should have won the war." [22]

During the day the men cleaned weapons and repaired those parts of the trench damaged during the night. Or they wrote letters, deloused themselves, or slept. The officers inspected, encouraged, and strolled about looking nonchalant to inspirit the men. They censored the men's letters and dealt with the quantities of official inquiries brought them daily by runner. How many pipe-fitters had they in their company? Reply immediately. How many hairdressers, chiropodists, bicycle repairmen? Daily "returns" of the amount of ammunition and the quantity of trench stores had to be made. Reports of the nightly casualties had to be sent back. And letters of condolence, which as the war went on became form-letters of condolence, had to be written to the relatives of the killed and wounded. Men went to and fro on sentry duty or working parties, but no one showed himself above the trench. After evening stand-to, the real work began.

Most of it was above ground. Wiring parties repaired the wire in front of the position. Digging parties extended saps toward the enemy. Carrying parties brought up not just rations and mail but the heavy engineering materials needed for the constant repair and improvement of the trenches: timbers, A-frames, duckboards, stakes and wire, corrugated iron, sandbags, tarpaulins, pumping equipment. Bombs and ammunition and flares were carried forward. All this ant-work was illuminated brightly from time to time by German flares and interrupted very frequently by machine gun or artillery fire. Meanwhile night patrols and raiding parties were busy in No Man's Land. As morning approached, there was a nervous bustle to get the jobs done in time, to finish fitting the timbers, filling the sandbags, pounding in the stakes, and then returning mauls and picks and shovels to the Quartermaster Sergeant. By the time of stand-to, nothing human was visible above ground anywhere, but every day each side scrutinized the look of the other's line for significant changes wrought by night.

Flanders and Picardy have always been notorious for dampness. It is not the least of the ironies of the war for the British that their trenches should have been dug where the water-table was the highest and the annual rainfall the most copious. Their trenches were always wet and often flooded several feet deep. Thigh-boots or waders were issued as standard

Duckboards going forward over flooded trenches near Pilckem, Ypres Salient, October 1917. (Imperial War Museum)

articles of uniform. Wilfred Owen writes his mother from the Somme at the beginning of 1917: "The waders are of course indispensable. In 2½ miles of trench which I waded yesterday there was not one inch of dry ground. There is a mean depth of two feet of water." [23] Pumps worked day and night but to little effect. Rumor held that the Germans not only could make it rain when they wanted it to—that is, all the time—but had contrived some shrewd technical method for conducting the water in their lines into the British positions—perhaps piping it underground. Ultimately there was no defense against the water but humor. "Water knee deep and up to the waist in places," one soldier notes in his diary. "Rumors of being relieved by the Grand Fleet." [24] One doesn't want to dwell excessively on such discomforts, but here it will do no harm to try to imagine what, in these conditions, going to the latrine was like.

The men were not the only live things in the line. They were accompanied everywhere by their lice, which the professional delousers in rest positions behind the lines, with their steam vats for clothes and hot baths for troops, could do little to eliminate. The entry *lousy* in Eric Partridge's *Dictionary of Slang and Unconventional English* speaks volumes: "Contemptible; mean; filthy. . . . Standard English till 20th C, when, especially after the Great War, colloquial and used as a mere pejorative." *Lousy with*, meaning *full of*, was "originally military" and entered the colloquial word-hoard around 1915: "That ridge is lousy with Fritz."

The famous rats also gave constant trouble. They were big and black, with wet, muddy hair. They fed largely on the flesh of cadavers and on dead horses. One shot them with revolvers or coshed them to death with pick-handles. Their hunger, vigor, intelligence, and courage are recalled in numerous anecdotes. One officer notes from the Ypres Salient: "We are fairly plagued with rats. They have eaten nearly everything in the mess, including the table-cloth and the operations orders! We borrowed a large cat and shut it up at night to exterminate them, and found the place empty next morning. The rats must have eaten it up, bones, fur, and all, and dragged it to their holes." [25]

One can understand rats eating heartily there. It is harder to understand men doing so. The stench of rotten flesh was over everything, hardly repressed by the chloride of lime sprinkled on particularly offensive sites. Dead horses and dead men—and parts of both—were sometimes not buried for months and often simply became an element of parapets and trench walls. You could smell the front line miles before you could see it. Lingering pockets of gas added to the unappetizing atmosphere. Yet men ate three times a day, although what they ate reflected the usual gulf between the ideal and the actual. The propagandist George Adam announced with satisfaction that "the food of the army is based upon the conclusions of a committee, upon which sat several eminent scientists." The result, he asserted, is that the troops are "better fed than they are at home." [26] Officially, each man got daily:

1¼ pounds fresh meat (or 1 pound preserved meat),
1¼ pounds bread,
4 ounces bacon,
3 ounces cheese,
½ pound fresh vegetables (or 2 ounces dried),

together with small amounts of tea, sugar, and jam. But in the trenches there was very seldom fresh meat, not for eating, anyway; instead there was "Bully" (tinned corned-beef) or "Maconochie" (ma-con'-o-chie), a tinned meat-and-vegetable stew named after its manufacturer. If they did tend to grow tedious in the long run, both products were surprisingly good. The troops seemed to like the Maconochie best, but the Germans favored the British corned beef, seldom returning from a raid on the British lines without taking back as much as they could carry. On trench duty the British had as little fresh bread as fresh meat. "Pearl Biscuits" were the substitute. They reminded the men of dog biscuits, although, together with the Bully beef, they were popular with the French and Belgian urchins, who ran (or more often strolled) alongside the railway trains bringing troops up to the front, soliciting gifts by shouting, "Tommee! Bull-ee! Bee-skee!" When a company was out of the line, it fed better. It was then serviced by its company cookers—stoves on wheels—and often got something approaching the official ration, as it might also in a particularly somnolent part of the line, when hot food might come up at night in the large covered containers known as Dixies.

Clothing and equipment improved as the war went on, although at the outset there was a terrible dearth and improvisation. During the retreat from Mons, as Frank Richards testifies, "A lot of us had no caps: I was wearing a handkerchief knotted at the four corners—the only headgear I was to wear for some time." Crucial supplies had been omitted: "We had plenty of small-arm ammunition but no rifle-oil or rifle-rag to clean our rifles with. We used to cut pieces off our shirts . . . and some of us who had bought small tins of vaseline . . . for use on sore heels or chafed legs, used to grease our rifles with that." [27] At the beginning line officers dressed very differently from the men. They wore riding-boots or leather puttees; melodramatically cut riding breeches; and flare-skirted tunics with Sam Browne belts. Discovering that this costume made them special targets in attacks (German gunners were instructed to fire first at the people with the thin knees), by the end they were dressing like the troops, wearing wrap puttees; straight trousers bloused below the knee; Other Ranks' tunics with inconspicuous insignia, no longer on the cuffs but on the shoulders; and Other Ranks' web belts and haversacks. In 1914 both officers and men wore peaked caps, and it was rakish for officers to remove the grommet for a "Gorblimey" effect. Steel helmets were introduced at the end of 1915, giving the troops, as Sassoon observed, "a Chinese look." [28] Herbert Read found the helmets "the only poetic thing in the British Army, for they are primeval in design and effect, like iron mushrooms." [29] A perceptive observer could date corpses and skeletons lying on disused battlefields by their evolving dress. A month before the end of the war, Major P. H. Pilditch recalls, he

spent some time in the old No Man's Land of four years' duration. . . . It was a morbid but intensely interesting occupation tracing the various battles amongst the hundreds of skulls, bones and remains scattered thickly about. The progress of our successive attacks could be clearly seen from the types of equipment on the skeletons, soft cloth caps denoting the 1914 and early 1915 fighting, then respirators, then steel helmets marking attack in 1916. Also Australian slouch hats, used in the costly and abortive attack in 1916.[30]

To be in the trenches was to experience an unreal, unforgettable enclosure and constraint, as well as a sense of being unoriented and lost. One saw two things only: the walls of an unlocalized, undifferentiated earth and the sky above. Fourteen years after the war J. R. Ackerley was wandering through an unfrequented part of a town in India. "The streets became narrower and narrower as I turned and turned," he writes, "until I felt I was back in the trenches, the houses upon either side being so much of the same color and substance as the rough ground between." [31] That lost feeling is what struck Major Frank Isherwood, who writes his wife in December, 1914: "The trenches are a labyrinth, I have already lost myself repeatedly. . . . you can't get out of them and walk about the country or see anything at all but two muddy walls on each side of you." [32] What a survivor of the Salient remembers fifty years later are the walls of dirt and the ceiling of sky, and his eloquent optative cry rises as if he were still imprisoned there: "To be out of this present, ever-present, eternally present misery, this stinking world of sticky, trickling earth ceilinged by a strip of threatening sky." [33] As the only visible theater of variety, the sky becomes all-important. It was the sight of the sky, almost alone, that had the power to persuade a man that he was not already lost in a common grave.

SUNRISE AND SUNSET

Morning and evening stand-to's were the occasions when the sky especially offered itself for observation and interpretation. Stand-to was a solemn moment. Twice a day everyone stared silently across the wasteland at the enemy's hiding places and considered how to act if a field-gray line suddenly appeared and grew larger and larger through the mist and the half-light. Twice a day everyone enacted this ritual of alert defense that served to dramatize what he was in the trench for and that couldn't help emphasizing the impossibility of escape. The "peculiar significance" of stand-to is one thing David Jones particularly remembers:

> [The] procedure was strict and binding anywhere in the forward zone, under any circumstances whatever. . . . So that the hour occurring twice in the twenty-four . . . was one of peculiar significance and there was attaching to it a degree of solemnity, in that one was conscious that from the sea dunes to the mountains, everywhere, on the whole front the two opposing lines stood alertly, waiting any eventuality.[34]

The hour of stand-to constituted a highly ritualized distillation of the state of anxious stalemate and the apparently absolute equivalence of force that had led to the stasis of mutual entrenchment. The hour was memorable as an emblem of the political essence of the war itself, for morning and evening stand-to's evened out the advantages possessed by either side. In the morning, the advantage lay with the British side as the rising sun ahead silhouetted belated German patrols and working parties. In the evening, as the sun went down behind the British lines, the Germans had the advantage: they could catch the silhouettes of the nightly ration and working parties starting out too early.

What one stared at twice daily for years on end, thus, was sunrise and sunset. "We have a unique opportunity of viewing sunrise and sunset in the trenches," writes Hugh Quigley, "owing to the decree that men must stand-to at daybreak and nightfall." And if Flanders was flat and wet, it had the distinction of offering terrain and weather productive of superb dawns and dusks. "I have never seen such glorious vistas of sunset or sunrise as in this country," Quigley continues. "Its flatness gathers up everything to the world overhead and adds a great depth and shimmering distance to the gold, rose, and green, touching them finely to delicate gleam." [35] Quigley's vision is that of a painterly Nineties aesthete—the very first words in his *Passchendaele and the Somme* (1928) are "At sunset we passed by the lake"—but even so practical a countryman as Sassoon recalls thinking once at evening stand-to that "the sky was one of the redeeming features of the war." [36] Christopher Tietjens, in Ford Madox Ford's *A Man Could Stand Up—*, notices that observing from a trench makes the sky overhead and to the front especially brilliant:

> Twice he had stood up on a rifleman's step enforced by a bully-beef case to look over. . . . Each time, on stepping down again, he had been struck by that phenomenon: the light seen from the trench seemed, if not brighter, then more definite. So, from the bottom of a pit-shaft in broad day you can see the stars.[37]

It was a cruel reversal that sunrise and sunset, established by over a century of Romantic poetry and painting as the tokens of hope and peace and rural charm, should now be exactly the moments of heightened ritual anxiety. What literate young man could forget, during the hour of stand-to, the sunrises of Blake and Shelley, of Tennyson and Arnold, the sunsets of Wordsworth and Byron and Browning, and even of Sir Henry Newbolt? Those attentive to the history of taste know that sky-awareness is a fairly late development. There is little of it, for example, in the eighteenth century, which felt no pressing need for such emblems of infinity as sky or sea. The crucial document in the history of British awareness of the sky is Ruskin's *Modern Painters*, the five volumes of which appeared from 1843 to 1860. And the crucial place in *Modern Painters* is the chapter "Of the Open Sky" in the first volume.

Here Ruskin asserts that hitherto no one has paid sufficient attention to the sky, and that we have thus missed much of what Nature has to tell us for our moral benefit. The effects of the sky, he says, are "intended" by their "Maker and Doer" for our pleasure as well as our moral instruction. The sky speaks universally to the human heart, "soothing it and purifying it from its dross and dust." And the great exponent of the sky's moral work is the painter J. M. W. Turner. His renderings of sunrise and sunset are far superior to those of Claude Lorraine, and for Ruskin the painter of such effects is a master artist whose achievements in light and color imitate those of God himself in His most earnest pedagogic moments.

Modern Painters was immensely influential in establishing the visual taste of the Victorians and their successors. Charlotte Brontë wrote of it: "This book seems to give me eyes," [38] and she was followed by several generations uniquely devoted to observing sunrise and sunset with an almost mystical enthusiasm. Oscar Wilde's "The Decay of Lying" (1889) implies the continuing vogue of sunsets among the popularly educated:

> Yesterday evening Mrs. Arundel insisted on my going to the window and looking at the glorious sky, as she called it. Of course I had to look at it. She is one of those absurdly pretty Philistines to whom one can deny nothing. And what was it? It was simply a very second-rate Turner, a Turner of a bad period, with all the painter's worst faults exaggerated and overemphasized.

By 1930 so much had been made of sunrises and sunsets that the sophisticated were sick of them, as Evelyn Waugh reveals in his travel book *Labels:*

> I do not think I shall ever forget the sight of Etna at sunset; the mountain almost invisible in a blur of pastel grey, glowing on the top and then repeating its shape, as though reflected, in a wisp of grey smoke, with the whole horizon behind radiant with pink light, fading gently into a grey pastel sky. Nothing I have ever seen in Art or Nature was quite so revolting. [39]

But if Wilde and Waugh were tiring of sunrises and sunsets, most people were not, and the general devotion to these effects lasted through the Great War and well beyond. In July, 1931, the Southern Railway advertised a dawn excursion this way: "Experience the novel thrill of watching a summer dawn from the first streaks to the full sunrise!" As Robert Graves and Alan Hodge report, "Forty people were expected; 1,400 showed up, and filled four trains." [40]

That is evidence that the influence of Ruskin leaked down abundantly to those who had not read him. C. S. Lewis was a representative young man of the period. Writing about his life at Oxford before entering the army, he recalls that "the sky was, and still is, to me one of the principal elements in any landscape, and long before I had seen [skies of all types] named and sorted out in *Modern Painters* I was very attentive to [their] different qualities. . . ." [41] Max Plowman remembers, even if vaguely, his Ruskin:

> Was it Ruskin who said that the upper and more glorious half of Nature's pageant goes unseen by the majority of people? . . . Well, the trenches have altered that. Shutting off the landscape, they compel us to observe the sky; and when it is a canopy of blue flecked with white clouds . . . , and when the earth below is a shell-stricken waste, one looks up with delight, recalling perhaps the days when, as a small boy, one lay on the garden lawn at home counting the clouds as they passed. [42]

Here one might object that sunsets are sunsets, in all times and places. But no. It is impossible to think of anyone at any other time noticing a sunset in such pedantic, painterly, high-minded, self-satisfied terms as the young Basil Willey at Cambridge in 1916:

> There was a superb cirrus sunset tonight. Down low behind King's Chapel there was a perfect welter of flaming orange, seen in wonderful mosaic-fragments through the high screen of elms in the Backs. Thronging the upper heaven was a flight of winged shapes tinted with all hues of rose, salmon, damask, and lavender. [43]

Welter, thronging, a flight of winged shapes, tinted with all hues, rose, lavender: superbly "of the period." Before the war it was a commonplace that one of the duties of "a writer" was to study such effects. Poverty-stricken and shelterless on the Venice Lido in 1910, Frederick Rolfe ("Baron Corvo") spent whole nights walking on the beach. When the police questioned him, he explained himself to their satisfaction by proclaiming that he was a writer "studying the dawns." [44] The Evelyn Waugh who says of himself at fifteen, "I had not read much Ruskin, but I had in some way imbibed most of his opinions," [45] reveals that for all his boredom with the vulgar picturesque off Sicily he is squarely in the British tradition of paying lots of attention to sky effects and knowing about their thematic value for painting. On the first page of his novel *Officers and Gentlemen* (1955) we find this scene from a later war:

> The [night] sky over London was glorious, ochre and madder, as though a dozen tropical suns were simultaneously setting round the horizon; every-

where the searchlights clustered and hovered, then swept apart; here and there pitchy clouds drifted and billowed; now and then a huge flash momentarily froze the serene fireside glow. Everywhere the shells sparkled like Christmas baubles.

"Pure Turner," said Guy Crouchback enthusiastically. . . .

"John Martin, surely?" said Ian Kilbannock. . . .

S. S. Horsley, a soldier whose diary and pocket notebook are preserved in the Imperial War Museum, is another legatee of Ruskin. In his diary he very often takes the trouble to celebrate the sunrises at stand-to, and when he wants to copy out a poem to treasure up for consolation in his notebook, he chooses Yeats's "He Wishes for the Cloths of Heaven," a poem entertaining no doubts that imagery drawn from the kinetic skies is of very high value.

Thus, by the time the war began, sunrise and sunset had become fully freighted with implicit aesthetic and moral meaning. When a participant in the war wants an ironic effect, a conventional way to achieve one is simply to juxtapose a sunrise or sunset with the unlovely physical details of the war that man has made. This is the method of 2nd Lieutenant William Ratcliffe, destined to be killed on July 1, 1916. He writes his parents in June: "Everywhere the work of God is spoiled by the hand of man. One looks at a sunset and for a moment thinks that that at least is unsophisticated, but an aeroplane flies across, and puff! puff! and the whole scene is spoilt by clouds of shrapnel smoke." [46] Details like "one looks at a sunset" and "the whole scene" indicate the formal, painterly dimensions of these Great War sunsets. They are almost in gilded frames. The sad contrast between Man's works and God's is one of Blunden's perpetual ironic means in *Undertones of War*. One of his sunsets is even more specifically theological (even if more sardonically so) than Lt. Ratcliffe's. Of the evening following a particularly ghastly battle near Ypres in 1917 he writes: "The eastern sky that evening was all too brilliant with British rockets, appealing for artillery assistance. Westward, over blue hills, the sunset was all seraphim and cherubim." [47]

Emphasized as objects of close attention by the daily rituals of stand-to, sanctified as morally meaningful by the long tradition of Ruskin, these sunrises and sunsets, already a staple of prewar Georgian poetry and the literature of the Celtic "Twilight," move to the very center of English poetry of the Great War. They are its constant atmosphere and its special symbolic method. The most sophisticated poem of the war, Isaac Rosenberg's "Break of Day in the Trenches," undertakes its quiet meditative ironies at morning stand-to. The most "representative"—I suppose I mean the most frequently anthologized—is Wilfred Owen's "Anthem for Doomed Youth," which darkens to this "period" dying fall:

> And each slow dusk a drawing-down of blinds.

(Bernard Bergonzi's "Study of the Literature of the Great War" is titled perhaps even more suggestively than its author realizes: *Heroes' Twilight*.) The most popular poem for quotation, still, at memorial occasions (at the Cenotaph, for example) is Laurence Binyon's "For the Fallen," written with uncanny prescience of the appropriate imagery in September, 1914, when the war was a mere seven weeks along. Its memorable stanza begins,

> They shall not grow old, as we that are left grow old:
> Age shall not weary them, nor the years condemn.

And the stanza concludes with lines which many parents and wives chose as supremely appropriate for their soldiers' headstones:

> At the going down of the sun and in the morning
> We will remember them.

Equally popular for headstones were such conventional evocations of a morally significant sunrise as

> Till morning breaks

and

> Until the day breaks and the shadows flee.

And sunrise and sunset did not have to wait for solemn occasions to enter poems. They appeared everywhere, even in whimsical verse of the sort printed by the Army Service Corps' *Hangar Herald* in the days when officers were still wearing their "pips" on their tunic cuffs:

> Oh, deem it pride, not lack of skill,
>   That thus forbids my sleeve's increase:
> The morning and the evening still
>   Have but one star apiece.[48]

In working up scenes for *The Memoirs of George Sherston* between 1926 and 1936, Sassoon recovered the Great War feeling for symbolic sunsets and adjusted important emotional climaxes to make them take place at such rich but melancholy moments. For example, after enlisting Sherston says his ritual goodbye to his Kentish drawing room at sunset:

> The sun blinds . . . were drawn down the tall windows; I was alone in the twilight room, with the glowering red of sunset peering through the chinks and casting the shadows of leaves on a fiery patch of light which rested on the wall by the photograph of [George Frederic Watts's painting] "Love and Death." So I looked my last and rode away to war on my bicycle. . . . [49]

It is in twilight that George Sherston first beholds his new ideal friend, his young fellow officer Dick Tiltwood: "Twilight was falling and there was only one small window [in the army hut], but even in the half-light his face surprised me by its candor and freshness. He had the obvious good looks which go with fair hair and firm features, but it was the radiant integrity of his expression which astonished me." [50] Dick's burial in "a sack" a few months later occurs when "the sky was angry with a red smoky sunset." [51] Sunset is likewise the time for making crucial decisions, like Sherston's to return to the war after the months of safety his pacifistic mutiny has earned him. Snug in Craiglockhart Hospital, he says, "I argued it out with myself in the twilight." [52] The very talisman which Sassoon carried in his pocket at the front was a lump of fire opal which he used to call his "pocket sunset." [53]

This exploitation of moments of waxing or waning half-light is one of the distinct hallmarks of Great War rhetoric. It signals a constant reaching out towards traditional significance, very much like the system of "high" diction which dominated the early stages of the war. It reveals an attempt to make some sense of the war in relation to inherited tradition. We can see this clearly by contrasting some of the memorable poems of the Second War, when there were few static stand-to's—it is hard to stand-to at 20,000 feet—and when dawn and dusk had become largely de-ritualized, no longer automatically portentous. Where are dawn and dusk in Randall Jarrell's "Eighth Air Force," or "Losses," or "The Death of the Ball Turret Gunner"? Or in Henry Reed's "Lessons of the War"? Or in Keith Douglas's *Vergissmeinnicht*? In Gavin Ewart's compelling "When a Beau Goes In," typical of Second War poetry in its laconic refusal to reach out for any myth, the time of day is as irrelevant as praying:

> When a Beau goes in,
> Into the drink,
> It makes you think,
> Because, you see, they always sink
> But nobody says "Poor lad"
> Or goes about looking sad
> Because, you see, it's war,
> It's the unalterable law.

Although it's perfectly certain
The pilot's gone for a Burton
And the observer too
It's nothing to do with you
And if they both should go
To a land where falls no rain nor hail nor driven snow—
Here, there, or anywhere,
Do you suppose *they* care?

You shouldn't cry
Or say a prayer or sigh.
In the cold sea, in the dark
It isn't a lark
But it isn't Original Sin—
It's just a Beau going in.

Ewart's flagrantly un-innocent *just* exemplifies what Larkin means to point to by saying, "Never such innocence again." *Just* asserts the utter irrelevance of theological or any other connotations. The tone is like that of the opening of Jarrell's "Losses":

It was not dying: everybody died.

In a world where myth is of no avail and where traditional significance has long ago been given up for lost, time of day doesn't matter.

Indeed, in Second War poetry there is something like a refusal to notice that sunrise and sunset ever were mythicized. An example is Louis Simpson's long poem "The Runner," whose more than 800 lines narrate a series of infantry actions occurring over several weeks. Not once is there a significant description of a sunrise or sunset. When the poem requires an indication of these times of day, Simpson performs the description, or rather the announcement, in the most abstract, inert way: "Day turned to dusk," he will write, or

Day seemed less to rise
Than darkness to withdraw.

What he has done is to pare away traditional connotation. We are well past the moment when we can turn for comfort to the implications of the day breaking and the shadows fleeing. It is perhaps less that we have "lost hope" than that we have lost literary resources.

On the other hand, traditional ritual meaning is what the poems of the Great War are at pains to awaken. Especially the earliest poems of the war, thoroughly resolute when not enthusiastic, conceive of dawn as morally meaningful, as a moment fit for great beginnings. Hardy's famous early call to action, "Men Who March Away" (published in the *Times*, September 5, 1914), chooses earliest morning as the significant time for volunteers—contrasted with "dalliers"—to step out, animated by their "faith and fire":

> What of the faith and fire within us
>   Men who march away
>   Ere the barn-cocks say
>   Night is growing gray,
> Leaving all that here can win us;
> What of the faith and fire within us
>   Men who march away?

It was Hardy's pre-dawn moment that lodged in Edward Thomas's memory. When he wrote W. H. Hudson to praise Hardy's poem, the cocks waiting for the incipient dawn are what he finds memorable: "I thought Hardy's poem in *The Times* 'Ere the barn-cocks say/Night is growing gray,' the only good one concerned with the war." [54] When Thomas writes his own call to action, he remembers Hardy's dawn. Thomas's "The Trumpet" inverts the terms of the standard aubade in order to encourage enlistment. Dawn is conceived as a signal to action rather than as the last hurried moment for dalliance:

> . . . Rise up and scatter
> The dew that covers
> The print of last night's lovers—
> Scatter it, scatter it!
>
>   .   .   .
>
> Open your eyes to the air
> That has washed the eyes of the stars
> Through all the dewy night:
> Up with the light,
> To the old wars;
> Arise, arise!

One who might have heard Thomas's summons is the clerk of Herbert Asquith's "The Volunteer," who, toiling away in the City, has feared that

>   his days would drift away
> With no lance broken in life's tournament.

He enlisted and was killed, and

> . . . now those waiting dreams are satisfied.

From a boring office to the "field" he went; or as Asquith puts it, amalgamating Tennyson with the practical convention of the attack at dawn,

> From twilight to the halls of dawn he went.

Now,

> His lance is broken; but he lies content
> With that high hour, in which he lived and died.

It is "that high hour," dawn, that has made the difference.

It also makes the difference in two of Rupert Brooke's five famous innocent sonnets of 1914. Brooke perhaps did not know yet that attacks do actually take place at dawn; he certainly could not have known that stand-to would become a trench rite transforming the Georgian literary symbolism of vague high hopes into a daily routine of quiet terror. But there is something like prescience in Brooke's sensing that dawn is the right figure for the beginning of the major action which is the war: God's very hour—the irony to be sure is almost unbearable for us—is "dawn":

> Now God be thanked who has matched us with His Hour,
> And caught our youth, and wakened us from sleeping.
> (Sonnet I: "Peace")

Dawn also visits Brooke's Fourth Sonnet, "The Dead," where, together with sunset, it is one of the specific "kindnesses" vouchsafed the young before their death:

> Dawn was theirs,
> And sunset, and the colors of the earth.

If the dead are now fixed in frost, at least the delights of "waking" once were theirs. Golden half-light was Brooke's world, and had been ever since "The Old Vicarage: Grantchester" of 1912, a poem whose 140 lines manage to encompass four separate dawns and three sunsets. No wonder Wilfrid Gibson, lamenting Brooke's death in 1915, recalled him as an emanation of a "sunset glow":

> He's gone.
> I do not understand.
> I only know
> That, as he turned to go
> And waved his hand,
> In his young eyes a sudden glory shone,
> And I was dazzled by a sunset glow,
> And he was gone.

The process by which, as the war went on, these Georgian metaphoric sunrises and sunsets turned literal and ghastly can be observed in "Before Action," a poem written by William Noel Hodgson two days before he was killed in the Somme attack. Writing as if during the last sunset before the next day's attack, he begins his three-part prayer with a Ruskinian survey of benefits received:

> By all the glories of the day
>   And the cool evening's benison,
> By that last sunset touch that lay
>   Upon the hills when day was done,
> By beauty lavishly outpoured
>   And blessings carelessly received,
> By all the days that I have lived
>   Make me a soldier, Lord.

We are still in the purlieus of the Georgian understanding of "Beauty." But as he writes Hodgson knows that the forthcoming attack will take place in the early morning, and the final stanza, while maintaining its focus on traditionally significant times of day, signals an awareness of the actual situation a day hence. The "familiar hill" is both metaphoric and literal, or rather it is a metaphor caught in the act of turning literal: the hill is at once the lucky eminence of youth and the actual observation post where the speaker has watched with too little feeling "a hundred" sunsets preceding morning attacks like the one in which he' will now take part:

> I, that on my familiar hill
>   Saw with uncomprehending eyes
> A hundred of Thy sunsets spill
>   Their fresh and sanguine sacrifice,
> Ere the sun swings his noonday sword
>   Must say goodbye to all of this:
> By all delights that I shall miss,
>   Help me to die, O Lord.

Spill, fresh, sanguine: with those terms the two dimensions of the poem merge into one. Georgian figure and discovered actuality merge; the red of sunset is seen as identical with the red of freshly shed blood. With *sacrifice*, the poem, although still maintaining its pose of abstract literariness, turns to face actual facts, and we realize that it is no longer talking about sky effects and Literature but about people and action. Without rudeness or abruptness, Ruskin has been invited to squat in a jump-off trench on a hill near Albert.

A similar deployment of traditional literary terms in aid of the new actuality occurs in Wilfred Owen's "Exposure." After a night of trench watch in icy wind and rain, those standing-to and peering to the east perceive the half-light of dawn breaking, gray as in pastoral literary tradition, but gray now also like the shade of the German uniform:

> The poignant misery of dawn begins to grow . . .
> We only know war lasts, rain soaks, and clouds sag stormy.
> Dawn massing in the east her melancholy army
> Attacks once more in ranks on shivering ranks of gray,
>   But nothing happens.

The new alliance of the terms of "poetry" with those of experience is apparent likewise in Blunden's "Trench Raid Near Hooge." A ten-man raiding party goes out

> At an hour before the rosy-fingered
>   Morning should come,

and is caught by a prepared artillery fix. The sudden flashes of the guns over the horizon are like "false dawns" usurping the prerogatives of the real one, while the fiery tracer trails of the projectiles constitute a new kind of rosy fingers,

> All at one aim
> Protending and bending. . . .

When the real dawn arrives, she is shocked—indeed, shell-shocked—by what, fingers trembling now, she beholds at the point of impact:

> Nor rosy dawn at last appearing
>   Through the icy shade
> Might mark without trembling the new-deforming
> Of earth that had seemed past further storming.
>   Her fingers played
>
> One thought, with something of human pity
>   On six or seven
> Whose looks were hard to understand,
> But that they ceased to care what hand
>   Lit earth and heaven.

Another complicated dawn in Blunden is the one marking the direction of the attack in "Come On, My Lucky Lads." The poem begins with an apostrophe to dawn couched in the highest terms of traditional ode:

> O rosy red, O torrent splendor
>   Staining all the orient gloom,
> O celestial work of wonder—
>   A million mornings in one bloom!

"The artist of creation" is busy, as in Ruskin, about his twice-daily job of work. But as the attack moves out, the speaker finds such rhetoric impossible to sustain. The red dawn now modulates to a "blood burst," and before the poem is over we are forced to gaze at a less inspiring red:

> The dawn that hangs behind the goal,
>   What is that artist's joy to me?
> Here limps poor Jock with a gash in the poll,
>   His red blood now is the red I see,

—and suddenly, "literature" forgotten, the poem explodes in a line,

> The swooning white of him, and that red!

whose twisted meter and vernacular idiom measure the inadequacy of language like "orient gloom" to register what's going on.

Dawn has never recovered from what the Great War did to it. Writing four years after the Armistice, Eliot accumulates the new, modern associations of dawn: cold, the death of multitudes, insensate marching in files, battle, and corpses too shallowly interred:

> Under the brown fog of a winter dawn,
> A crowd flowed over London Bridge, so many,
> I had not thought death had undone so many.
> Sighs, short and infrequent, were exhaled,
> And each man fixed his eyes before his feet. . . .
> There I saw one I knew, and stopped him, crying: "Stetson!
> You who were with me in the ships at Mylae!
> That corpse you planted last year in your garden,
> Has it begun to sprout? Will it bloom this year?
> Or has the sudden frost disturbed its bed? . . ."

When he was touring the Somme battlefields in 1920, observing them in all lights, Stephen Graham found that full light did not properly illuminate them. "Sunlight and noonday," he says, "do not always show us truth." They suggest too insistently things like vitality and ecstasy and the soaring consciousness of complete life. "Only in the grey light of afternoon and evening, and looking with the empty eye-socket of night-darkness can one easily apprehend what is spread out here—the last landscape of tens of thousands who lie dead." [55]

A RIDICULOUS PROXIMITY

In the Second World War the common experience of soldiers was dire long-term exile at an unbridgeable distance from "home." One fought as far from home as Alsace is from California, or Guadalcanal from Manhattan, or Bengasi from Birmingham and Berlin. "Shipping out" is significantly a phrase belonging to the Second War, not the First. And once committed to the war, one stayed away until it should be over.

By contrast, what makes experience in the Great War unique and gives it a special freight of irony is the ridiculous proximity of the trenches to home. Just seventy miles from "this stinking world of sticky trickling earth" was the rich plush of London theater seats and the perfume, alcohol, and cigar smoke of the Café Royal. The avenue to these things was familiar and easy: on their two-week leaves from the front, the officers rode the same Channel boats they had known in peacetime, and the presence of the same porters and stewards ("Nice to serve you again, Sir") provided a ghastly pretence of normality. One officer on leave, observed by Arnold Bennett late in 1917, "had breakfasted in the trenches and dined in his club in London." [56]

The absurdity of it all became an obsession. One soldier spoke for everyone when he wrote home, "England is so absurdly near." [57] Another, bogged down in the Salient in September, 1917, devoted many passages of his diary to considering the anomaly: "I often think how strange it is that quiet home life is going on at Weybridge, and everywhere else in England, all the time that these terrific things are happening here." [58] Remembering the war fifty years afterwards and shaping his recollections into a novel, one participant discovers that the thing he finds especially "hard to believe" about the war is this farcical proximity. On the way back up the line on a train, one officer returning from leave remarks to others in his compartment: "Christ! . . . I was at *Chu Chin Chow* last night with my wife. Hard to believe, isn't it?" And the narrator observes:

> Hard to believe. Impossible to believe. That other life, so near in time and distance, was something led by different men. Two lives that bore no relation to each other. That was what they all felt, the bloody lot of them.[59]

And on a similar visit to the past, but this time a literal one, other observers are struck by the same "disquieting" anomaly. Visiting the battlefields forty-six years after the war, John Brophy finds it "fantastic" that so battered an area can have been restored to cultivation even after nearly half a century. This gross contrast reminds him of an even more striking one, and he goes on: "What is most disquieting on such a visit is

to realize how little space . . . separated the line, the soldier's troglodyte
world, the world which might have been another planet, from home,
from England, from sanity." [60]

There were constant reminders of just how close England was. Some
were sweetly well-intentioned, like Their Majesties' Christmas Card,
distributed to all hands in 1914. On the same occasion Princess Mary's
gifts of pipes and tobacco were much appreciated, as were the Christmas
puddings sent across by the *Daily Mirror* (as the war proceeded, known
increasingly to the troops as the *Daily Prevaricator*). At even the worst
moments, when apparently nothing could live through the shelling,
troops found the *Daily Mail* being peddled by French newsboys at the
entrances to communication trenches, and buying it was a way, if not of
finding out what was going on, at least of "feeling at home" and achiev-
ing a laugh as well. There were other anomalous apparitions of "London." During the winter of 1914–15 a familiar sight just behind the line
was a plethora of London heavy transport vehicles—brewer's trucks,
moving vans, London buses—often with their original signs intact. "You
meet them," says one observer, "toiling along the greasy road with heavy
loads of shells or food, their gay advertisements of Bass's Beer or Crosse
and Blackwell scarred and dirtied by the war." [61] Some traditional civilian comforts might find themselves in odd circumstances. Major Frank
Isherwood relied on Mothersill's seasickness preventative to mitigate the
Channel crossings,[62] and General J. L. Jack sniffed often at "a smelling
salts bottle given years ago by an old aunt without any thought of this
purpose" to disguise "the stench from the older corpses in our parapets." [63] Sassoon, perceiving that uncut wire was going to be one of the
problems on July 1, 1916, devoted his last leave just before the Somme
jump-off to shopping. At the Army and Navy Stores he bought two pair
of wire-cutters with rubber-covered handles, which he issued to his company on his return. These were, he says, "my private contribution to the
Great Offensive." [64]

The postal connection between home and the trenches was so rapid
and efficient that it constituted a further satire on the misery of the
troops in their ironic close exile. Writing on July 18, 1916, during some
of the worst moments on the Somme, one officer says:

> It is extraordinary how the post manages to reach one out here. The other
> day we were moving about all day, over trenches and back again, and when
> we had been sitting still about an hour, the post came in. We got the Lon-
> don papers only a day late, as we always do.[65]

Letters and parcels normally took about four days, sometimes only two.
Exotic foodstuffs could easily be sent across, not just standard non-

perishables like tinned kippers and oysters, tinned butter and fowl, pâté and chocolate, cheese and cherry brandy and wine (by the dozen bottles at once), but perishables like gingerbread, cakes, and tarts; fresh fruit and butter and eggs; and fresh flowers (primroses, violets) for the "table." Sometimes one took postal pot-luck. In hospital, Captain M. J. H. Drummond received a gift of eggs, on each of which was written "Wishing you good luck from Violet Palmer, Girls Council School, Thetford." [66] And only occasionally did things arrive spoilt, as once for Corporal J. L. Morgan, who sent "Annie and Art" this little report of a minor hope abridged:

> I . . . received your parcel quite safe. . . . I am sorry to say though the tart had gone bad. I was so mad as I just felt like a bit of tart then too. I think the tin box done it. I don't think a tin box is as good as a cardboard box or wood box for something in the tart line. [67]

A different kind of spoilage is what worried F. E. Smith, who had boxes of very good cigars sent from home. To prevent their being pillaged en route, he instructed his wife to

> tell the Stores not to print any indication that the boxes are cigars. Have printed yourself some gummed labels as follows:

> ┌─────────────────────────────────┐
> │   ARMY TEMPERANCE SOCIETY   │
> │   PUBLICATIONS SERIES 9   │
> └─────────────────────────────────┘

and put these and nothing else on the outside. [68]

Both Fortnum and Mason's and Harrod's specialized in gift assortments for the front, Fortnum's fruit cake being especially popular for lasting well. The wits of the *Wipers Times* thought it funny that these normalities—redolent of the well-to-do sending off hampers of treats to boys at the better public schools—should persist as if nothing had changed. They printed frequent satiric advertisements in which "Herod's, Universal Provider" and "Messrs. Shortone and Pastum" announced their Christmas boxes and described such desirable novelties as "Our Latest Improved Pattern Combination Umbrella and Wire Cutter." Sometimes excesses crept in. Brigadier F. P. Crozier recalls one officer who in 1916 "kept a sort of open mess for which he paid, but which was more like a society lunch than a serious effort to establish war hardness and endurance. Chickens, hams, jellies, wines and various rich and unwholesome foods were posted to this soldier daily by a devoted wife, or consigned from fashionable stores. It was not war." [69]

The postal service brought the troops their usual magazines: one sim-

ply indicated a change of address and the subscription continued unin-
terrupted. The standard officers' dugout required, as an index of swank,
current copies of the *Bystander*, the *Tatler*, and *Punch*, and as testimony to
upper-class interests in the pastoral and in connoiseurship, *Country Life*
and the *Burlington*. Major P. H. Pilditch received his copy of *The Wyke-
hamist* regularly. Geoffrey Keynes specialized in receiving antiquarian
booksellers' catalogs and buying books by return mail. He wrote in 1968:
"I still have a copy of Sir Thomas Browne's *Pseudodoxia Epidemica*, 1669,
containing the note: 'Sent off from Edgehill, Dernancourt, near Albert,
21 March 1918, the day of the great German offensive. It passed through
Albert on 22 March, the last day on which the post could be sent that
way.' " [70] The smallest memorials of a familiar beauty could sustain the
spirits. Committed to the Somme, Max Plowman receives a packet of
small reproductions of Renaissance paintings—Rubens, Botticelli—and
observes: "My dear in England has sent me half a dozen small Medici
prints, and I cannot describe the joy they are to me in this stricken
waste." [71] The most appropriate comment would seem to be the line
from Frost's "Directive" about children's playthings:

> Weep for what little things could make them glad.

We hear of gramophones and boxing gloves sent by mail, of fly-sprayers
and "metal waistcoats," of Zeiss field-glasses and wading boots and the
popular Harrison's Anti-Lice Pomade. One man wanted engraved visiting cards ("Oh reason not the need"), and V. de S. Pinto asked his father
to send "an indelible pencil, candles, and the works of Petronius in the
Loeb edition." [72] Twice a year regularly one artillery officer received a
pair of Abbott's Phiteezi Boots from a London shop. [73] Once when
Wilfred Owen inadvertently left two towels at home on leave, he simply
had them posted back to him in his trench. [74] The efficiency of the postal
arrangements meant that there was parcel traffic in both directions and
that souvenirs could be dispatched readily, so readily, in fact, that one
had to be careful to eschew the obvious. "Best love," Oliver Lyttelton
writes his mother: "I am sending you a German forage cap which is a
little more interesting, I think, than the hackneyed helmet." [75]

In the trenches one was absurdly near not just to England but to normal social usages. Discoursing on the effectiveness of overhead German
harassing machine-gun fire, General Sir Richard Gale notes that one
bullet came down out of nowhere and hit his orderly:

> I got him into my shelter and could see no mark, but when he turned round
> I could see the back of his tunic was dripping with blood. The bullet had
> gone right through him. . . . The bullet passed very near his lung; he was
> lucky not to have been killed.

But thanks to the social norms and the postal service,

> After he got home he sent me a lovely cake, which kind thought I appreciated very much.[76]

If the troops were constantly reminded by transmissions like these how ironically close England was, those at home were kept equally mindful of how near the trenches were. And not just by assertions to that effect by editors and politicians. They could literally hear the war, at least if they lived in Surrey, Sussex, or Kent, where the artillery was not only audible but, with the wind in the right direction, quite plainly audible. When the mines went off at Messines, not merely was the blast heard in Kent: the light flashes were visible too. The guns were heard especially during preparation for a major assault, when they would fire unremittingly for a week or ten days, day and night. Thus Edmund Blunden recalls that in late June, 1916, as the artillery strove to cut the German wire for the Somme attack, "in Southdown villages the schoolchildren sat wondering at that incessant drumming and rattling of the windows." [77] The preparation at Passchendaele was "distinctly" audible to Kipling at Burwash, Sussex, 100 miles from the guns.[78]

One reason modern English poetry can be said to begin with Hardy is that he is the first to invite into poems the sound of ominous gunfire heard across the water. There is not just "Channel Firing"; there is "The Discovery," written well before the war and published in *Satires of Circumstance* in November, 1914. The speaker remembers a love affair which occurred in an unlikely coastal setting sinister with a mysterious menace:

> I wandered to a crude coast
>   Like a ghost;
> Upon the hills I saw fires—
>   Funeral pyres
> Seemingly—and heard breaking
> Waves . . .

—waves which break like

> distant cannonades that set the land shaking.

So sinister and unpromising was the setting, so utterly brutal and "modern," that he could never have guessed that there he would find

> A Love-nest,
> Bowered and candle-lit.

The poem is a condensed redaction of Arnold's "Dover Beach," but with certain adjustments in the idea of the modern since Arnold's day. Arnold's ignorant armies are guessed at but unheard. Those implicit even in Hardy's simile are palpable, audible. They are fully imaginable, as they are in this passage too, where two people are walking on a beach very like that at Dover:

> The boots of the two . . . crunch through to sand or shingle. The very bottom of the year. They can hear the guns in Flanders today, all the way across the Channel on the wind.

We recognize it as distinctly in the tradition. Yet it was written not in 1914, nor as a recollection in 1929, but in 1973. The war is not the First but the Second; and the two who walk on the beach and listen there to the guns of Flanders are Roger Pointsman and Tyrone Slothrop, of the Special Operations Executive and of Thomas Pynchon's brilliant *Gravity's Rainbow*,[79] in whose pages persists the Great War theme—already mastered by Hardy even before the war broke out—of the ironic proximity of violence and disaster to safety, to meaning, and to love.

THE BONEYARD

Writing his sister in August, 1916, one soldier marvels at the fantastic holes and ditches which scar the whole landscape and wonders, "How ever they will get it smoothed out again is more than I can imagine."[80] The work of smoothing it out continues to this day. At first, some thought restoration of the area impossible and advised that it simply be abandoned. In 1919 the battlefields were still much as 1918 had left them, but a tourist could visit them, assisted by a series of *Illustrated Michelin Guides to the Battlefields (1914–1918)*, written in English and printed in England. Gradually the road network was re-established; craters and trenches were filled in; duds were collected and exploded; nasty things were collected and buried; and villagers began returning and rebuilding, often exactly reproducing a leveled town on its original site.

Today the Somme is a peaceful but sullen place, unforgetting and unforgiving. The people, who work largely at raising vegetables and grains, are "correct" but not friendly. To wander now over the fields destined to extrude their rusty metal fragments for centuries is to appreciate in the most intimate way the permanent reverberations of July, 1916. When the air is damp you can smell rusted iron everywhere, even though you see only wheat and barley. The farmers work the fields without joy. They collect the duds, shell-casings, fuses, and shards of old barbed wire as the plow unearths them and stack them in the corners of their fields. Some of the old barbed wire, both British and German, is used for fencing. Many of the shell craters are still there, though smoothed out and grown over. The mine craters are too deep to be filled and remain much as they were. When the sun is low in the afternoon, on the gradual slopes of the low hills you see the traces of the zig-zag of trenches. Many farmhouses have out in back one of the little British wooden huts that used to house soldiers well behind the lines; they make handy toolsheds. Lurking in every spot of undergrowth just off the beaten track are eloquent little things: rusted buckles, rounds of corroded small-arms ammunition, metal tabs from ammunition boxes, bits of Bully tin, buttons.

Albert today is one of the saddest places in France. It has all been restored to its original ugliness. The red-brick Basilica is as it was before the war, with the gilded virgin back up on top of the tower, quite erect. But despite an appearance of adequacy, everything human in Albert seems to have been permanently defeated. The inhabitants are dour. Everywhere there is an air of bitterness about being passed over by the modernity, sophistication, and affluence of modern France. Everywhere one senses a quiet fury at being condemned to live in this boneyard and backwater, where even the crops contend with soil once ruined by gas.

And a boneyard it is. Every week bones come to light. Depending on one's mood one either quietly buries them again, or flings them into the nearby brush, or saves them to turn over to the employees of the Commonwealth (formerly "Imperial") War Graves Commission, which supervises the 2,500 British military cemeteries from offices in the main cities. The cemeteries are both pretty and bizarre, fertile with roses, projecting an almost unendurably ironic peacefulness. They memorialize not just the men buried in them, but the talents for weighty public rhetoric of Rudyard Kipling. He was called on to devise almost all the verbal formulas employed by the Imperial War Graves Commission, from "Their Name Liveth For Evermore," carved on the large "Stone of Remembrance" in each cemetery, to the words incised on headstones over the bodies of the unidentified: "A Soldier of the Great War/Known unto God." The unforgettable, infinitely pathetic inscriptions are not Kipling's but those which the families of the dead were allowed—after long debate within the Commission about "uniformity"—to place on their stones. In addition to the still hopeful ones about dawn and fleeing shadows we find some which are more "modern," that is, more personal, particular, and hopeless:

Our dear Ted. He died for us.

.   .   .

Our Dick.

.   .   .

If love could have saved him he would not have died.

And some read as if refusing to play the game of memorial language at all:

A sorrow too deep for words.

The notorious Butte of Warlencourt, a fifty-foot knoll on the road to Bapaume from which the Germans strenuously held the British advance in the autumn and winter of 1916, is overgrown and silent. Crops grow right to its foot, dipping here and there to betray the persistent shapes of shell holes and mine craters. Tens of thousands of men simply disappeared here. The sticky Somme mud makes large unwieldy spheres of your shoes as you climb to the top through the thick undergrowth. At the top you can picnic, if you have the heart for it, and inspect the large weathered wooden cross erected by the Germans at the summit and apparently renewed at the end of the Second World War. On it is carved the word *Friede*.

WILL IT EVER END?

But the likelihood that peace would ever come again was often in serious doubt during the war. One did not have to be a lunatic or a particularly despondent visionary to conceive quite seriously that the war would literally never end and would become the permanent condition of mankind. The stalemate and the attrition would go on infinitely, becoming, like the telephone and the internal combustion engine, a part of the accepted atmosphere of the modern experience. Why indeed not, given the palpable irrationality of the new world? Why not, given the vociferous contempt with which peace plans were received by the patriotic majorities on both sides?

The possibility that the war might be endless began to tease the mind near the end of 1916. In November, Queen Mary hinted in a letter to Lady Mount Stephen that even her expectation of an expeditious and rational conclusion was beginning to weaken: "The length of this horrible war is most depressing. I really think it gets worse the longer it lasts." [81] In its "Review of the Year [1916]," published on New Year's Day, 1917, the *Times*, whose optimism usually made it sound like merely a government gazette, had to admit that "the year closes, as its two predecessors closed, in blood and destruction; and after 29 months of fighting, which has involved nearly all the States of Europe, anything like a definite decision seems far distant."

At the front, as might be expected, views were considerably darker. It was there, in dugouts and funk-holes, that the bulk of what were called the Neverendians could be found. R. H. Mottram remembers one pessimistic officer who, in the summer of 1917,

> roughed out the area between the "front" of that date and the Rhine, . . . and divided this by the area gained, on the average, at the Somme, Vimy and Messines. The result he multiplied by the time taken to prepare and fight those offensives, averaged again. The result he got was that, allowing for no setbacks, and providing the pace could be maintained, we should arrive at the Rhine in one hundred and eighty years.[82]

The always intelligent—and here brilliantly prophetic—Major Pilditch came to a similar conclusion with less suggestion of frivolity. Observing events on the Somme in August, 1917, he notes: "Both sides are too strong for a finish yet. God knows how long it will be at this rate. None of us will ever see its end and children still at school will have to take it over." The French and Germans on the spot thought so too. After a dinner party one night with some French officer colleagues, at which everyone "talked a bit on . . . the never failing topic, 'Quand sera-ce fini?' " Pilditch found that his companions "seemed to think the war might go on forever." [83] German prisoners interviewed by Philip Gibbs after the Somme battles agreed:

> "How will it end?" I asked [a German doctor].
> "I see no end to it," he answered. "It is the suicide of nations. . . ."
> I met other prisoners then and a year afterward who could see no end of the massacre.[84]

One way of dealing with the intolerable suspicion that the war would last forever was to make it tolerable by satire. Some chose to link the apparent infiniteness of the war with the rigidities of Haig's character. After visiting the front in 1916 George Bernard Shaw wrote of Haig: "He was, I should say, a man of chivalrous and scrupulous character. He made me feel that the war would last thirty years, and that he would carry it on irreproachably until he was superannuated." [85] Simpler kinds of humor were a help, too. A well-known Bruce Bairnsfather cartoon depicted this front-line conversation about lengths of enlistment:

" 'Ow long are you up for, Bill?"

"Seven years."

"Yer lucky—I'm duration."

This sort of fiction could always be equalled in "real life." The middle-aged batman Private Hale reports that one Porter, a colleague,

> always went about with a bent head and with a constant air of acute depression. In fact it was quite in the nature of things that he should come up to me one morning [in 1917] . . . and in his usual slow depressed voice remark to me that it was a "life-sentence." I asked him what was a life-sentence? He replied, "the War." [86]

Some humorists imagined the distant future with the war still proceeding. A poem in the *Wipers Times* for September 8, 1917, presents Tommy aged and sere some years hence, and still in the line:

> The Summer had been long and cold,
> And Intha Pink was growing old;
> He stroked his hoary snow-white beard,
> And gazed with eyes now long since bleared.

One Divisional entertainment featured a sketch titled "The Trenches, 1950," in which "A fed-up Tommy confronts an equally fed-up . . . Hun across a 'No Man's Land' shrunk in width to a couple of yards or so. All hopes of 'relief' have long since vanished." Reporting this, Major Pilditch says: "I laughed more than I have for years." [87]

All three of the great English memoirists of the war, Graves, Blunden, and Sassoon, confess to entertaining the idea or the image of the war's literally continuing forever. Graves recalls with accustomed jauntiness: "We held two irreconcilable beliefs: that the war would never end and that we would win it." [88] Newly arrived in the line, Blunden recognizes that the first question put to him, "Got any peace talk?" is entirely facetious and rhetorical. From that moment, he says, "One of the first ideas that established themselves in my inquiring mind was the prevailing sense of the endlessness of the war. No one here appeared to conceive any end of it." [89] And the persistence of the war (at least through 1936) is the theme of one of Sassoon's recurring anxiety nightmares. He reports that he undergoes this one "every two or three months":

> The War is still going on and I have got to return to the Front. I complain bitterly to myself because it hasn't stopped yet. I am worried because I can't find my active-service kit. I am worried because I have forgotten how to be an officer. I feel that I can't face it again, and sometimes I burst into tears

and say "It's no good. I can't do it." But I know that I can't escape going back, and search frantically for my lost equipment.[90]

The poet Ivor Gurney is one in whose mind such dreams finally filled all the space. He had fought with the Gloucester Regiment, and was wounded and gassed in 1917. He died in a mental hospital in 1937, where he had continued to write "war poetry," convinced that the war was still going on. And looking back at the war from an awareness provided by fifty-two years of subsequent history, William Leonard Marshall, a young "war novelist" not born until 1944, has found it appropriate that one of his Great War characters should say, "The war'll go on forever—what's the difference?" By the end of the same novel, the continuities of modern history have brought us to Auschwitz in the forties, where the theme of infinite war is again replayed. One prisoner asks another: "When do you think the war will end?" The other shakes his head. "I don't know. . . . About 1950, I suppose—perhaps never." [91]

The idea of endless war as an inevitable condition of modern life would seem to have become seriously available to the imagination around 1916. Events, never far behindhand in fleshing out the nightmares of imagination, obliged with the Spanish War, the Second World War, the Greek War, the Korean War, the Arab-Israeli War, and the Vietnam War. It was not long after the Second World War, says Alfred Kazin, that even most liberal intellectuals abandoned the hope that that war had really put an end to something. As he says, there were

> so many uncovered horrors, so many new wars on the horizon, such a continued general ominousness, that "*the* war" [that is, the Second] soon became War anywhere, anytime—War that has never ended, War as the continued experience of twentieth-century man.[92]

The 1916 image of never-ending war has about it, to be sure, a trace of the consciously whimsical and the witty hyperbolic. But there is nothing but the literal in this headline from the *New York Times* for September 1, 1972:

U. S. AIDES IN VIETNAM
SEE AN UNENDING WAR

Thus the drift of modern history domesticates the fantastic and normalizes the unspeakable. And the catastrophe that begins it is the Great War.

III
Adversary Proceedings

THE ENEMY

What we can call gross dichotomizing is a persisting imaginative habit of modern times, traceable, it would seem, to the actualities of the Great War. "We" are all here on this side; "the enemy" is over there. "We" are individuals with names and personal identities; "he" is a mere collective entity. We are visible; he is invisible. We are normal; he is grotesque. Our appurtenances are natural; his, bizarre. He is not as good as we are. Indeed, he may be like "the Turk" on the Gallipoli Peninsula, characterized by a staff officer before the British landings there as "an enemy who has never shown himself as good a fighter as the white man." [1] Nevertheless, he threatens us and must be destroyed, or, if not destroyed, contained and disarmed. Or at least patronized. "He" is the Communist's "Capitalist," Hitler's Jew, Pound's Usurer, Wyndham Lewis's Philistine, the Capitalist's Communist. He is Yeats's "rough beast" and the social menace of

> . . . unremembering hearts and heads,
> Base-born products of base beds,

as well as the collective enemy who plans the murder of Yeats's infant son,

> . . . for they know
> Of some most haughty deed or thought
> That waits upon his future days,

> And would through hatred of the bays
> Bring that to nought.

He is Faulkner's Snopeses, Auden's "trespasser" and "ragged urchin," Eliot's Sweeney and young man carbuncular, Lawrence's nice sexless Englishman, Roy Fuller's barbarian, and Anthony Burgess's Alex and his droogs. Prolonged trench warfare, with its collective isolation, its "defensiveness," and its nervous obsession with what "the other side" is up to, establishes a model of modern political, social, artistic, and psychological polarization. Prolonged trench warfare, whether enacted or remembered, fosters paranoid melodrama, which I take to be a primary mode in modern writing. Mailer, Joseph Heller, and Thomas Pynchon are examples of what I mean. The most indispensable concept underlying the energies of modern writing is that of "the enemy."

"He's got a fixed rifle on the road," warns a sergeant guide in Jones's *In Parenthesis*, and Jones notes: "He, him, his—used by us of the enemy at all times." [2] The inferred, threatening presence of "him" across the way is what seems to give significant dimension to a modern landscape. "It is curious," T. E. Hulme writes home in 1915, "how the mere fact that in a certain direction there are the German lines, seems to alter the feeling of a landscape." The presence of the enemy off on the borders of awareness feeds anxiety in the manner of the dropping-off places of medieval maps:

> You unconsciously orient things in reference to it. In peacetime, each direction on the road is as it were indifferent, it all goes on ad infinitum. But now you know that certain roads lead as it were, up to an abyss. [3]

The German line and the space behind are so remote and mysterious that actually to see any of its occupants is a shock. H. H. Cooper recalls seeing some Germans—his first—who have surrendered:

> Germans! How often we had talked of them in camp; imagined them! Even in the tumult of a few hours ago they had been distant and such very "unknown," mysterious, invisible beings. It seemed now that we were looking upon creatures who had suddenly descended from the moon. One felt "So this is the Enemy. These are the firers of those invisible shots, those venomous machine guns, all the way from Germany and here at last we meet . . . ."

Cooper closes his meditation by finding words for the ineradicable and paradoxical "otherness" of enemy terrain: it is "the 'other' mysterious, vacant yet impenetrable land." [4]

It was eerie never to see the Germans, or almost never. "One might be

weeks in the trenches," says Charles Carrington, "without seeing or
hearing anything of the enemy except his shells and bullets." One saw
the blue smoke of his breakfast bacon frying and occasionally "a flitting
figure in the distance" or "a head and shoulders, seen from the sniper's
loophole, leaping past a gap in the enemy's parapet." [5] Looking through
binoculars over the sandbags once, General Jack saw "the keen, hard face
of a German, capless and with iron-grey hair, staring in my direction
from a cleft in their sandbags 150 yards away." But by the time Jack had
turned to pick up a periscope, the German had vanished.[6] The sighting
was rare enough to record.

Living in the "other" land, "the strange land that we could not enter,
the 'garden over the wall' of the nightmare," as one soldier remembers, it
is no wonder that the enemy took on attributes of the monstrous and
grotesque. After all, this soldier goes on, polarizing vigorously, the other
world was "peopled by men whose way of thinking was totally and absolutely distinct from our own." [7] *Totally* and *absolutely*: the words are the
equivalent in rhetoric of the binary deadlock, the gross physical polarization, of the trench predicament. Sometimes the shadowy enemy resembled the vilest animals. Guy Chapman recalls: "The Boche were . . . invisible by daylight. . . . Sometimes in the valley on the right, a grey
shadow would stand for a few seconds, and then slide from sight, like a
water-rat into his hole." [8] John Easton likewise remembers the enemy
first as a shade and then as a light-colored slimy thing. During a night attack, "wraiths in spiked helmets—the cloth covers shining white in the
moonlight—were dashing for safety on all sides. . . , like disturbed
earwigs under a rotten tree stump." [9] To David Jones, the field-grey of
the German uniforms "seemed always to call up the grey wolf of Nordic
literature," although while he polarizes thus he is also aware that the
Germans must be doing so too: "It would be interesting to know," he
says, "what myth-conception our own ochre coats and saucer hats suggested to our antagonists." [10] Ernest Jünger answers his question. To the
Germans, the ochre of the British tunic appears as "a brownish-yellow
fleeting shadow," constituting one of the "signs that we puzzled over as
though they were the runes of a secret book or the spoor of some mighty
and unknown beast that came nightly to drink." [11] All this animal
figuration is the more artistic "line" equivalent of the home-front hate
promulgated by journalists like Horatio Bottomley, whose periodical
John Bull coined the term *Germ-huns* and was fond of speculating whether
the enemy was human.

Even the German dead were different from "ours." They were por-
cine. Blunden remembers the body of one forty-year-old German officer

as "pig-nosed . . . and still seeming hostile," [12] while Sassoon recalls the dead enemy's "look of butchered hostility." [13] Naturally gigantism was ascribed to the enemy, as Blunden indicates: when a German grenade-thrower was seen for an instant stretching above his trench, "he was always reported to be of gigantic stature." [14] It was this psychological illusion that lent force to Graves's early poem "Goliath and David," which ironically depicts the new German Goliath winning the encounter because, by the standards of the age, he deserves to win—he is larger and stronger.

"There's no doubt about it, they're a dirty lot of bastards," said one of George Coppard's mates after seeing a snapshot taken from a dead German:

> The picture showed a row of a dozen Jerry soldiers with their backs to the camera, sitting on a long latrine pole above a pit. Each man had his shirt pulled up, exposing his backside and genitals. All wore big grins as they looked round over their shoulders, and as if to crown the ugly sight, the *pickelhaubes* on their heads made them look a leering bunch of devils engaged on some hellish prank. . . .[15]

Everything attaching to such creatures was grotesque and inhuman. Their steel helmets were like coal scuttles and covered so much of the head that every German looked alike. "In battle, this anonymity was chilling," one survivor reports.[16] If the British helmet could be said to resemble a bowler, with its familiar associations of normality and domesticity, the German helmet looked like a "helm": it was indisputably Gothic. Where the British helmet was somehow comic, the German was "serious." As Ford's Tietjens says, "When you saw a Hun sideways he looked something: a serious proposition. Full of ferocity. A Hun up against a Tommie looked like a Holbein *landsknecht* fighting a music-hall turn. It made you feel that you were indeed a rag-time army. Rubbed it in!" [17] The German anti-gas respirators were of leather instead of fabric. Their haversacks had animal hair on them. Their trenches and dugouts had a special "heavy and clothy" smell,[18] and—no surprise—their bread was black instead of white. Their barbed wire was different, "more barbs in it and foreign-looking." [19] Even the sandbags with which they reinforced their parapets were bizarre. While the British sandbags were a dignified and uniform gunnysack color and were aligned on the parapet with some attention to the discipline of headers and stretchers, "The Bosche appeared to economize in the matter of sandbags by ransacking the countryside . . . and confiscating pillow slips, etc., anything which would hold earth." The speaker, "Uncle Harry," whose memoir is in the

Imperial War Museum, continues: "These in all sizes and colors from black to light grey were piled higglety pigglety to form the outside of his parapet." There was perhaps a rationale behind this motley: the multicolored sandbags made the German steel loopholes hard to spot. Against the uniform British sandbags the loopholes showed up as conspicuous dark ovals. Young Hugh Britling, in Wells's *Mr. Britling Sees It Through* (1916), is shot in the head through such an egregious loophole, and his father has ample cause to asperse "the stupidity of the uniform sandbags." [20]

THE *VERSUS* HABIT

The physical confrontation between "us" and "them" is an obvious figure of gross dichotomy. But less predictably the mode of gross dichotomy came to dominate perception and expression elsewhere, encouraging finally what we can call the modern *versus* habit: one thing opposed to another, not with some Hegelian hope of synthesis involving a dissolution of both extremes (that would suggest "a negotiated peace," which is anathema), but with a sense that one of the poles embodies so wicked a deficiency or flaw or perversion that its total submission is called for. When Wordsworth said of his mind in "seed-time" that it was "fostered alike by beauty and by fear," he was, as Herbert Read says, thinking of "some dialectic of beauty and fear, a process leading to a higher synthesis." [21] But with the landscape, the former domain of "beauty," ravaged and torn, and with "fear" no longer the thrill of the old Sublime but a persistent physical terror, the time-honored nineteenth-century synthesis is no longer thinkable. Such subtle overlaps and associative connections as Wordsworth had in mind will now seem hopelessly sentimental and archaic.

The sharp dividing of landscape into known and unknown, safe and hostile, is a habit no one who has fought ever entirely loses. In 1965 the veteran Charles Carrington testifies to its staying power:

> In fifty years I have never been able to rid myself of the obsession with No Man's Land and the unknown world beyond it. This side of our wire everything is familiar and every man a friend; over there, beyond their wire, is the unknown, the uncanny.

What sharpens this consciousness in 1965 is that Carrington has been confronting the Berlin Wall.[22] One of the legacies of the war is just this habit of simple distinction, simplification, and opposition. If truth is the main casualty in war, ambiguity is another. The minority fear that a civilized ambiguity and nuance were in serious jeopardy lay behind the promulgation and eager reception of Cyril Connolly's *Horizon* during the Second War; the same fear helped promote the revival of Henry James's fiction at the same time. Nuance has a hard time surviving in a world whose antitheses are as stark and whose oppositions are as absolute as they are in, say, Read's poem "Leidholz":

> We met in the night at half-past one
> between the lines.
> Leidholz shot at me
> And I at him.

There, typically, even such details as *half-past one* and *between the lines* insist upon the theme of unambiguous division. The *versus* habit derived from the war had the power to shape later recall not just of the whole war but of its tiniest incidents. Looking back on the war from 1953, Duff Cooper says,

> I was very proud of having possibly saved my life by my rudimentary knowledge of astronomy. . . . [On a night patrol] we were walking straight into the German trenches when I happened to look up at the heavens, and ordered an immediate about-turn.

This, at least, is the way he remembers it. "I had convinced myself it was so," he says, "until I came to look into [my] diary, which makes no mention of such a dramatic incident." [23] The drama of the binary here resides not in the actual event but in the postwar tendency of imagination that has imposed a dramatic binary action upon it.

The image of strict division clearly dominates the Great War conception of Time Before and Time After, especially when the mind dwells on the contrast between the prewar idyll and the wartime nastiness. Thus Blunden's dichotomous "Report on Experience," a poem whose contrasts he manages to redeem from facility by the wry mock-puzzlement of the final lines of the stanzas, as in the second and third:

> I have seen a green country, useful to the race,
> Knocked silly with guns and mines, its villages vanished,
> Even the last rat and last kestrel banished—
>   God bless us all, this was peculiar grace.
>
> I knew Seraphina; nature gave her hue,
> Glance, sympathy, note, like one from Eden.
> I saw her smile warp, heard her lyric deaden;
>   She turned to harlotry;—this I took to be new.

(Those who suspect, by the way, that the poetry of John Crowe Ransom has something to do with the Great War and its legacies can study with profit what he has derived from Blunden.)

Of course August 4, 1914, is the classical "dividing line," but the image of time abruptly divided emerges everywhere. "Passchendaele drew an abrupt dividing line across my experience," says Sir Mortimer Wheeler.[24] In Stuart Cloete's novel *How Young They Died* Jim Hilton, on his way up the line for the first time, pauses at a railway canteen. He naturally wishes that this brief respite might last forever, but "the engine blew its whistle, a sad, wailing sound that broke the night in two. It ended the canteen interlude; it divided time into the before-the-whistle and after it."[25] The dividing line is as unshaded in the experience of Wilfred Owen. Before seeing action at the Somme in January, 1917, he writes his mother, "There is a fine heroic feeling about being in France, and I am in perfect spirits." But sixteen days later everything has changed:

> I can see no excuse for deceiving you about these 4 days. I have suffered seventh hell.
> I have not been at the front.
> I have been in front of it.[26]

There is no dialectic capable of synthesizing those two moments in Owen's experience.

Another division trench warfare magnifies is that between day and night. By day, a deserted landscape; by night, frenzied activity everywhere. Cecil Lewis recalls the look of things just before the attack on the Somme:

> By day the roads were deserted; but as soon as dusk fell they were thick with transport, guns, ammunition trains, and troops, all moving up through Albert to take their positions in or behind the lines. . . . Endlessly, night after night, it went on. . . . Yet when dawn came, all signs of it were gone. There was the deserted road, the tumble-down farmhouses, the serene and silent summer mornings. Never do I remember a time when night so contradicted day.[27]

"It is curious," notices T. E. Hulme typically: "It is curious to think of the ground between the trenches, a bank which is practically never seen by anyone in the daylight, as it is only safe to move through it at dark." This odd state of things, with its curious divisions and distinctions, prompts him to project yet another distinction, one between time-now and time-later: "It is curious," he goes on, "to think of it later on in the war, when it will again be seen in the daylight."[28]

Other unambiguous distinctions encouraging the *versus* habit were of course rife in the Army itself. There was the wide, indeed gaping distinction between officers and men, emphasized not merely by separate quarters and messes and different uniforms and weapons but by different accents and dictions and syntaxes and allusions. In London an officer was forbidden to carry a parcel or ride a bus, and even in mufti—dark suit, white collar, bowler, stick—he looked identifiably different from the men. When a ten-minute break was signalled on the march, officers invariably fell out to the left side of the road, Other Ranks to the right. Army contrasts were bold, and the polarizations were always ready to flower into melodrama. An example is the practice, which became standard as the war grew worse, of leaving behind ten per cent of a battalion destined for attack. The function of the ten per cent was to rebuild the battalion by training the reinforcements if the ninety per cent should be unlucky. One can imagine the scenes of greeting after the attack between the spruce ten per cent and the others. There were many such scenes on the night of July 1, 1916, as Martin Middlebrook notes, and "these meetings were tense occasions." To depict a sample reunion, Middlebrook must rely on the standard spatial metaphor of polarization: "On the one side, the shabby survivors; on the other, the fresh smart soldiers who had escaped the holocaust." As we can imagine, "there was a gulf between the two parties. . . ." [29]

Simple antithesis everywhere. That is the atmosphere in which most poems of the Great War take place, and that is the reason for the failure of most of them as durable art. Graves has declined to reprint more than a handful of his. The title of his wartime volume of poems suggests in its melodramatic dichotomy a fatal lack of subtlety in its contents: *Fairies and Fusiliers*—pastoral and anti-pastoral, the "home" of Spenser and Shakespeare and Herrick *versus* the "France" of Haig and Ludendorff. *Ardours and Endurances*, the title of Robert Nichols's volume of 1917, is another which points in two opposite directions. Ivor Gurney's title, *Severn and Somme*, is even more explicitly polar.[30] And in a volume titled, like Sassoon's, *Counter-Attack and Other Poems*, we would be surprised to find the proceedings anything but remorselessly binary.

THE ENEMY TO THE REAR

In Sassoon's poems the oppositions become the more extreme the more he allows his focus to linger on the Staff and its gross physical, moral, and imaginative remove from the world of the troops. "Base Details" is an example, with its utterly unbridgeable distinctions between "the scar-

let Majors at the base" and the "glum heroes" of the line sacrificed to
their inept commands. The distinction Sassoon draws between the two
"sides" is fully as wide as that between "us" and "him." Looking towards
the enemy from a hilltop near Albert, Max Plowman devised this formulation: "Heaven on one side: hell on the other." [31] The two domains normally occupied by the troops and the staff invited a similar dichotomy.
Major Pilditch reports: "I saw that one humorist had painted an arrow
on the notice board marking the beginning of [the] road [leading up to
the line], and the words 'To the War,' and the reverse way, 'To
Heaven.' " [32]

Staff officers were separated from line officers the same way officers in
general were separated from Other Ranks, by uniform. Their cap-bands
and lapel-tabs were bright scarlet. The flamboyant unsuitability of this
color as camouflage served to emphasize that the place of the Staff was
distinctly not in the line. "The Red Badge of Funk" was a phrase often
on the lips of the troops. The line officers' deficiency of scarlet marked
them as a sort of high proletariat carrying always the visible stigma of
deprivation. Red tabs connoted intellectual work performed at chairs and
tables; khaki tabs, regardless of rank, connoted the work of command,
cajolement, and negotiation of the sort performed by shop foremen.
Anyone who had ever seen a factory in operation could sense what was
implied by these tabs. Anyone acquainted with the social topography of
London could arrive at the appropriate analogy: "G.H.Q. . . . had
heard of the trenches, yes, but as the West End hears of the East End—a
nasty place where common people lived." [33] The sharp division between
Staff and troops was as visible in London as at the front. Major General
Herbert Essame remembers the nightly spectacle of the leave trains in
Victoria Station waiting to start the journey back to France:

> There were six of them side by side at the departure platforms. Into five of
> them piled a great crowd of men with bulging packs on their backs to sit five
> a side in badly lit compartments: these were the regimental officers and men
> returning to the trenches.

Essame can't help noticing—and then magnifying (with "obsequious
myrmidons")—a gross dichotomy:

> In sharp contrast the sixth train was brightly lit: it had two dining cars and
> all the carriages were first class. Obsequious myrmidons . . . guided red-
> hatted and red-tabbed officers to their reserved seats. It was nearly 6:30 and
> the waiters in the dining cars were already taking orders for drinks.

Essame correctly interprets his vignette of the polarized leave trains:

The irony of this nightly demonstration at Victoria Station of the great gap between the leaders and the led, this blatant display of privilege was to rankle in the minds of the soldiers in the front line and to survive in the national memory for the next half century.[34]

It certainly survived to help sustain the General Strike of 1926, and the national memory probably retained a sufficient trace of it to help account for the "surprising" electoral reverse of 1945, which, before the Second War was even over, deposed the Tories in favor of Attlee's Labor Party.

The standard indictment brought by the troops against the Staff was that it was innocent of actual conditions on the line, and that the reason was scandalous: its funk, it was said, prevented its even approaching the front. The classic ironic narrative is the one included in almost all accounts of the protracted mud-sodden attack on Passchendaele in 1917. After weeks of frustration, the attack finally (and literally) bogged down in early November. Lieutenant General Sir Launcelot Kiggell, of the Staff, "paid his first visit to the fighting zone":

> As his staff car lurched through the swampland and neared the battleground he became more and more agitated. Finally he burst into tears and muttered, "Good God, did we really send men to fight in that?"
> The man beside him, who had been through the campaign, replied tonelessly, "It's worse further on up." [35]

As narrative that is certainly very good, although with its tears and its economical irony bringing neat retributive self-knowledge to hubris, it does sound too literary to be quite true, as if originally either conceived or noted down by someone who knew his Greek tragedy and perhaps Shakespeare's history plays as well. But it is true in spirit. During his eight months in France as a subaltern in a machine-gun company, Alec Waugh saw no officers above the rank of lieutenant colonel,[36] and Oliver Lyttelton saw only one corps commander—Sir Julian Byng—who ventured closer to the line than Brigade Headquarters.[37] Even the imaginative and humane General Plumer didn't always get close enough to know what was going on. One day in October, 1918, says General Jack, "Sir Herbert Plumer arrived in a large car with his chief staff officer . . . and asked heartily, 'Well, Jack, how are your men?' He seemed astonished to be told that they are pretty tired, but will be fit to attack again with a few days' rest." It is the old story, says Jack: " 'Plum' is most human, but . . . those who live right away from the troops engaged cannot possibly understand the strain and weariness affecting fighting troops at the front. . . ." [38]

No soldier who has fought ever entirely overcomes his disrespect for the Staff. David Jones is one in whom forty-five years after the war that disrespect is still vital and fructive. In his essay "The Utile," in *Epoch and Artist* (1959), his point is that to make art one must hurl oneself into it, get down into one's material, roll in it, snuff it up: know it, in fact, the way troops know fighting, rather than the way the Staff conjectures about it:

> Ars is adamant about one thing: she compels you to do an infantryman's job. She insists on the tactile. The artist in man is the infantryman in man. . . . all men are aboriginally *of* this infantry, though not all serve *with* this infantry. To pursue the analogy, this continued employment "away from the unit" has made habitual and widespread a "staff mentality."

Which is to say that the artist is overweighed by critics, reviewers, discussants, conjecturers, manipulators. "Today," Jones concludes, "most of us are staff-wallahs of one sort or another." [39] Graves is another who has never softened his views of the Staff and of institutions like it, as the parodic canting conclusion of "The Persian Version," with its mimicry of complacent Staff rhetoric, suggests:

> Despite a strong defence and adverse weather
> All arms combined magnificently together.

And it is tempting to believe that the oppositions between Graves's Ogres and Pygmies—the Ogres with their stinking armpits and great earthworks, the Pygmies with their cupid lips and little tassel-yards— owe something to the earlier impress on Graves of the polar opposition of troops and Staff. Representatives of the two sides are still seldom seen in each other's company. In Duff Cooper's immense biography *Haig* (London, 1935; 2 vols.), neither Graves, Blunden, nor Sassoon makes a single appearance in either text or index, which would suggest that fighting men and their reports on what they have done and seen are as irrelevant to the career of the Staff officer as is literature itself.

Here is a scene from a novel:

> The [training-camp] C.O. spoke to us on passing-out parade, neat, heavy, anonymous, standing immobile in the shade while we sweated immobile in the sun to listen to him.
> ". . . You have acquitted yourselves well. You must be proud to know that you are now fully-trained fighting machines. Your training . . . has not been wasted. It has given you experience of any conditions you are likely to meet. . . . You have been a splendid body of men to train, well worthy of the division to which you belong, well worthy of the objective for which you have been trained. . . . And I only wish I was coming with you. . . ."
> "You can have my fucking place, for one." That was Dusty Miller.
> "No talking in the ranks," [Sergeant] Meadows said.

The action is pure "Great War," say about autumn of 1916. But the
scene is from Brian W. Aldiss's *A Soldier Erect*, published in 1971; [40] the
troops are on their way not to the line in Flanders but to the Second War
fighting in Burma; and the enemy is Japanese. The Great War paradigm
proves adequate to any succeeding confrontation.

It was not just from their staffs that the troops felt estranged: it was
from everyone back in England. That division was as severe and un-
compromising as the others generating the adversary atmosphere. The
visiting of violent and if possible painful death upon the complacent, patriotic, uncomprehending, fatuous civilians at home was a favorite fantasy indulged by the troops. Sassoon, as he indicates in "Blighters,"
would like to see them crushed to death by a tank in one of their silly patriotic music halls, and in "Fight to a Finish" he enacts a similar fantasy.
The war over, the army is marching through London in a Victory
Parade, cheered by the "Yellow-Pressmen" along the way. Suddenly,
the soldiers fix bayonets and turn on the crowd:

> At last the boys had found a cushy job.

The poem ends with a further fantasy of violent retribution, this time
with Sassoon himself in the role of commander of a very modern kind of
"guerrilla" grenade attack:

> I heard the Yellow-Pressmen grunt and squeal;
> And with my trusty bombers turned and went
> To clear those Junkers out of Parliament.

The poet Charles Sorley is not joking when he writes in a letter as early
as November, 1914: "I should like so much to kill whoever was primarily
responsible for the war." [41] Philip Gibbs recalls the deep hatred of civilian England experienced by soldiers returning from leave: "They hated
the smiling women in the streets. They loathed the old men. . . . They
desired that profiteers should die by poison-gas. They prayed God to get
the Germans to send Zeppelins to England—to make the people know
what war meant." [42]

Men on leave from the front were not comforted to hear of things like
Philip and Lady Ottoline Morrell's farm, Garsington Manor, near Oxford, where, after conscription was introduced in January, 1916, numerous Bloomsbury essayists harbored as conscientious objectors performing "agricultural work." [43] Nor could they help noticing phenomena
like this newspaper poster of March, 1917:

> BATTLE RAGING AT YPRES
> GATWICK RACING—LATE WIRE [44]

But even if those at home had wanted to know the realities of the war, they couldn't have without experiencing them: its conditions were too novel, its industrialized ghastliness too unprecedented. The war would have been simply unbelievable. From the very beginning a fissure was opening between the Army and civilians. Witness the *Times* of September 29, 1914, which seriously printed for the use of the troops a collection of uplifting and noble "soldiers' songs" written by Arthur Campbell Ainger, who appeared wholly ignorant of the actual tastes in music and rhetoric of the Regular Army recently sent to France.[45]

The causes of civilian incomprehension were numerous. Few soldiers wrote the truth in letters home for fear of causing needless uneasiness. If they did ever write the truth, it was excised by company officers, who censored all outgoing mail. The press was under rigid censorship throughout the war. Only correspondents willing to file wholesome, optimistic copy were permitted to visit France, and even these were seldom allowed near the line. A typical kept correspondent was Hilaire Belloc, a favorite butt of the troops. The *Wipers Times* satirized him as "Belary Helloc," whose regular column advanced idiot suggestions for winning the war without any trouble. Another was George Adam, Paris correspondent of the *Times*. His *Behind the Scenes at the Front*, published in 1915, exudes cheer, as well as warm condescension toward the "Tommy," whom he depicts as well-fed, warm, safe, and happy—better off, indeed, than at home. One of the best-known apologists was John Masefield, whose *Gallipoli* (1918) argued not just the thrilling "adventure" but—like Dunkirk later—the triumph (at least moral triumph) of the campaign. To assist this argument Masefield prefaced each chapter with a quotation of heroic tendency from *The Song of Roland*.

The official view of the soldiers' good luck in being on the line was roughly that promulgated by Lord Northcliffe, the publisher of the *Times*, who finally assumed full charge of government propaganda. In an essay titled "What to Send 'Your Soldier,' " addressed to Americans now facing this question, he counsels the transmission of—among other comforts—peppermint bulls' eyes:

> The bulls' eyes ought to have plenty of peppermint in them, for it is the peppermint which keeps those who suck them warm on a cold night. It also has a digestive effect, though that is of small account at the front, where health is so good and indigestion hardly ever even heard of. The open-air life, the regular and plenteous feeding, the exercise, and the freedom from care and responsibility, keep the soldiers extraordinarily fit and contented.[46]

Raw data makes a sufficient comment on this. General Sir Richard Gale on the winter of 1916–17, Ypres sector: "During this winter I lost several men who were literally frozen to death." [47] Leonard Thompson, who fought at Gallipoli: "We were all lousy and we couldn't stop shitting because we had caught dysentery. We wept, not because we were frightened but because we were so dirty." [48]

One of Northcliffe's techniques was the shrewd use of domestic similes which anchored the novelties of modern war to the world of the familiar, the comfortable, and the safe. Writing late in 1917, for example, about an advanced dressing station just behind the lines (where he finds splendid work going on), he says: "Now and then we heard the brisk note of a machine-gun, which sounds for all the world like a boy rasping a stick along palings or the rattle which policemen carried in Mid-Victorian days." [49] It is no surprise to find Northcliffe's *Times* on July 3, 1916, reporting the first day's attack on the Somme with an airy confidence which could not help deepen the division between those on the spot and those at home. "Sir Douglas Haig telephoned last night," says the *Times*, "that the general situation was favorable," and the account goes on to speak of "effective progress," nay, "substantial progress." It soon ascends to the rhetoric of heroic romance: "There is a fair field and no favor, and [at the Somme] we have elected to fight out our quarrel with the Germans and to give them as much battle as they want." In short, "everything has gone well": "we got our first thrust well home, and there is every reason to be most sanguine as to the result." No wonder communication failed between the troops and those who could credit prose like that as factual testimony. The opposition between the world which could believe such an account and the real world is nicely set forth in miniature by David Jones, who at one point in *In Parenthesis* sandwiches the meaty idiom of the front between two doughy layers of upper-class home-idiom:

> Nothing is impossible nowadays my dear if only we can get the poor bleeder through the barrage and they take just as much trouble with the ordinary soldiers you know. . . .

A sensitive Other Rank, Gunner W. R. Price, came to feel that the front and home were environments so totally different that to transport the usages of the one into the other was an indecency. Price had been doing a lot of amateur photographing, but during Passchendaele he decided to abjure this hobby:

> I took a photograph of B gun and its damage [four men had been wounded] . . . , and as I did so I said to myself that this should be the end of my photography, which seemed to me was a pastime belonging to another world

. . . and that things like this . . . did not need photographing by me. So I contrived soon after to get rid of my little camera. . . .[50]

Nor did the Armistice do much to unite the vision of those who had fought with that of those who had (as they imagined) watched. With an air of having learned on the line itself how "distinct" oppositions can be, Graves and Alan Hodge put it quite unequivocally: "It must . . . be emphasized," they say, "that by the end of 1918 there were two distinct Britains: . . . the Fighting Forces, meaning literally the soldiers and sailors who had fought, as opposed to garrison and line-of-communication troops, and the Rest, including the Government." [51] But not everybody understood. Reviewing Wilfred Owen's *Poems* on January 6, 1921, the anonymous *Times Literary Supplement* writer, clearly among "the Rest," finds himself puzzled by the other side's continuing insistence on the convention of the two Britains. He says, "The suggestion [in Owen's angrier poems] is that a nation is divided into two parts, one of which talks of war and ordains it, while the other acts and suffers. We can understand how such a thought might arise but not how it can persist and find sustenance." But it readily persisted as a reaction to impudent romancing like the following, an account for American readers of the first use of British tanks at Combles on the Somme, published a year after the war was over and the survivors able to give the lie to it had all returned:

Beyond doubt the presence of the "tanks" added greatly to the zeal and confidence of the assaulting infantry. An element of sheer comedy had been introduced into the grim business of war, and comedy was dear to the heart of the British soldier. The crews of the "tanks" seemed to acquire some of the light-heartedness of the British sailor. Penned up in narrow and stuffy quarters, condemned to a form of motion compared with which that of the queasiest vessel was steady, and at the mercy of unknown perils, these adventurers faced their task with the zest of boys on holiday. With infinite humor they described how the enemy sometimes surrounded them when they were stuck, and tried in vain to crack their shell, while they themselves sat inside laughing.[52]

That is the historiographical equivalent of Sassoon's "jokes in Music-halls" which, making merry on the subject of the tank, "mock the riddled corpses round Bapaume."

The opposition between the troops and the enemy to the rear is like an odd resumption of the collision between Arnold's Educated and his Philistines, or between the Vulgarians and Aesthetes of the Nineties. Only here the role of the Sensitive is assumed by those who have been brutalized on the line, while those who have remained unscarred are the

Brutes. With fury D. H. Lawrence designated the adversary in *Kangaroo* (1923):

> We hear so much of the bravery and horrors at the front. . . . It was at home the world was lost. . . . At home stayed all the jackals, middle-aged, male and female jackals. And they bit us all. And blood-poisoning and mortification set in.
>
> We should never have let the jackals loose, and patted them on the head. They were feeding on our death all the while.[53]

THE BINARY VISION OF SIEGFRIED SASSOON

Sassoon is thus hardly unique when he polarizes this way: "The man who really endured the War at its worst was everlastingly differentiated from everyone except his fellow soldiers." [54] What is unique in Sassoon is the brilliance with which he exploits the dichotomies forced to his attention by his wartime experience and refines them until they become the very fiber of his superb memoir of the war.

One of the most flagrant of dichotomies is that between prewar and postwar Sassoon, between the "nice" unquestioning youth of good family, alternately athlete and dreamer, and the fierce moralist of 1917, surging with outrage and disdain. He was born in Kent in 1886, one of three brothers, of whom one was killed at Gallipoli in 1915. His Jewish father, who was pleased to trace the family's origins to ancient Persia, left Siegfried's mother when the boy was five, and like Wilfred Owen he grew up protected and encouraged by his mother. She had artistic and literary relatives—one of Siegfried's eccentric aunts edited the *Sunday Times*—and she knew people like Edmund Gosse and Edward Marsh, who interested themselves in the boy's poetry. The local society in which he matured was pastoral and traditional: his boyhood drawings were of hop-kilns and gardening implements, and one nearby old lady was said to have been kissed by King George IV. It was, naturally, a horsy world. Siegfried got his first horse when he was twelve.

He was sent to Marlborough College and later to Clare College, Cambridge, where he began by reading law but shifted to history. At Cambridge he was not an impressive student, and he left without taking a degree, consoling himself with the thought that Tennyson had once done the same. He returned to the quiet, orderly pleasures of well-to-do rural life—an inherited income gave him about 500 pounds a year—which did not include very strenuous intellectual operations. He spent his time in cricket, fox-hunting, book-collecting, and poeticizing. Between the ages of nineteen and twenty-six he published privately nine volumes of dreamy Keatsian and Tennysonian verse, and it is this body of work that Blunden has in mind when he observes that, oddly, "no poet of twentieth-century England, to be sure, was originally more romantic and floral than young Siegfried Sassoon from Kent." [55]

The war changed all that. Siegfried was healthy and unthinkingly patriotic, and he loved anything to do with horses. By the morning of August 5, 1914, he was in uniform as a twenty-eight-year-old cavalry trooper. In a short time he transferred to the Royal Welch Fusiliers as a Second Lieutenant of Infantry, and before long he was in action in France. Initially enthusiastic, he was very soon appalled by what he saw there. But throughout he was an extremely brave and able officer, nicknamed "Mad Jack" by his men. He was wounded in the shoulder, but after convalescence returned to the line, where, as Frank Richards remembers, he was greatly admired by the troops: "It was only once in a blue moon," says Richards, "that we had an officer like Mr. Sassoon." [56]

What Blunden calls Sassoon's "splendid war on the war" [57] took place initially in the pages of the *Cambridge Magazine*, which published many of the anti-war—or better, anti-home-front—poems collected in the volume *The Old Huntsman* in May, 1917. Two months later he issued his notorious *non serviam*, "A Soldier's Declaration." He expected to be court-martialed for this, but his friend Robert Graves managed to arrange that he face a medical board instead. It found that he needed rest, and he was sent off to a mental sanitarium in Scotland, where one of his admiring readers, Wilfred Owen, sought him out and praised his poems. Owen described him in a letter (August 22, 1917) to his friend Leslie Gunston: "He is very tall and stately, with a fine firm chisel'd (how's that?) head, ordinary short brown hair. The general expression of his face is one of boredom." Sassoon, says Owen, "is 30! Looks under 25!" And, to Owen's wonder, "Sassoon admires Thos. Hardy more than anybody living." [58] "Cured" at last and convinced now that he was abandoning his men at the front, Sassoon returned to action. In June, 1918, he brought out his final blast against the war, *Counter-Attack and Other Poems.* In July he was wounded in the head and invalided home.

By the time of the Armistice he was exhausted and trembly, sleepless and overwrought, fit for no literary work. He found peace and quiet again in Kent, but nightmares kept intruding. By 1926, however, he had recovered sufficiently to begin work on the obsessive enterprise which occupied most of the rest of his life, the re-visiting of the war and the contrasting world before the war in a series of six volumes of artful memoirs. The writing took him from one war to another: he finished the job in 1945. The first three volumes (later collected and titled *The Memoirs of*

George Sherston) comprise *Memoirs of a Fox-Hunting Man* (1928), *Memoirs of an Infantry Officer* (1930), and *Sherston's Progress* (1936). This trilogy deals with elements of his life from about 1895 to 1918. In the next volume, *The Old Century and Seven More Years* (1938), he went back to his infancy and boyhood and with less fictional concealment covered the years 1886 to 1909. In *The Weald of Youth* (1942), he focused on that part of his prewar life, the "indoor" and "London" part, which for the sake of artistic coherence he had largely suppressed in the "outdoor" *Memoirs of a Fox-Hunting Man*. In *The Weald of Youth* he depicts his hobbledehoy country self in embarrassed contact with literary and stylish London from 1909 to 1914. The final volume, *Siegfried's Journey* (1945), deals with his literary activity during the Great War and with his first two postwar years, from 1918 to 1920, from the viewpoint of the Second War. The whole autobiographical labor consumed almost twenty years of nervous concentration, during which he finally married (in 1933, at the age of forty-seven) and sired a son, George. In 1957 he became a Roman Catholic, and in 1967 he died at the age of 80. Exactly half his life he had spent plowing and re-plowing the earlier half, motivated by what—dichotomizing to the end—he calls "my queer craving to revisit the past and give the modern world the slip." [59] The life he cared to consider ran from 1895 to 1920 only.

Of all those for whom remembering the war became something like a life work, Sassoon is the one whose method of recall, selection, and expression seems to derive most directly from the polarities which the war pressed into the recesses of his mind. I am not thinking merely of the notable leisure and amplitude and luck of his prewar life—his circumstances were more comfortable than Graves's or Blunden's—although these things did provide a background against which the war could be appraised as especially monstrous. As Arthur E. Lane has observed about the method of "pointed juxtaposition" in Sassoon's poems, "Since his prewar background had been leisured and pleasant, Sassoon was all the more alert to the contrasts provided by the conditions of life at the front." [60] That is true. But I am thinking more of the grinding daily contrasts which no line-officer ever forgets: those between "his" ground and ours; the enemy and "us"; invisibility and visibility; his dead and ours; day rest and night labor; the knowledge born of the line and the ignorant innocence at home; the life on the line and the life of the Staff. To see what Sassoon does with such insistent polarities, to see the way they determine his whole lifetime mental set and become the matrices of his memory, we must turn to the three works constituting *The Memoirs of George Sherston*.

Siegfried Sassoon. (Courtesy Faber & Faber)

In *Memoirs of a Fox-Hunting Man* Sassoon has rearranged events to emphasize a polarity between his "Aunt Evelyn," with whom he lives in "Butley," Kent, and Tom Dixon, her groom. To simplify and clear the ground for this operation, he gets rid of his own parents, saying, "My father and mother died before I was capable of remembering them" (1),[61] although in *Siegfried's Journey* his mother is still living when he's thirty-four. Being raised by "Aunt Evelyn" is a way of not writing about one's real mother, a way of turning a deeply emotional actuality into something close to romantic comedy. Aunt Evelyn is comic and innocent and likable, fearful of "George" 's breaking his neck one day on those horses. The groom Dixon, on the other hand, finds his main pleasure in shrewdly slipping round Aunt Evelyn's wishes and introducing George to serious, risky horsemanship. Sherston tells us of his loneliness and his yearning for a "dream friend," whose failure to materialize made his early life "not altogether happy" (1). Although we hear of no girls at all, we hear a lot about the mare Sheila, the chestnut gelding Harkaway, and the bay gelding Cockbird.

Instructed by Dixon, George gradually masters the techniques and social polity of the local hunt meetings. Much of the time he is away at "Ballboro" School, but we never see him there, only in Kent hunting or playing cricket. His attendance at and return from Cambridge are indicated in a sentence or so, and he is back on the Weald, pursuing sport in one heady meet after another, hunting foxes and stags, and steeplechasing as well. Two young men almost assume the form of real "dream friends." There is the brilliant horseman Stephen Colwood, with whom the hero-worshipping George stays in Sussex. And there is Denis Milden, an equestrian whom George envies from a greater social distance.

Four-fifths of *Memoirs of a Fox-Hunting Man* is pastoral romance, but pastoral romance complicated by criticism from the forty-year-old Sassoon. Looking back with affectionate pity at young Sherston's innocence, and with some sorrow at his social snobbery, Sassoon manages to assay him while summarizing one particularly exciting day of hunting: "All the sanguine guesswork of youth is there, and the silliness; all the novelty of being alive and impressed by the urgency of tremendous trifles" (194).

The final fifth of the book begins in the autumn of 1914. Sherston is in the army, finding his cavalry exercises a "picnic in perfect weather" (290). But he is soon bored by the garrison life of a trooper. Through a neighbor, "Captain Huxtable," he wangles a commission in the "Royal Flintshire Fusiliers," with whom he trains at "Clitherland Camp," near Liverpool. And here his dream friend finally appears in the shape of his hutment roommate, "Dick Tiltwood," another hero—blond this time—like Colwood and Milden. Dick Tiltwood was actually Lt. David Thomas, from South Wales, described by Graves in *Good-bye to All That* as "simple, gentle, fond of reading." [62] It is while Sherston is standing with Dick that he hears by telegram of Colwood's death at the front.

Finally, in November, 1915, George and Dick's battalion makes the Channel crossing from Folkestone to Calais and undertakes the standard ironic pilgrimage up the line, starting from the infamous combat refresher course (nicknamed "the Bull Ring") at Etaples and then proceeding by train to Béthune. In trenches near Morlancourt, George hears that Sergeant Tom Dixon, formerly of Butley but now of the Veterinary Corps, has died in France—of pneumonia. The only two friends uneliminated are Dick and Denis Milden. While on a wiring party, Dick is shot through the throat and dies, leaving George devastated. And what of Milden? George hears that he's behind the lines, trying—the polarities of the memoir demand it—to get on the Staff. At this stage, says George, "I had more or less made up my mind to die" (373). The book has begun with fox-hunting: it ends with Boche-hunting. Seeking revenge for Dick, George takes out a night patrol to track down a prisoner, but returns with only a flesh-filled boot, and one belonging to a long-dead French soldier at that. The last remembered day is Easter Sunday, 1916. The last remembered moment of that day is morning stand-to.

The reader sensitive to the thematic architecture of the book will do his own remembering at this point, not least of the numerous prewar "stand-to's," delighted anticipatory dawn-watches, of young Sherston back in Kent. One is especially memorable, on the morning of his triumphant performance at the Flower-Show Cricket Match, on a day destined to bring him nothing but joy:

> I loved the early morning; it was luxurious to lie there, half-awake, and half-aware that there was a pleasantly eventful day in front of me. . . . Soon I was up and staring at the tree-tops which loomed motionless against a flushed and brightening sky. . . . There was no sound except the first chirping of the sparrows in the ivy. [Downstairs] there was the familiar photograph of "Love and Death," by Watts, with its secret meaning which I could never quite formulate in a thought, though it often touched me with a vague emotion of pathos. When I unlocked the door into the garden the early morning air met me with its cold purity. . . . The scarlet disc of the sun had climbed an inch above the hills. Thrushes and blackbirds hopped and pecked busily on the dew-soaked lawn, and a pigeon was cooing monotonously from the belt of woodland which sloped from the garden toward the Weald. . . . How little I knew of the enormous world beyond that valley and those low green hills (54–56).

If the theme of the "peaceful" part of the book had to be expressed as an infinitive, it would be *to be alive*, or *to be glad*, that is, not to know the "secret meaning" of Love and Death. That idyllic morning is typical: the first four-fifths of the book is a gathering of such separate single days, rigorously selected by Sassoon from a running diary and arranged to indicate the ways both fox-hunting and cricket constitute forms of mimic war.

Countless smaller ironies are seeded into the tiniest recesses of the book. Innocent prewar George is depicted "mooning, more or less moodily, about the looming landscape, with its creaking-cowled hop-kilns and whirring flocks of starlings and hop-poles piled in pyramids like soldiers' tents" (100). Preparing for the Flower-Show Cricket Match, George uses Blanco to clean his cricket-pads, not suspecting that he is going to become all too familiar with that product in cleaning his army web equipment. There is subtle irony in his recalling a moment when the Master of the Ringwell Hunt declared that barbed wire is "the most dangerous enemy of the hunting-man" (120); this hints the viciousness which it will later assume. One of Sherston's memorable prewar falls is occasioned by a strand of barbed wire concealed in a luxuriant hedge. We finally understand why all this barbed wire is here when we learn that it was on a wiring party that Dick Tiltwood was killed.

One large prevailing irony never overtly surfaces, but the whole course of events implies it. This is the fate of cavalry in the unanticipated war of static confrontation. The whole first part of *Memoirs of a Fox-Hunting Man* prepares us, as it prepares George, to expect for him a brilliant career in the cavalry, which was still, in the days of his youth, virtually the equivalent of "the Army." What is going to happen instead is just hinted in the little participial phrase about motoring here: "Dixon had taken [George's favorite horse] Cockbird to Downfield the day after mobilization, and had returned home just in time to interview some self-important persons who were motoring about the country requisitioning horses for the Army" (294). The joyous comings and goings of horses and riders which make the early part of the book a celebration of utterly unrestrained movement stop abruptly, to be replaced by their opposite, stasis in fixed trenches.

Memoirs of an Infantry Officer, the second volume of the trilogy, exhibits a similar overall ironic structure. But here the irony underlines not only the futility of the war, but—surprisingly—the futility of polarizing one's opposition to it. If the target of hatred in the first volume was the war itself, the target of a wonder verging on ridicule in the second is George Sherston himself, and especially his habit (from which the first volume has so notably benefited) of facile antithesis.

The dynamics of *Memoirs of an Infantry Officer* are penetration and withdrawal: repeated entrances into the center of trench experience, repeated returns to the world of "home." "So it will go on, I thought; in and out, in and out, till something happens to me" (116). From the repeated ins and outs and the contrasts they emphasize, George learns "the truth about the war": that it is ruining England and has no good reason for continuing.

At the outset of the book, in spring, 1916, George finds he must leave the trenches for four weeks and thus interrupt what he calls, because of Dick's death, "my personal grievance against the Germans" (9). He is ordered to attend a preposterous, optimistic "open-warfare school" at Flixécourt and take part in "field-day make-believe" (11). All the while Dick remains unavenged. In April George goes "in" again, rejoining the battalion near Mametz. He takes part in a raid, and, playing "the happy warrior," goes out unasked to bring in wounded, a feat for which he receives the Military Cross.

The next thing he knows, he is "out" again, back in pastoral Butley on an incongruous seven-day leave. Before returning to the line, he stops at the Army and Navy Stores and outfits himself for the Somme offensive, which everyone knows is coming. He buys the two pairs of wire-cutters as well as some salmon to take back to C Company mess, but decides that his ears can get along without a set of Mortleman's Patent Sound Absorbers. Back in again, he finds preparations for the July 1 attack well advanced. On the night of June 30 he goes out and cuts the German wire with his private wire-cutters. The day of the attack his brigade is in reserve, but he watches and records it all in his diary. It seems to him that the attack is going very well. He learns the truth later. After an encounter at Mametz Wood, where he captures a German trench quite alone, the battalion is withdrawn to rest twelve miles behind the line. There is time for George to renew acquaintance with "David Cromlech" (Robert Graves), an officer of the 2nd Battalion well known for his abrasiveness and untidiness and suspected and feared for his intelligence and originality. A few days later Cromlech's death in High Wood is reported, and it is not until two weeks later that George learns that the report is wrong—Cromlech has merely been seriously wounded.

Now George goes out of the line for seven months. He finds that he has a temperature of 105°. "Enteritis" (typhoid fever?) is diagnosed, and he moves from one hospital to another, always further back, from Amiens to Rouen to Southampton, until he discovers that he is in a hospital at Somerville College, Oxford. Visits from an elderly friend of Aunt Evelyn's, the amiable but unaware Mr. Farrell, do nothing to alter his feeling that he is becoming seriously estranged from life anywhere out of the line. Recovering slowly through the month of August, 1916, he enjoys two months of sick leave at Butley and its Trollopian environs.

The dead, especially Dixon and Stephen Colwood, are much on his mind. He does not dare think of Dick. The Somme affair is proceeding: letters from his battalion inform him of a disastrous attack on Ginchy. His friend Dottrell writes: "Only two officers got back without being hit. C. S. M. Miles and Danby both killed. The Battn. is *not now* over strength for rations!" And George says: "I walked about the room, whistling and putting the pictures [including "Love and Death"] straight. Then the gong rang for luncheon. Aunt Evelyn drew my attention to the figs, which were the best we'd had off the old tree that autumn" (135).

In early November a Medical Board in London pronounces him fit for garrison duty. He rejoins Cromlech, now recovered from his lung wound, at Clitherland Camp, where, observing training, he learns a little more about the drift of the war:

> The raw material to be trained was growing steadily worse. Most of those who came in now had joined the Army unwillingly, and there was no reason why they should find military service tolerable. The War had become undisguisedly mechanical and inhuman. What in earlier days had been drafts of volunteers were now droves of victims. I was just beginning to be aware of this (143).

In nearby Liverpool George observes the profiteers dining out, and wonders whether the war is justified. Was Dick's death justified?

Passed fit for general service in February, 1917, he rejoins his regiment but is posted to the 2nd Battalion this time. It harbors in a hateful muddy camp, where it is preparing to move up toward Arras. George notes everywhere a new cynicism about the war and observes in himself an increasing confusion of attitudes about it. His own personal poles are widening daily. As Graves says of Sassoon, "He varied between happy warrior and bitter pacifist." [63] In April George's battalion is committed to the line near Arras, and his intellectual and political awareness begins to sharpen. The sight of "an English soldier lying by the road with a horribly smashed head" triggers one of his first political perceptions. "In 1917," George says, "I was only beginning to learn that life, for the majority of the population, is an unlovely struggle against unfair odds, culminating in a cheap funeral" (201). Although he is not moved quite to Lear's

> O, I have ta'en
> Too little care of this!

he is now able to understand how for some the war is simply a continuation of "life" by other means, a mere intensification of the gross state of nature which is their normal lot. They are always engaged in what George recognizes in *Sherston's Progress* as "the 'battle of life' " (22).

He is about to be wounded and to go out of the line for what seems the last time. Sassoon prepares for this climax by setting George and a fellow officer to reminiscing about the Kentish Weald just before George's bombing party goes forward:

> Thus, while working-parties and machine-gunners filed past the door with hollow grumbling voices, our private recess in the Hindenburg Tunnel was precariously infused with evocations of rural England and we challenged our surroundings with remembrances of parish names and farmhouses with friendly faces (222).

A few minutes later George is clearing German trenches with grenades. Unwisely looking out over a parapet, he is shot through the shoulder and goes out again, through the usual sequence of dressing stations and hospitals, all the way to London. "At Charing Cross a woman handed me a bunch of flowers and a leaflet by the Bishop of London who earnestly advised me to lead a clean life and attend Holy Communion" (234). In hospital, he stokes his bitterness by gobbling up liberal reading matter like "Markingtons' " *The Unconservative Weekly* (H. W. Massingham's *The Nation*). Sent to convalesce at "Nutwood Manor," Lord and Lady Asterisk's country seat (actually Lord and Lady Brassey's, in Sussex), he grows even angrier as he contrasts the tenor of letters from his battalion with his hostess's highminded but vague Theosophical optimism.

On June 7, at the very moment the attack on Messines is in progress, he lunches—a gross polarity!—with Markington at his London club and hears him hint that Britain's war aims have now become crudely acquisitive. In consultation with "Thornton Tyrrell" (Bertrand Russell) and now brimming with zeal and self-righteousness, he composes his statement of mutiny and defiance. He launches his gesture from the recesses of Butley, where he carefully overstays his month's leave. He mails copies to newspapers and sends one direct to Clitherland Camp to "explain" his non-appearance there for re-posting. The statement itself perceptively asserts the hopeless division that has opened between the two Britains, but Sassoon's presentation of George writing it is satirical and ironic; he speaks of "working myself up into a tantrum" (276) and of convincing himself that, through his agency, "truth would be revealed" (278). The rhetoric of the statement, which busily divides all things into twos, admits of few shadings or qualifications:

> . . . I believe that the war is being deliberately prolonged by those who have the power to end it.

.   .   .

I believe that this war, upon which I entered as a war of defense and libera-
tion, has now become a war of aggression and conquest.

.   .   .

On behalf of those who are suffering now I make this protest against the
deception which is being practiced on them; also I believe that I may help to
destroy the callous complacence with which the majority of those at home
regard the continuance of agonies which they do not share, and which they
have not sufficient imagination to realize (297–98).

But what he discovers when he finally presents himself at Clitherland,
rigid with self-consciousness, is not at all the adversary proceedings he
has been soliciting. To his astonishment, everyone there is very pleasant
to him, if a bit embarrassed by his imprudence. He has expected to encounter the enemy in a melodramatic court-martial. Ironically, the authorities decline to accept the role of enemy that George insists they assume. Slipping out of the binary trap that George has contrived for
them, they convene a Special Medical Board, pronounce him shell-
shocked, and send him, with all good wishes, to a benign sanitarium.
(The convalescent David Cromlech has been their shrewd intermediary.)
Their solution is one which George has not even entertained as a possibility; it is a gentle rebuke not so much to his mutinous pacifism as to his
habit of simpleminded binary vision, and Sassoon's treatment of that
habit from a distance of thirteen years bespeaks his awareness of its intellectual and emotional and artistic limitations, even if he has invited it to
dominate the structure and texture of his memoir. Sassoon is being critical of his art as well as of his wartime character. He is pointing to the
defect both owe to their origin in something so exaggerated, mechanical,
and overwrought as the Great War.

In the final volume, *Sherston's Progress*, George makes, as he says, a
"definite approach to mental maturity" (32). He learns that the world-
wide disaster implicating him is beyond his understanding and thus
surely beyond his influence or control. A severe critic would find the
texture of *Sherston's Progress* the least artistic among the three volumes:
there are signs of impatience and fatigue. One such sign is the frequent
and sometimes embarrassingly self-conscious auctorial intrusions which
shatter the verisimilitude, like this chapter opening: "Sitting myself
down at the table to resume this laborious task after twenty-four hours'
rest, I told myself that I was 'really feeling fairly fresh again' " (54).
Another index of fatigue is Sassoon's leaving seventy-three pages in
something close to their original form as diary, without troubling to
transmute them into something closer to fiction. The book was written

in four months at the end of 1935, and Sassoon was doubtless distracted
and disheartened both by the Depression and by signs in Europe that the
Great War was about to come alight again. But whatever its weaknesses
in its smaller elements, the book in its overall structure is as tellingly
ironic as its predecessors.

It is summer, 1917. Sherston has arrived at "Slateford War Hospital"
(actually Craiglockhart War Hospital) near Edinburgh, a former lavish
hydrotherapy center with ample gardens and golf courses. At Slateford
Sherston expects to undergo a melodramatic martyrdom. What he meets
instead is the rare sympathy and humanity of Dr. (currently Captain)
W. H. R. Rivers, a distinguished fifty-year-old Royal Army Medical
Corps neurologist. From the start Sherston the pastoralist is prepared to
like him: "His name had obvious free associations with pleasant landscapes and unruffled estuaries" (3). Before Rivers has finished his sessions with George, he has become to him almost a middle-aged version
of the "dream friend." Golf, quiet, and decent food are Rivers's allies,
and by autumn, George is beginning to feel ashamed that his pacifistic
convictions have so conveniently installed him in comfort and safety.
"Reality was on the other side of the Channel, surely" (16), he begins to
believe. He is haunted by a set-piece of imagery, almost a genre picture
in the manner of Hardy's *Dynasts:*

> I visualized an endless column of marching soldiers, singing "Tipperary" on
> their way up from the back areas; I saw them filing silently along ruined
> roads, and lugging their bad boots through mud until they came to some
> shell hole and pillar-box line in a landscape where trees were stumps and
> skeletons . . . (43).

George feels at once an unbearable shame at not being there and a pressing need for a personal *coup de théâtre:* namely, his sacrificial death on
the line, which will conclusively validate the nobility of his pacifism. He
concludes that "going back to the War as soon as possible was my only
chance of peace" (44). And thirteen years after the event he is not sorry
he went back: "Much as [Rivers] disliked speeding me back to the
trenches, he realized that it was my only way out. And the longer I live
the more right I know him to have been" (67). Encouraged by Rivers,
George appears in November before a Medical Board which restores his
former status. But now that he is ready for his act of sacrifice, George is
of two minds about it: he is full one moment of self-congratulation, the
next of anger and bitterness. In May he is shipped back to France, where
as a result of the German successes of March he finds an appropriately
depressing theater for his sacrifice. "It doesn't cheer me to read [in the

Daily Mail] that 'we advanced our line a little near Morlancourt, a position of great tactical importance.' Two years ago we were living there, and it was five miles behind our Front Line" (165).

He is now a captain and company commander, and thus in a better position than as a platoon leader to father his men, to be elaborately decent to them. He is an elderly thirty-one; most of them are much younger. His affection for them surfaces especially in his notice of one of his subalterns, "Handsome boy Howitt" (actually one Lieutenant Jowitt), a "lad" who is "dark-eyed and lover-like and thoughtful" (196). It is as if the troublesome spirit of Dick Tiltwood had never been laid to rest. (This may be the place to observe that *The Memoirs of George Sherston* could easily be subtitled *Men Without Women*. We have Aunt Evelyn and Lady Asterisk, of course, but only two young women of George's age are noticed in the trilogy. Each is made the strident mouthpiece of political fatuity. Both are despised by Sherston.)

Captain Sherston is assisted by a new second-in-command, the scholarly "Velmore" (that is, Vivian de Sola Pinto, later Professor of English at the University of Nottingham). In trenches near St. Hilaire, Sherston becomes increasingly foolhardy. On July 18 he takes out an unnecessary two-man night patrol to bomb a German machine gun, persuading himself that he is "having fun." He returns at dawn, whereupon one of his own sergeants, on sentry duty, mistakes him for a German and shoots him in the head.

Despite the pain and confusion, Sherston is almost happy: he assumes that he has been killed by a German bullet and that his action of nobly returning to the front has been most artistically rounded off. But he finds that he has merely been wounded in the scalp, and ignominiously; and a few days later he is (in his view, shamefully) back in London. From a hospital bed he torments himself over the anticlimax and sets himself to despise his existence. He wants to be killed. While he is agonizing in self-contempt, Rivers saves his sanity by walking into the room:

> Without a word he sat down by the bed; and his smile was benediction enough for all I'd been through. "Oh Rivers, I've had such a funny time since I saw you last!" I exclaimed. And I understood that this was what I'd been waiting for (245).

That sounds so exactly like the last words of "a novel" that we must wonder how much of the *The Memoirs of George Sherston* is fiction.

As I have indicated, the trilogy is elaborately structured to enact the ironic redemption of a shallow fox-hunting man by terrible events. If the

delights of mindless sport are warmly and straightforwardly celebrated
in the first volume and recalled with pleasure even in the last, the intervening events make George at the end an illustration of William Shenstone's mid-eighteenth-century dictum: "The world may be divided into
people that read, people that write, people that think, and fox-hunt-
ers." [64] It is to reveal George's new capacity for understanding this—and
the cost of that understanding—that the trilogy is written. The trilogy,
that is, enforces a "moral" point; Sassoon's own life at the time makes no
moral point at all.

For during the years covered by the *Memoirs* Sassoon was most conspicuous not as a fox-hunter but as a publishing poet excited by his literary career and careful always to advance it. This was the center of his
life, but we hear nothing of it whatever in the *Memoirs*. To find out about
it we must go to *Siegfried's Journey*, where he confesses that "Sherston was
a simplified version of my 'outdoor self.' He was denied the complex advantage of being a soldier poet." [65] Although "Sherston" allows that he
may read a poem from time to time, he never discloses that Sassoon's
publishing a great many, that they are important to him, and that they
are being constantly read and discussed. The young Sassoon, in fact,
became the talk of literary London, and his *Counter-Attack* was reviewed
in the *Times Literary Supplement* by no less a figure than Virginia Woolf.
He suppresses entirely the poetic friendship with Wilfred Owen at
Craiglockhart. Owen edited *The Hydra*, the hospital magazine there, and
both published poems in it. Sherston's only mention of the magazine is
as a very occasional, desultory reader able to remember one bit of its
facetious humor. V. de S. Pinto testifies that the reason he was over-
joyed to be assigned as second-in-command to Sassoon was that he was
"the author of *The Old Huntsman*." Meeting his new company commander, the first thing Pinto said was: ". . . er, are you the poet Siegfried Sassoon?" *Counter-Attack* had just been published, and Sassoon apparently had copies with him at the front, for he gave Pinto an inscribed
copy just before they went into the line in July, 1918.[66] About all this
Sherston is silent. At one point in *Sherston's Progress* he says that he is
tempted to give the names of two novelists (Bennett and Wells, as Sassoon reveals in *Siegfried's Journey* [67]) who wrote him about his pacifist
stand while he was at Craiglockhart. But he decides not to mention the
names, explaining, ". . . somehow I feel that if I were to put them on
the page my neatly contrived little narrative would come sprawling out
of its frame" (7). I think we will not be wrong to take *frame* to suggest the
ironic structure that requires Sherston not to be a poet and not even to
be literary nor to have any access to important literary people. And I

think the reason is that Sassoon wants his contrasts to be as bold and dramatic as possible. A poet is a contrast to an infantry officer, all right; but there is an even greater contrast between a fox-hunter brimming with health and a puling neurasthenic.

I must emphatically disagree, then, with Arthur E. Lane, who says: "*The Memoirs of George Sherston* is in no way fictional, unless one balks at the use of pseudonyms such as Thornton Tyrrell (for Bertrand Russell) and David Cromlech (for Robert Graves)." [68] I would say that the *Memoirs* is in every way fictional and that it would be impossible to specify how it differs from any other novel written in the first person and based on the author's own experience—that is, from any other "first novel." The conventions of first novels disguised as memoirs are apparently not well understood: witness the innocent literalism of Wintringham N. Stable's letter to the *Daily Telegraph* of September 13, 1967. Stable, a former major in the Royal Welch Fusiliers, is puzzled by Sassoon's many egregious departures from "fact." Not expecting fiction, he can impute such departures only to lapse of memory caused by some sort of "brain disturbance":

> I rather believe . . . that [Sassoon's] head wound did set up some brain disturbance. [*Sherston's Progress*] . . . contains a number of inaccuracies in relation to events of that particular time from which I infer that his head wound in some measure must have impaired his memory.

Even if we allow that in Sassoon's *Memoirs* historical events and personal fiction do walk hand-in-hand, it is clear that what is presiding throughout is not fidelity to fact but workmanship.

And for Sassoon what workmanship means is very largely the bodying forth of contrasts in every conceivable way. It means the application of binary vision everywhere, even in the smallest details. We have encountered such a detail already in a passage from *Memoirs of an Infantry Officer* contrasting earlier and later, where the contrast is ironically supported by orderly symmetrical alliteration: "What in earlier days had been *d*rafts of *v*olunteers were now *d*roves of *v*ictims" (143). A similar if more complicated effect attends George's meditation on his future after composing his Statement: "I compared the *pr*ospect of *b*eing in a *pr*ison *c*ell with the *pr*osy *s*erenity of this *b*uzzing *s*ummer afternoon" (298). Merely to flip through the pages of the *Infantry Officer* is to encounter contrast repeatedly:

> That evening my temperature had been normal, which reminded me that this change from active service to invalidism was an acute psychological experience (118).

. . .

Parkins was an obvious contrast to this modest youth (166).

. . .

The contrast between the Front Line and the Base was an old story . . . (75).

. . .

Everything in the pretty room was an antithesis to ugliness and discomfort (249).

Sometimes a whole scene will derive its dynamics from successive revelations of things that are "funny" or "queer," that is, from a contrast between the expected and the actual. This scene offers three such revelations in rapid succession:

> One Sunday in January [1916] I got leave to go into Amiens. . . . Dick went with me. After a good lunch we inspected the Cathedral, which was a contrast to the life we had been leading. But it was crowded with sight-seeing British soldiers: the kilted "Jocks" walked up and down the nave as if they had conquered France, and I remember seeing a Japanese officer flit in with curious eyes (343).

Even theory of narrative technique comes under the binary vision, as we learn in *Sherston's Progress:* "There are two ways of telling a good story well—the quick way and the slow way" (61). And of course there are the larger structural contrasts dominating the trilogy: sane, innocent Aunt Evelyn at home, mad Europe abroad; Butley at home, Bapaume abroad; the past Sherston acting, the present Sassoon writing. For Sassoon "the line" divides everything and always will. His persisting imaginative habit stems from this memory in *Memoirs of a Fox-Hunting Man:* "The landscape was in front of us; similar in character to the one behind us, but mysterious with its unknown quality of being 'behind the Boche line' " (372).

THE PERSISTENT ENEMY

In looking at writing and at rhetorical formulations and conventions following the war, it is easy to see division everywhere, and one must be careful not to impute all of it to the impact of four years of dichotomizing. But some special ways the modern world chooses to put things do appear profoundly affected by the sense of adversary proceedings to which the war accustomed both those who had fought and those who had not. I am thinking, for example, of something so apparently remote from the war as Eliot's perception in 1921 that the concept of a "dissociation of sensibility" might now be an effective way of suggesting what happened to English poetry during the seventeenth century. In his view, thought and feeling, intellect and reflection, allied in the best work of the Metaphysicals, separated afterwards, retreating, as it were, into two camps. There would seem to be something distinctly "of the period" in Eliot's binary analytical operation, as well as in the avidity with which it was embraced. It will remind us of Hugh Selwyn Mauberley's "consciousness disjunct," a psychological phenomenon perceivable in 1920 as never before. And Auden's formulation in "It's No Use Raising a Shout" seems in its imagery to hint the origin of such disjunction:

> In my spine there was a base;
> And I knew the general's face:
> But they've severed all the wires,
> And I can't tell what the general desires.

Another phenomenon implying a special sensitivity to "division" is the post-war popularity, perhaps especially at the University of Cambridge, of the famous injunction on the title page of Forster's *Howards End,* which was published four years before the war: "Only connect." To become enthusiastic about connecting it is first necessary to perceive things as regrettably disjoined if not actively opposed and polarized. "Only connect": "It could be said," Goronwy Rees observes, "that those two words, so seductive in their simplicity, so misleading in their ambiguity, had more influence in shaping the emotional attitudes of the English governing class between the two world wars than any other single phrase in the English language." [69]

These are, one would have to admit, subtle manifestations of the staying power of the adversary habit. There are more obvious evidences of its postwar persistence and of its power long after the war to bend the mind and to determine its intellectual set. The mind of Pound is an example, with its constant truffle-snuffing for enemies and its delight in exhibiting its adversary capacities. The mind of Wyndham Lewis likewise. He was not only happy to refer to himself as "The Enemy"; he also chose to title his magazine, begun in 1927, *The Enemy: A Review of Art and Literature.* (A precursor, in its way, of the later American *Fuck You! A Magazine of the Arts.*) Conspicuous during 1914 and 1915 were the antithetical manifestos of Lewis's *Blast,* "Blasting" one set of phenomena, "Blessing" another. "BLAST HUMOR—Quack ENGLISH drug for stupidity and sleepiness . . . BLAST SPORT . . . BLAST—years 1837 to 1900." On the other hand, "BLESS the HAIRDRESSER . . . correcting the grotesque anachronisms of our physique . . . BLESS SWIFT for his solemn bleak wisdom of laughter." Adversary listings of this sort grew so stylish that after the war they were taken up by the Bright Young People, as Waugh recalls. At a chic party in *Vile Bodies* (1930) Miss Mary Mouse says, "Wasn't the invitation clever? Johnny Hoop wrote it." And Waugh appends this footnote:

> Perhaps it should be explained—there were at this time three sorts of formal invitation card; there was the nice sensible copybook hand sort with a name and *At Home* and a date and time and address; then there was the sort that came from Chelsea, *Noel and Audrey are having a little whoopee on Saturday evening: do please come and bring a bottle too, if you can;* and finally there was the sort that Johnny Hoop used to adapt from *Blast*. . . . These had two columns of close print; in one was a list of all the things Johnny hated, and in the other all the things he thought he liked. Most of the parties which Miss Mouse financed had invitations written by Johnny Hoop.[70]

Of course the war provided literary criticism with a whole armory of military images fit to suggest what it conceived as the crucial importance of its adversary proceedings. Writing *Science and Poetry* in 1926, I. A. Richards argued the apparent vulnerability of humanistic studies menaced by the enemy, "Science," which is beginning to "come into action." He conceived a vivid image of the German trenches blown up by a shocking mine, as at Messines on June 7, 1917, and—this is the interesting thing—imagined those elaborate trenches to be peopled not by the apparent enemy, scientists, but by humanists of his own persuasion:

> The most dangerous of the sciences is only now beginning to come into action. I am thinking less of Psychoanalysis or of Behaviorism than of the whole subject which includes them. It is very probable that the Hindenburg Line to which the defence of our traditions retired as a result of the onslaughts of the last century will be blown up in the near future. If this should happen a mental chaos such as man has never experienced may be expected. We shall then be thrown back, as Matthew Arnold foresaw, upon poetry. Poetry is capable [like a reserve?] of saving us. . . .[71]

Revising this passage nine years later, Richards permits himself to regard the issues as not quite so stark and vivid and momentous. The imagery of violent explosion has disappeared, and the relation now between defenders and attackers suggests a more civilized scene. Nothing is "blown up" now, and the attack results in no "chaos." Although "officially held"—marked as "ours" on staff maps—the humanistic "line" is simply abandoned and ignored by the troops on the scene. The retreating survivors of the former mine explosion are no longer "thrown back . . . upon" a new defensive position to the rear; rather, they fade away like experienced and very tired soldiers:

The most dangerous of the sciences is only now beginning to come into action. I am thinking not so much of psychoanalysis or of behaviorism as of
the whole subject which includes them.

   The Hindenburg Line to which the defence of our traditions retired as the
result of the onslaughts of the last century may still be officially held (in the
schools, for example), but it is really abandoned as worth neither defence
nor attack. The struggle is elsewhere. . . .[72]

Nine years have apparently shown Richards that the confrontation is
not so melodramatically adversary as, closer to the war, he was imagining.

But the full exaggerated imagery of battlefield confrontation has been
found indispensable by a later generation which likes to think of itself as
highly politicized, especially when questions of literature are under discussion. Thus in 1973 Anthony Thwaite delivers a radio talk on the
"Two Poetries," that is, the traditional complex kind designed to be
apprehended by a single reader in an act of special concentration and the
Pop kind designed to be heard once in coffeehouse or pub. In deploring
the adversary relation between these two kinds, Thwaite deploys the full
imagery of 1914–18: he speaks of "forces" and "the enemy"; he insists
(by using the phrase twice) that the opposition between the two poetries
constitutes "a state of deadlock," as well it might, since the practitioners
of Pop poetry think of their enemy as "entrenched authority." We hear
of "the dangerous minefield of ambiguity" and of venturing out into
"unfriendly territory." At the moment, however, the best poems are
"poems that stand above the battle." As an example of the overheated
spirit of opposition he's exposing, Thwaite quotes from the polemical
preface to *Children of Albion*, Michael Horovitz's anthology of poetry by
children. One of Horovitz's passages—perhaps the trigger of Thwaite's
imagery in rebuttal—undertakes an amalgam of older "heroic" military
figures ("Legions," "brandishing standards") with Great War images of
mine explosions and trenches hastily dug in muddy fields during re-
treats:

   At Oxford I saw budding talents buried alive, most elegantly taught to lie,
   still most persuasively cast in Eliot's calligraphy of dry bones. Legions of
   professional hollow men—brandishing standards of the New Criticism and
   New Lines—re-laid their trenches, held the muddied field and apportioned
   the spoils.[73]

The hysteria of Horovitz's first sentence, with its strained, all-but-illi-
terate pseudo-figure, "cast in . . . calligraphy of dry bones," is of course
common in critical discussion animated by convictions of political recti-

tude. As it grows politically conscious, literary discourse naturally takes
on the rhetorical characteristics of postwar political adversary proceedings. And that is fatal for it. As Correlli Barnett says of popular attitudes
toward modern international politics, "The pattern established during
the Great War is still with us: Great international questions are seen in
simple moralistic and idealistic terms, as emotional crusades." [74] It was
this habit that made the Spanish War so depressing for George Orwell.
As he says in "Inside the Whale,"

> The thing that, to me, was truly frightening about the war in Spain was not
> such violence as I witnessed, nor even the party feuds behind the lines, but
> the immediate reappearance in left-wing circles of the mental atmosphere of
> the Great War. The very people who for twenty years had sniggered over
> their own superiority to war hysteria were the ones who rushed straight
> back into the mental slum of 1915. All the familiar war-time idiocies, spy-
> hunting, orthodoxy-sniffing (Sniff, sniff. Are you a good anti-Fascist?), the
> retailing of incredible atrocity stories, came back into vogue as though the
> intervening years had never happened.

Which is to say that "by 1937 the whole of the intelligensia was mentally
at war." [75]

After such strenuousness and hazards, such sticky mud and loud reports as Thwaite's and Horovitz's, it is refreshing to turn to a wittier
tradition. It is a fact of literary history not always noticed that the year
1928, a decade after the war, is notable for two unique kinds of books:
on the one hand, the first of the war memoirs setting themselves the task
of remembering "the truth about the war"; on the other, clever novels
exhibiting a generation of bright young men at war with their elders. In
1928 we have Remarque's *All Quiet on the Western Front* and the first performance of Sheriff's *Journey's End*, as well as Sassoon's *Memoirs of a Fox-
Hunting Man*, Blunden's *Undertones of War*, Max Plowman's *A Subaltern
on the Somme*, and Hugh Quigley's *Passchendaele and the Somme*. At the
same time and on the same bookshop counters we find Aldous Huxley's
Point Counter Point and Waugh's *Decline and Fall*.

The psychology and dynamics of this interesting concurrence are suggested in Christopher Isherwood's memoir *Lions and Shadows: An Education in the Twenties* (1938), which chronicles the persistence of the binary
sense and the prevalence of the conception of "the enemy," first at school
and later at university just after the war. "The Sixth [Form]," says
Isherwood, who was fourteen years old in 1918, "was still composed of
boys who had only just missed being conscripted, potential infantry
officers trained to expect the brief violent career of the trenches: they had

outgrown their school life long before they left it." The Armistice found them fully prepared for adversary proceedings, but the accustomed outlet was now cut off: "Now, suddenly, the universal profession of soldiering was closed to them; and the alternatives seemed vague and dull" (11).[76]

A result of the subliminal persistence of "the war" was the enthusiasm with which the sensitive and the intelligent conceived of their relations with the rest of the university as a form of fighting. The lines were drawn: on one side, aesthetes, wits, subversives, and winners; on the other, dons, hearties, "stooges," and losers. It was a war very like that between troops and Staff a year or so earlier. Young Isherwood's ally in his war against "the Staff" at Cambridge was Edward Upward, whom he calls Chalmers. "The whole establishment," he says, "seemed to offer an enormous tacit bribe. We fortified ourselves against it as best we could, in the privacy of our rooms; swearing never to betray each other, never to forget the existence of 'the two sides' and their eternal, necessary state of war" (24). "We were going to open," he announces, "a monster offensive against the dons" (99).

With impressive perception and honesty Isherwood analyzes what was really motivating him: "We young writers of the middle 'twenties were all suffering, more or less subconsciously, from a feeling of shame that we hadn't been old enough to take part in the European war" (74). And as he goes on, we may catch the substance and the tone of such a Second War writer as Mailer:

> Like most of my generation, I was obsessed by a complex of terrors and longings connected with the idea "War." "War," in this purely neurotic sense, meant The Test. The test of your courage, your maturity, of your sexual prowess. "Are you really a Man?" Subconsciously, I believe, I longed to be subjected to this test; but I also dreaded failure. I dreaded failure so much—indeed, I was so certain that I *should* fail—that, consciously, I denied my longing to be tested altogether. I denied my all-consuming morbid interest in the idea of "war" (75–76).

Hence the frantically clever displacement of the idea of war onto such whimsical fantasies as Isherwood and Upward devised. The university establishment and those who believed in it became "The Combine," or "the 'Other Side,' " a conspiracy against all wit and decency; the opposed world of Isherwood and his friends was denominated—with appropriate trench imagery—"The Rats' Hostel." This fantastic confrontation ("Everywhere we encountered enemy agents" [67]) will perhaps strike us as a foreshadowing of some of the paranoid oppositions, likewise displacing the excessively raw facts of a later war, in Thomas Pynchon's *V.* and *Gravity's Rainbow*.

Isherwood's youthful literary interests sufficiently indicate the obsessive source of his adversary playfulness. Among his and "Chalmers' " favorite poets is Wilfred Owen. The title of Isherwood's early attempt at a
novel, "Lions and Shadows," derives, as he reveals, "from a passage in
Fiery Particles, by C. E. Montague (himself pre-eminently a writer of war
stories)" (75). Doubtless Isherwood also knew Montague's *Disenchantment*
(1922), one of the earliest reprehensions of the conduct of the war, of the
government's contribution to the rapid shift during the war from ideal-
ism to cynicism, and of the postwar neglect of the veterans. When Isherwood offers samples of the texture of "Lions and Shadows," the landscape of the war in France seems never very far away. For example,
"Leonard Merrows," thinking of his childhood home in England, remembers that "tiles had fallen from the coach-house roof"; we hear of
rats, and that "down by the river, where there was a gap in the garden
wall, the meadow had flowed through like a tide, in wave after wave of
poppies" (80). Striking here is the application to domestic description of
metaphors associated during the war with infantry attacks: *gap, tide,
wave*. Of these details Isherwood recalls: "The passages I have quoted
seemed to me, privately, so beautiful that, in re-reading them, I sometimes found my eyes were full of tears" (81).

Isherwood finally tells how he wanted to register his feelings about the
persisting war in a novel, to be titled "A War Memorial" or "The War
Memorial." "It was to be about war," he says. "Not the War itself, but
the effect of the idea of 'War' on my generation" (296). When he finally
published *The Memorial* in 1932, he allowed his young postwar hero Eric
to fantasize about what being in the War would have been like:

> Eric saw their life together in the training-camp, watched himself and
> Maurice drilling, being taught how to fight with bayonets, embarking on the
> troopship, cheering from French trains—Are we downhearted?—arriving in
> billets, going up along miles of communication trenches to the front line,
> waiting for the zero hour at dawn, in thin rain (144).

Like Anthony Burgess in *The Wanting Seed* (1962), Isherwood has packed
in the maximum number of Great War clichés. One way he learned
them was from a young veteran he met in Cornwall, where the disused
shafts of abandoned tin mines "looked like shell craters, surrounded with
barbed wire" (173). This veteran, "Lester," turns all conversation back to
his memories of the war:

> He never suspected, I think, how violently his quietly told horribly
> matter-of-fact anecdotes affected me. I had heard plenty of war stories be-
> fore, from older men, and the war novel was just coming into fashion; but
> Lester alone had the knack of making all those remote obscenities and hor-

rors seem real. Always, as I listened, I asked myself the same question; always I tried to picture myself in his place (255–56).

But the only war available to Isherwood is that against the dons. He sets himself to win it and to be shut of "the 'Other Side' " forever. His strategy is to be sent down by failing his Tripos; but he must fail so cleverly and scandalously that losing will be like winning. "My gesture must be in connection with the Tripos itself," he says. "That was it! I must actually go into the examination-room and write my insults as answers to the questions." Lest he somehow pass, he fills in "any possible funk-holes"—the figure of course is from the trenches—by burning his notes and disposing of his textbooks (126–27). His assault on "The Combine" must be ravishingly clever, and it is: his weapon is "concealed verses." Answering a question about the causes of the Restoration, for example, he writes in his examination book:

> When Charles the Second had, at length, returned, most Englishmen received him with relief rather than real affection. Bonfires burned to welcome back the Stuarts; and no grief was then expressed for Puritanical customs and fasts and sabbaths. Men were tired of martial law and major-general. A monarchy was all that they desired.
> And, besides this, now Oliver was dead and Richard had been tried with ill-success as Lord Protector; now that all the slow tide of opinion, gathering to a head, demanded Charles; now they were leaderless—the Roundheads were resigned to lying low (132–33).

"The 'Other Side' " is not so dull as not to recognize the effrontery of an anomalous sonnet, and Isherwood is soon having a terminal interview with his tutor. "What was there to say? . . . How was I to tell the tutor that we had often plotted to blow him sky-high with a bomb?" (134). Isherwood's adversaries will call to mind others in works continuing the tradition: the fatuous complacent elderly in Anthony Powell's novels of the thirties, for example, or figures like Professor Welch in Kingsley Amis's *Lucky Jim* (1953). The Great War Staff and "home-front" merge to assume the shape of the common enemy, persisting as club-man, don, divine, editor, industrialist, and politician. Staff-wallahs all, and hence the enemy.

In 1926 C. G. Jung had a troubling but instructive dream. He dreamed that he was "driving back from the front line with a little man, a peasant, in his horse-drawn wagon. All around us shells were exploding, and I knew that we had to push on as quickly as possible, for it was very dangerous." Then he adds this note:

The shells falling from the sky were, interpreted psychologically, missiles coming from the "other side." They were, therefore, effects emanating from the unconscious, from the shadow side of the mind.

And he goes on to posit what inspection of the writing of the twenties and thirties makes clear:

The happenings in the dream suggested that the war, which in the outer world had taken place some years before, was not yet over, but was continuing to be fought within the psyche.[77]

IV

Myth, Ritual, and Romance

A NEW WORLD OF MYTH

It was in December, 1916, that Second Lieutenant Francis Foster underwent what he later recognized as "the most momentous experience of my life." As a new and timid subaltern, he had just joined C Company of the East Lancashire Regiment on the line. His company commander sensed that his new young officer needed some steadying if he was to be of any use. Consequently, at sunset he took him on an outrageously risky impromptu stroll right out into No Man's Land, culminating in the swank of smoking cigarettes in a notoriously dangerous willow copse between the lines. Throughout, the captain is wholly phlegmatic, and Foster suddenly sloughs off his fear. "Because I was no longer fearful," he remembers, "elation filled me. But I could not understand what had caused the transformation. It was as though I had become another person altogether, or, rather as though I had entered another life." [1] A similar and equally portentous image is the resort of Max Plowman as he describes his feelings upon being relieved from a hazardous position: "It is marvellous to be out of the trenches: it is like being born again." [2] To Henry Williamson, those who "passed through the estranging remoteness of battle" were "not broken, but reborn," [3] and a similar rhetoric of Conversion dominates Ernest Parker's recall of moving up the line for the first time: "What effect this experience would have on our lives we could not imagine, but at least it was unlikely that we should survive without some sort of inner change. Towards this transmutation of our personalities we now marched." [4] The personal issues at stake in infantry warfare are so momentous that it is natural to speak of "baptism by fire." "We are still an initiate generation," says Charles Carrington, a generation possessing "a secret that can never be communicated." [5] And after the initiatory rite of baptism, there is the possibility of resurrection. Returning from an apparently hopeless patrol, Blunden says, "We were received as Lazarus was." [6] "I had been feeling much more cheerful lately," says Sassoon, "for my friend Cromlech had risen again from the dead." [7]

A world of such "secrets," conversions, metamorphoses, and rebirths is a world of reinvigorated myth. In many ways it will seem to imply a throwback way across the nineteenth and eighteenth centuries to Renaissance and medieval modes of thought and feeling. That such a myth-ridden world could take shape in the midst of a war representing a triumph of modern industrialism, materialism, and mechanism is an anomaly worth considering. The result of inexpressible terror long and inexplicably endured is not merely what Northrop Frye would call "displaced" Christianity. The result is also a plethora of very un-modern superstitions, talismans, wonders, miracles, relics, legends, and rumors.

RUMOR, FICTION, BELIEF

Rumor, "painted full of tongues," is in attendance, as Shakespeare knew, at every war. Yet the Great War seems especially fertile in rumor and legend. It was as if the general human impulse to make fictions had been dramatically unleashed by the novelty, immensity, and grotesqueness of the proceedings. The war itself was clearly a terrible invention, and any number, it seemed, could play. What Marc Bloch recalls about inverse skepticism is from his experience of the French trenches, but it is true of the British scene as well. "The prevailing opinion in the trenches," he notes, "was that anything might be true, except what was printed." From this skepticism about anything official there arose, he says, "a prodigious renewal of oral tradition, the ancient mother of myths and legends." Thus, ironically, "governments reduced the front-line soldier to the means of information and the mental state of olden times before journals, before news sheets, before books." The result was an approximation of the popular psychological atmosphere of the Middle Ages, where rumor was borne not as now by ration-parties but by itinerant "peddlars, jugglers, pilgrims, beggars." [8]

Two of the earliest and best-known legends have known originators. The Angels of Mons, reputed to have appeared in the sky during the British retreat from Mons in August, 1914, and to have safeguarded the withdrawal, developed from a short story which mentioned no angels at all. On September 29, 1914, Arthur Machen published in the *Evening News* an openly fictional romantic story, "The Bowmen," in which the ghosts of the English bowmen dead at Agincourt came to the assistance of their hard-pressed countrymen by discharging arrows which killed Germans without leaving visible wounds. Machen described these bowmen, who appeared between the two armies, as "a long line of shapes, with a shining about them." It was the *shining* that did it: within a week Machen's fictional bowmen had been transformed into real angels, and what he had written as palpable fiction was soon credited as fact. He was embarrassed and distressed at the misapprehension, but he was assured, especially by the clergy, that he was wrong: the angels—in some versions, angel bowmen—were real and had appeared in the sky near Mons. It became unpatriotic, almost treasonable, to doubt it.

The second famous early legend also derives from the inventive power of one man; this *canard* also involves the Mons retreat. The Kaiser, it was said, had referred to the British troops as "a contemptible little army." It is now known that the phrase emanated not from the German side but from the closets of British propagandists, who needed something memorable and incisive to inspirit the troops. The phrase was actually devised at the War Office by Sir Frederick Maurice and fathered upon the Kaiser.

But the other main legends and rumors are quite anonymous. No one knows who it was who contrived the German Corpse-Rendering Works, or Tallow Factory. This legend held that fats were so scarce in Germany because of the naval blockade that battlefield corpses were customarily taken back by the Germans to be rendered at special installations. The fats produced from this operation were then utilized in the manufacture of nitroglycerine, as well as candles, industrial lubricants, and boot dubbing. The legend probably originated in an intentional British mistranslation of the phrase *Kadaver Anstalt* on a captured but routine German administrative order about sending all available *cadavers*—in German, animal remains—to an installation in the rear to be reduced to tallow. An analogous legend locates the sinister Reducer or Destructor on the British side. The notorious training center at Etaples was selected as the most appropriate site of this manifestation of rampant industrialism. The rumor, like most of those circulated by the troops, has high literary quality: it concentrates in the image of The Destructor—brilliant term!—the essence of the whole war. The version encountered by Alfred M. Hale is representative:

O'Rorke . . . had said that [in Etaples] was the largest Destructor the British Army possessed. Everything that could come under the head of refuse
was brought here from over a wide area, to be reduced to ashes—even, according to a sinister report, the arms and legs of human beings. It was also
said that military executions took place here.[9]

What better place?

Like the Tallow Works, many other atrocity rumors were devised to
blacken the enemy. One is the story that the Germans used bayonets
with a saw-like edge, the better to rip open the British belly. Actually
there were such bayonets—they are to be seen in museums today—but
they were not supposed to be sadistic instruments: they were issued to
German pioneer units for sawing tree-branches, and they were carried in
addition to the regular "anti-personnel" bayonet. Such is the desire for
these bayonets to bespeak nastiness in the German character that to this
day the rumor persists that they were a specific instrument of Hun
sadism.[10] Every war gives rise to rumors of dum-dum bullets, and in this
one the Germans were said to have had recourse to such ammunition
quite instinctively. But anyone who has seen the damage done by quite
ordinary bullets fired from high-velocity military rifles—what happened
to John F. Kennedy's head is a case in point—will realize that there's no
need for special bullets. The same can be said of what ordinary bayonets
do.

Another well-known rumor imputing unique vileness to the Germans
is that of the Crucified Canadian. The usual version relates that the Germans captured a Canadian soldier and in full view of his mates exhibited
him in the open spread-eagled on a cross, his hands and feet pierced by
bayonets. He is said to have died slowly. Maple Copse, near Sanctuary
Wood in the Ypres sector, was the favorite setting. The victim was not
always a Canadian. Ian Hay, who places the incident as early as spring,
1915, maintains that the victim was British, that he was wounded when
captured, and that he was crucified on a tree by German cavalrymen,
who then "stood round him till he died." [11] A version popular in
America retains the element of the Canadian victim but—typically, some
will say—magnifies it twofold. Dalton Trumbo thus registers a moment
in the American popular consciousness: "The Los Angeles newspapers
carried a story of two young Canadian soldiers who had been crucified
by the Germans in full view of their comrades across Nomansland. That
made the Germans nothing better than animals and naturally you got interested and wanted Germany to get the tar kicked out of her." [12]

The Crucified Canadian is an especially interesting fiction both be-

cause of its original context in the insistent visual realities of the front
and because of its special symbolic suggestiveness. The image of crucifixion was always accessible at the front because of the numerous real
physical calvaries visible at French and Belgian crossroads, many of
them named Crucifix Corner. One of the most familiar terrain features
on the Somme was called Crucifix Valley after a large metal calvary that
once stood there. Perhaps the best-known calvary was the large wooden
one standing in the town cemetery at Ypres. It was famous—and to
some, miraculous—because a dud shell had lodged between the wood of
the cross and the figure of Christ (it stayed there until 1969, when the
excessively weathered crucifix was replaced). Stephen Graham was one
of many who wanted to behold miraculous power in the arrest of the
dud in that place. Writing of the cemetery in 1921, he says,

> In this acre of death the high wooden crucifix still stands, with its riven
> agonized Lord looking down. Of the hundreds of thousands of shells which
> fell in Ypres all spared Him—all but one which came direct and actually hit
> the Cross. That one did not explode but instead half-buried itself in the
> wood and remains stuck in the upright to this day—an accidental symbol of
> the power of the Cross.[13]

Roadside calvaries were not likely to go unnoticed by British pas-
sersby, not least because there was nothing like them on the Protestant
rural roads at home. Rupert Brooke reports one private saying, "What I
don't like about this 'ere Bloody Europe is all these Bloody pictures of
Jesus Christ an' 'is Relatives, be'ind Bloody bits of glawss."[14] But another reason the image of crucifixion came naturally to soldiers was that
behind the lines almost daily they could see some Other Ranks undergoing "Field Punishment No. 1" for minor infractions. This consisted of
being strapped or tied spread-eagled to some immobile object: a favorite
was the large spoked wheel of a General Service Wagon. Max Plowman
once inquires, "Wouldn't the army do well to avoid punishments which
remind men of the Crucifixion?"[15]

Reminded of the Crucifixion all the time by the ubiquitous foreign calvaries and by the spectacle of uniformed miscreants immobilized and
shamed with their arms extended, the troops readily embraced the image
as quintessentially symbolic of their own suffering and "sacrifice." Forty
years after the war Graves recalls the one-time popularity of George
Moore's *The Brook Kerith* (1916), and observes:

> It is in wartime that books about Jesus have most appeal, and *The Brook
> Kerith* first appeared forty years ago during the Battle of the Somme, when
> Christ was being invoked alike by the Germans and the Allies for victory.

. . . This paradox made most of the English soldiers serving in the purgatorial trenches lose all respect for organized Pauline religion, though still feeling a sympathetic reverence for Jesus as our fellow-sufferer. Cross-road Calvaries emphasized this relationship.[16]

The sacrificial theme, in which each soldier becomes a type of the crucified Christ, is at the heart of countless Great War poems like Robert Nichols's "Battery Moving Up to a New Position from Rest Camp: Dawn," in which the men passing a church congregation at Mass silently solicit their intercessory prayers:

> Entreat you for such hearts as break
> With the premonitory ache
> Of bodies whose feet, hands, and side
> Must soon be torn, pierced, crucified.

In Sassoon's "The Redeemer," the speaker, directing a party working with planks in a soaking trench, is struck by the resemblance of one of his men, both arms supporting his heavy planks, to Christ at Golgotha:

> He faced me, reeling in his weariness,
> Shouldering his load of planks, so hard to bear.
> I say that He was Christ. . . .
> Then the flame sank, and all grew black as pitch,
> While we began to struggle along the ditch;
> And someone flung his burden in the muck,
> Mumbling, "O Christ Almighty, now I'm stuck!"

Wilfred Owen draws the same equation between his soldiers and the Christ who approaches His Crucifixion. In a letter to Osbert Sitwell written in early July, 1918, he speaks of training new troops in England:

For 14 hours yesterday I was at work—teaching Christ to lift his cross by numbers, and how to adjust his crown; and not to imagine he thirst until after the last halt. I attended his Supper to see that there were no complaints; and inspected his feet that they should be worthy of the nails. I see to it that he is dumb, and stands at attention before his accusers. With a piece of silver I buy him every day, and with maps I make him familiar with the topography of Golgotha.[17]

(Better prose, by the way, is hardly to be found in the war, except perhaps from the hand of Blunden.) The idea of sacrifice urged some imaginations—Owen's among them—to homoeroticize the Christ-soldier analogy. Leonard Green, in his short story "In Hospital" (1920), depicts a handsome boy dying of his wounds. "His blood poured out in sacrifice," says Green, "made possible the hazardous success of the more fortunate. He was the pattern of all suffering. He was Christ. . . . He was my God, and I worshipped him." [18] All this considered, the rumor of the Crucified Canadian seems to assume an origin and a locus, as well as a meaning. He is "the pattern of all suffering." His suffering could be conceived to represent the sacrifice of all, at the same time that it was turned by propaganda into an instrument of hate. No wonder that, serving both purposes, it was a popular legend.

Most of the rumors originating in the Great War have become standard for succeeding wars. On the American line in January, 1945, it was believed that the Germans across the way had crucified an American. In 1944 and 1945 it was also fervently believed that the Germans would shoot any prisoner caught with German objects on his person. This was a replay of a rumor originating in the Great War. Once H. H. Cooper, in danger of capture, emptied his pockets: "Before going further I threw away any souvenirs I had in my pockets, German buttons and badges, coins and bullets, for if I was caught and had these things I had visions of being 'done in' by irate Jerries." [19] In both wars alike a perennial rumor was that the enemy had women in his entrenchments. The women's underwear sometimes found in dugouts was assumed to belong to the residents rather than to represent gifts destined for home by soldiers hoping for leave. And in both wars it was widely believed but never, so far as I know, proved that the French, Belgians, or Alsatians living just behind the line signaled the distant German artillery by fantastically elaborate, shrewd, and accurate means. In 1972 Stuart Cloete is still convinced that one Belgian near Ypres in 1916 signaled to the Germans by changing from time to time the position at the plow of his grey horse and his brown. Cloete believes that this Belgian was caught and shot.[20] So much plowing was always going on behind the line that it suggested a host of variations on signaling technique. Some French farmers, says Blunden, were reputed to plow arrows pointing to crucial British emplacements, as well as, in the classical way, "ploughing in view of the Germans with white or black horses on different occasions." [21] In the same way, back-of-the-line laundresses were believed to send signals by arranging their drying bedsheets in various significant patterns on the ground.

For sheer inventiveness in this line the prize would have to go to a Canadian artillery sergeant, Reginald Grant, whose fatuous and self-congratulatory *S.O.S. Stand To!* appeared while the war was going on and thus can be assumed to contain nothing at odds with official views. We can now see that the book is a virtual anthology of fables, lies, superstitions, and legends, all offered as a sober report. Sergeant Grant's problem is simple: he simply can't believe that Huns can be skilled at counter-battery location through sound and flash calculations. Seeing his own battery constantly hit by accurate counter-battery fire no matter how cleverly it moves or hides itself, he must posit some explanation. This he does by conceiving of the Belgian landscape as swarming with disloyal farmers who signal the Canadian artillery locations to the Germans by the following means: (1) windmills which suddenly turn the wrong way; (2) four white cows positioned briefly in front of the guns of the battery; (3) manipulation into anomalous positions of the hands of the clock in the village steeple; and (4) heliographic apparatus concealed in farmhouse attics. All this fantasy of folk espionage Grant projects in a frantic search for some way of explaining the disasters suffered by the Canadian artillery which will not have to acknowledge the enemy's skill in observation, mathematics, and deduction.

The Belgian station-master at Poperinghe is as subtle as his treasonable countrymen in the field. He was shot, Grant reports, when it was discovered that he had "a wire running from the station depot straight to the German lines, together with some other signaling apparatus." This he employed on one occasion to notify the German artillery that the baths at Poperinghe were at that moment full of British troops. Thirty were killed and forty wounded.[22] Again, two Belgian women lurking about British installations were seen to release a pigeon from a basket: one hour later a German shell hit a hospital, killing every single man in one of the wards.[23]

These rumors resemble much of the more formal literature of the war in that their purpose is to "make sense" of events which otherwise would seem merely accidental or calamitous. Other rumors were consolatory in function, like the popular one hinting the imminent transfer of a unit to Egypt, or later, to Italy. Or the rumor Stanley Casson remembers, that "no shell ever bursts in a hole made by one of its predecessors." He adds: "For so we fondly believed." [24] Some rumors were witty and sardonic, like the one maintaining that the British Army paid the French Government rent for the use of the trenches; or the famous one—a dramatization of the conviction that the war would continue forever—that the end of the war would be signaled by four black or dark blue Very lights shot up into the night sky.

Still other rumors developed into fully fleshed narrative fictions, almost short stories. One of the best of these, bred by anxiety as well as by the need to find a simple cause for the failure of British attacks, is the legend of the ghostly German officer-spy who appears in the British trenches just before an attack. He is most frequently depicted as a major.

No one sees him come or go. He is never captured, although no one ever sees him return to the German lines. The mystery is never solved. The version Blunden retails in *Undertones of War* includes all the classic details, not omitting the significant giveaway deviation in uniform (at this time the British were all wearing steel helmets in the line):

> A stranger in a soft cap and a trench coat approached, and asked me the way to the German lines. This visitor facing the east was white-faced as a ghost, and I liked neither his soft cap nor the mackintosh nor the right hand concealed under his coat. I, too, felt myself grow pale, and I thought it as well to direct him down the communication trench, . . . at that juncture deserted; he scanned me, deliberately, and quickly went on. Who he was, I have never explained to myself; but in two minutes the barrage was due, and his chances of doing us harm (I thought he must be a spy) were all gone.[25]

George Coppard's odd major also arouses suspicion by an irregularity of uniform:

> I remember during the Loos battle seeing a very military-looking major complete with monocle, and wearing a white collar. He asked me the way to Hay Alley and spoke good English. I never suspected that anything was wrong, though I was puzzled about his collar, as all our officers were then wearing khaki collars. Shortly after there was a scare, and officers dashed about trying to find the gallant major, but he had vanished.[26]

Since Reginald Grant is in the artillery rather than the infantry, his wraithlike spy must appear at a battery position instead of in a trench, asking pointed questions about the guns and their ammunition. He is of course a major, accompanied by a captain.

For the student of folklore there is much interest in watching the officer-spy-in-the-trenches legend mutate into other narrative types. An example is the legend of the lunatic inventor nicknamed The Admiral, which retains two elements of its officer-spy original or analog: the irregularities of uniform and the odd total disappearance of the protagonist. As Henry Williamson tells the story, a familiar but mysterious figure behind the lines was an eccentric officer called The Admiral because he wore both an Army captain's insignia and the wavy cuff stripes of an officer in the Royal Naval Volunteer Reserve. His passion was inventions. He devised a steel body-shield which did not work, and died finally by blundering into his own cunning booby-trap. Says Williamson: "His body was found (according to rumor) half a mile away. . . . Half a mile! It seems a long way to be wafted. Anyhow, the Admiral and his fearful toys were never seen again. Some said he was a spy, and had gone through the German lines again when he had got all the information he required. . . ." [27]

The finest legend of the war, the most brilliant in literary invention and execution as well as the richest in symbolic suggestion, has something of this fantastic quality. It is a masterpiece. The rumor was that somewhere between the lines a battalion-sized (some said regiment-sized) group of half-crazed deserters from all the armies, friend and enemy alike, harbored underground in abandoned trenches and dugouts and caves, living in amity and emerging at night to pillage corpses and gather food and drink. This horde of wild men lived underground for years and finally grew so large and rapacious and unredeemable that it had to be exterminated. Osbert Sitwell was well acquainted with the story. He says that the deserters included French, Italians, Germans, Austrians, Australians, Englishmen, and Canadians; they lived

> —at least they *lived*—in caves and grottoes under certain parts of the front line. . . . They would issue forth, it was said, from their secret lairs, after each of the interminable checkmate battles, to rob the dying of their few possessions. . . . Were these bearded figures, shambling in rags and patched uniforms . . . were they a myth created by suffering among the wounded, as a result of pain, privation, and exposure, or did they exist? . . . It is difficult to tell. At any rate, the story was widely believed among the troops, who maintained that the General Staff could find no way of dealing with these bandits until the war was over, and that in the end they had to be gassed. [28]

In some versions the ghouls are even wilder. Ardern Beaman tells of meeting a salvage company at work on the battlefields of the Somme, where the "warren of trenches and dugouts extended for untold miles":

> They warned us, if we insisted on going further in, not to let any man go singly, but only in strong parties, as the Golgotha was peopled with wild men, British, French, Australian, German deserters, who lived there underground, like ghouls among the mouldering dead, and who came out at nights to plunder and kill.

Beaman's details are telling:

> In the night, an officer said, mingled with the snarling of carrion dogs, they often heard inhuman cries and rifle-shots coming from that awful wilderness.

And he concludes with a vignette whose tone seems to evoke the style less of the First World War than of the Second:

> Once . . . the Salvage Company had put out, as a trap, a basket containing
> food, tobacco, and a bottle of whisky. But the following morning they found
> the bait untouched, and a note in the basket, "Nothing doing!" [29]

That sounds like Joseph Heller or Thomas Pynchon. Pynchon, indeed,
has refracted the legend of the wild deserters in *Gravity's Rainbow*, adapt-
ing it to the special plausibilities of the Second War. The ghoulish wild
men have now metamorphosed into wild dogs, who, immediately after
the war, occupy a German village:

> One village in Mecklenburg has been taken over by army dogs, . . . each
> one conditioned to kill on sight any human except the one who trained him.
> But the trainers are dead men now, or lost. The dogs have gone out in
> packs. . . . They've broken into supply depots Rin-Tin-Tin style and
> looted K-rations, frozen hamburger, cartons of candy bars. . . . Someday
> G-5 might send in troops. [30]

One reason the legend of the wild deserters is so rich is that it gathers
and unifies the maximum number of meaningful emotional motifs. For
one thing, it offers a virtual mirror image, and a highly sardonic one, of
real, orderly trench life, in which, for example, night was the time for
"work." For another, it projects the universal feeling of shame about
abandoning the wounded to spend nights suffering alone between the
lines. It embodies in objectified dramatic images the universal fantasy—
the Huckleberry Finn daydream—of flagrant disobedience to authority.
It conveys the point that German and British are not enemies: the enemy
of both is the War. And finally, it enacts in unforgettable terms a feeling
inescapable in the trenches—that "normal" life there was equal to out-
right bestiality and madness.

One would have to be mad, or close to it, to credit talismans, in the
first quarter of the twentieth century, with the power to deflect bullets
and shell fragments. And yet no front-line soldier or officer was without
his amulet, and every tunic pocket became a reliquary. Lucky coins, buttons, dried flowers, hair cuttings, New Testaments, pebbles from home,
medals of St. Christopher and St. George, childhood dolls and teddy
bears, poems or Scripture verses written out and worn in a small bag
around the neck like a phylactery, Sassoon's fire-opal—so urgent was the
need that no talisman was too absurd. And sometimes luck depended not
on what one carried but on what one did, or refrained from doing. Robert Graves asserts that he regarded the preservation of his virginity essential to his survival at the front, and imputes to his continence his remarkable good fortune in surviving for many months when the average
front-line life of a "wart" was only six weeks. [31] Philip Gibbs met a colo-

nel of the North Staffordshires who believed that it was his will power
that warded off flying metal. He told Gibbs:

> "I have a mystical power. Nothing will ever hit me as long as I keep that
> power which comes from faith. It is a question of absolute belief in the dom-
> ination of mind over matter. I go through any barrage unscathed because
> my will is strong enough to turn aside explosive shells and machine-gun
> bullets. As matter they must obey my intelligence. They are powerless to
> resist the mind of a man in touch with the Universal Spirit. . . ."

"He spoke quietly and soberly," says Gibbs, "in a matter-of-fact way. I
decided that he was mad." [32]

THREES

As we have seen, there are three separate lines of trenches: front, support, reserve. A battalion normally spent a third of its duty time in each,
and the routine in each line was similar: the unit was divided into three
groups, one of which kept alert while the other two stood down. "Day
and night," says Max Plowman, "we have three men to every bay: one
on sentry while the other two rest or sleep." [33] The daily pattern was
similar in the artillery: an officer would spend one day, the most dangerous, as forward observing officer; one day, the next most dangerous, firing with the battery; and one day, the least dangerous, in the rear,
supervising the transport of ammunition and supplies. Even the U-boats
observed the ubiquitous pattern of threes: one-third were actively pa-
trolling while a third were moving to or from duty stations and a third
underwent repair or refitting. After endless months and years of participating in such tripartite ways of dividing things, it was natural to see everything as divisible by threes. "I had become very sensitive to atmosphere, to the three zones of war," says Stuart Cloete, projecting the
pattern of front, support, and reserve lines onto areas well behind the
line. The three zones he distinguishes are

> the line where there were only fighting men; the next zone that was semi-
> immune to shellfire, where there were the ancillary services, Army Service
> Corps, casualty clearing stations, horse lines, and possibly some heavy guns.
> There were also some civilians and one could buy food, wine, and women
> . . . ; and finally the back areas peopled by old men, cripples, children and
> virtuous women. This arrangement must have applied equally to the Ger-
> man side.[34]

In Cloete's mind "the sequence of desires" in those just relieved from the
front was also triadic: "First sleep. . . . Then food. And only then a

woman—when they had been rested and fed. That was something the bloody civilians never knew. The sequence." [35]

Even colors formalized themselves into threes. One observer described the attack on Gommecourt on July 1, 1916, this way: "Everything stood still for a second, as a panorama painted with three colors—the white of the smoke, the red of the shrapnel and blood, the green of the grass." [36] White, red, and green can turn (sardonically) to red, white, and blue once the element of temporal sequence is added. A facetious poem by "R.W.M." titled "Tricolor" in the *Wipers Times* for December, 1918, finds that when wounded the soldier sees red; in hospital, he sees white; and released as ambulatory, he sees blue (the color of the official invalid bathrobes). When John Ball, the just-wounded soldier of David Jones's *In Parenthesis*, entertains a fantasy of the hospital at home, it comes out in a similar tripartite way, and the vision of the three colors seems then to trigger a further triadic division, this time of the setting:

> Mrs. Willy Hartington has learned to draw sheets and so has
> Miss Melpomené; and on the south lawns,
> men walk in red white and blue
> under the cedars
> and by every green tree
> and beside comfortable waters.

For the poet Charles Sorley the transformation of man into corpse is a three-part action. First man; then, when hit, animal, writhing and thrashing in articulate agony or making horrible snoring noises; then a "thing." Bringing in wounded, Sorley observes in himself and his fellows "the horrible thankfulness when one sees that the next man is dead: 'We won't have to *carry* him in under fire, thank God; dragging will do.' " And he notes "the relief that the thing has ceased to groan: that the bullet or bomb that made the man an animal has now made the animal a corpse." [37]

It could be said that it was the habit of thinking strictly by threes, and of considering the three land combat arms, infantry, artillery, and cavalry, as entirely distinct from each other, that contributed to the frustration of Haig's and General Sir Henry Rawlinson's plan of attack on the Somme. The order of events was planned to be as precisely sequential as Cloete understands men's desires to be once out of the line. First the artillery was to perform while the other two elements did nothing; then the infantry was to take over; once it penetrated, it was to stand fast and allow the cavalry to pass through it for the pursuit. It was not until two weeks after the initial attack that any major change was made in this rigorous three-part conception: only on July 14 was artillery used simultaneously with an advance by the infantry, which was allowed to follow a creeping barrage.

Counting off by threes: no soldier ever forgets it or its often portentous implications. "At night, we numbered off," says Frank Richards, "one, two, three, one, two, three—ones up on sentry, twos and threes working." [38] That was in 1914. In 1943 Guy Sajer, an Alsatian fighting with the German Army on the Eastern Front, was in a company facing Russian infantry defending a factory building. By telephone came down the order: "One-third of the men forward. Count off by threes":

> One, two, three . . . One, two, three. Like a miracle from heaven I drew a "one," and could stay in that splendid cement hole. . . . I cut off a smile, in case the sergeant should notice and send me onto the field. . . .
>
> The fellow beside me had number three. He was looking at me with a long desperate face, but I kept my eyes turned front, so he wouldn't notice my joy and relief, and stared at the factory as if it were I who was going to leap forward, as if I were number three. . . . The sergeant made his fatal gesture, and the . . . soldier beside me sprang from his shelter with a hundred others.
>
> Immediately, we heard the sound of Russian automatic weapons. Before vanishing to the bottom of my hole I saw the impact of the bullets raising little fountains of dust all along the route of my recent companion, who would never again contemplate the implications of number three. [39]

What we must consider now is the relation between this practical, *ad hoc*, empirical principle of three in military procedure and the magical or mystical threes of myth, epic, drama, ritual, romance, folklore, prophecy, and religion. In the prevailing atmosphere of anxiety, the military threes take on a quality of the mythical or prophetic. The well-known triads of traditional myth and ritual donate, as it were, some of their meanings and implications to the military threes. The result is that military action becomes elevated to the level of myth and imbued with much of its portent.

The tripartite vision is so ancient in Indo-European myth, religion, and folklore that there is no tracking it to its origins. [40] By the time of written documents it already has an infinite history. To Pythagoras, three is the perfect number, implying beginning, middle, and end. It is an attribute of deity, and ultimately of the Trinity. Graves's primordial White Goddess is a triple deity, presiding over sky, earth, and underworld. Greek religion recognizes the tripartite rule of the universe and assigns a deity to each part: Jupiter to heaven, Neptune to the sea,

Pluto to Hades. Each is equipped with his triadic emblem: Jupiter wields his three-forked lightning bolts and Neptune his trident, while Pluto is accompanied by his three-headed dog. It is on a tripod that an oracle sits above her gas-vent, and those who share her world and credit her visions, like Oedipus, are dominated by threes: the Sphinx's riddle solved by Oedipus divides the stages of human life into three, and the crucial place where Oedipus slays Laius is one where three roads meet. In early Christianity the enemies are three: the World, the Flesh, and the Devil, just as the virtues are three: Faith, Hope, and Charity. And in adjacent mythologies there are three Furies, three Graces, and three Harpies, Norns, or Weird Sisters.

As Northrop Frye points out, "A threefold structure is repeated in many features of romance—in the frequency, for instance, with which the successful hero is the third son, or the third to undertake the quest, or successful on his third attempt. It is shown more directly in the three-day rhythm of death, disappearance and revival which is found in the myth of Attis and other dying gods, and has been incorporated in our Easter." In the First Book of *The Faerie Queene*, "the battle with the dragon lasts, of course, three days." [41] And drama which is close to folklore and romance sources seems to behave in a similar triadic way. Observing that Shakespeare's plays generally exhibit distinct beginnings, middles, and ends, Maurice Charney finds that in such traditional drama "there is an implicit assumption that human experience, which supplies the plots for plays, also has beginnings, middles, and ends, and is causative, rational, progressive, and triadic in structure." [42] Traditional in pre-romantic thought—which would extend from 1789 all the way back to the fifth or sixth millennium B.C.—is the understanding of human life as meaningfully tripartite: each of the three stages, youth, maturity, and old age, imposes unique, untransferable duties, and each offers unique privileges and pleasures. Bent in a Christian direction, the classical concept generates the three stages of Christian experience: Innocence, Fall, Redemption. Laid "laterally," as it were, over the experience of infantry soldiers, in the Great or any other War, it produces, as we shall see, the structure of the paradigmatic war memoir: training, "combat," recovery. Or innocence, death, rebirth.

We can get a sense of the way traditional triadic meanings visit the practical threes of military experience and attach special import to them by looking at one of Ivor Gurney's poems, "Ballad of the Three Spectres." It is about what a folklorist would recognize as the Weird Sisters, and just as in Guy Sajer's ominous counting-off, the number three is the sinister, magical number:

> As I went up by Ovillers,
>   In mud and water cold to the knee,
> There went three jeering, fleering spectres,
>   That walked abreast and talked of me.
>
> The first said, "Here's a right brave soldier
>   That walks the dark unfearingly;
> Soon he'll come back on a fine stretcher,
>   And laughing for a nice Blighty."
>
> The second, "Read his face, old comrade,
>   No kind of lucky chance I see;
> One day he'll freeze in mud to the marrow,
>   Then look his last on Picardie."
>
> Though bitter the word of these first twain,
>   Curses the third spat venomously:
> "He'll stay untouched till the war's last dawning,
>   Then live one hour of agony."

Typically, the last hour of dawn, the last stand-to of the war, is the portentous, ironic moment chosen by the Third Spectre for the "right brave soldier's" fatal wound. The final stanza assures us that

> Liars the first two were . . .

and invites us to contemplate the soldier's year-long agonizing wait for the prophecy of the third to fulfill itself. While waiting, the soldier constantly recalls the prophecy while performing armsdrill "by the numbers," that is, while literally counting to three:

> Behold me
> At sloping arms by one—two—three;
> Waiting the time I shall discover
> Whether the third spoke verity.

The daily counting to three used to be only practical, mnemonic. It is now prophetic, or both practical and prophetic at once. It has become invested with myth.

One rumor circulating during the war suggests the naturalness with which the traditional threes of folktale could be aligned with a war having so much of the tripartite about it. This rumor, Stephen Southwold reports.

> told of a babe born into a Welsh farm up in the hills. Twenty-four hours
> after the child's birth, the nurse, in washing up, dropped a plate, and the

child cried out, "One." . . . Some few seconds later a second crock was dropped, whereat the infant cried gravely, "Two." Finally a third article fell crashing, and the babe in a great voice that filled the house with a rushing sound cried, "Three; in three days the war will end," and incontinently expired. Weeks, months, years were variants, according to the pessimism of the teller, but the three was invariable.[43]

The threefold actions of Gurney's spectres and Southwold's "babe" resemble in miniature the structure of threes in larger mythic and folk narratives. As Frye reminds us, a standard "quest" has three stages: first, "the stage of the perilous journey and the preliminary minor adventures"; second, "the crucial struggle, usually some kind of battle in which either the hero or his foe, or both, must die"; and third, "the exaltation of the hero, who has clearly proved himself to be a hero even if he does not survive the conflict." [44] It is impossible not be be struck by the similarity between this conventional "romance" pattern and the standard experience re-enacted and formalized in memoirs of the war. First the perilous journey, by both water and land, through the Bull Ring at Etaples and up to the ever more menacing line itself; second, the "crucial struggle" of attack or defense or attrition in the trenches; and third, the apotheosis of the soldier turned literary rememberer, whose survival—not to mention his ability to order his unbelievable, mad materials into proportion and serial coherence—constitutes his "victory," and thus his heroism. Every total experience of the war is "romantic" in the strict sense of the word. Every successful memoir of that experience shares something with traditional literary "romance," and indeed, regardless of its "truth" or accuracy of documentary fact, in its "plot" could be said to lean towards that generic category.

We must notice too that training in military maneuver and technique is governed by a simplified three-part conception of the elements of the process: first, preparation; then execution; and finally, critique. What war memoirs do is replicate this process, drummed during training into the head of every recruit and every new officer until he wants to scream with boredom. The "paradigm" war memoir can be seen to comprise three elements: first, the sinister or absurd or even farcical preparation (comic experiences on the French railway moving up the line are conventional; so are humorous accounts of the training at Etaples: "We have just returned from a lecture on sandbags, not bad" [45]); second, the unmanning experience of battle; and third, the retirement from the line to a contrasting (usually pastoral) scene, where there is time and quiet for consideration, meditation, and reconstruction. The middle stage is always characterized by disenchantment and loss of innocence, which tends to

make the whole tripartite experience resemble the psychological scheme of the lost and regained Paradise posited by traditional literary Christianity. Another way of putting it would be to say that war experience and its recall take the form of the deepest, most universal kind of allegory. Movement up the line, battle, and recovery become emblems of quest, death, and rebirth.

Because simplification is a characteristic of all ritual, ritual is likely to flourish where experience is simplified to essentials. Stand-to is easily conceived as a "ritual" in an environment that knows only two essential actions—attack and defense. Rituals come readily to those whose experience of life and dread of death have undergone such drastic simplification. Thus one of the survivors of the first day's attack on the Somme, Private H. C. Bloor, says in 1971: "I first went back to the Somme on a motorbike in 1935. I have been back twelve times since then and I intend to keep going as long as I can. I try to be there on 1 July. I go out and, at 7:30 AM, I stand at the exact spot where we went over the top in 1916." [46] Ernest Parker, "miraculously" spared while his battalion was all but wiped out on September 16, 1916, says in 1964: "One day . . . I shall revisit that little undulation in the fields between Gueudecourt and Delville Wood on an early morning in mid-September. There I will give thanks for being spared another fifty years of happy and fruitful life. . . ." [47] Such leanings towards ritual, such needs for significant journeys and divisions and returns and sacramental moments, must make us skeptical of Bernard Bergonzi's conclusion: "The dominant movement in the literature of the Great War was . . . from a myth-dominated to a demythologized world." [48] No: almost the opposite. In one sense the movement was towards myth, towards a revival of the cultic, the mystical, the sacrificial, the prophetic, the sacramental, and the universally significant. In short, towards fiction.

THE GOLDEN VIRGIN

A memorable instance of the prevailing urge towards myth is the desire felt by everyone to make something significant of the famous leaning Virgin and Child atop the ruined Basilica at Albert. No one wanted it to remain what it literally was, merely an accidentally damaged third-rate gilded metal statue now so tenuously fixed to its tower that it might fall any moment. Myth busily attached portentous meaning to it.

Mystical prophecy was first. The war would end, the rumor went, when the statue finally fell to the street. Germans and British shared this belief, and both tried to knock the statue down with artillery. When this proved harder than it looked, the Germans promulgated the belief that the side that shot down the Virgin would lose the war. This is the prophecy recalled by Stephen Southwold, who associates the wonders attaching to the leaning Virgin with those ascribed to miraculously preserved front-line crucifixes:

> There were dozens of miracle-rumors of crucifixes and Madonnas left standing amid chaos. In a few cases the image dripped blood or spoke words of prophecy concerning the duration of the war. Around the hanging Virgin of Albert Cathedral there gathered a host of these rumored prophecies, wonders and marvels, the chief one being that whichever side should bring her down was destined to lose the war.[49]

The statue remained hanging until April, 1918, after the British had given up Albert to the Germans. Determined that the Germans not use the tower for an artillery observation post, the British turned heavy guns on it and brought it down, statue and all. Frank Richards was there:

> The Germans were now in possession of Albert and were dug in some distance in front of it, and we were in trenches opposite them. The upside-down statue on the ruined church was still hanging. Every morning our bombing planes were going over and bombing the town and our artillery were constantly shelling it, but the statue seemed to be bearing a charmed existence. We were watching the statue one morning. Our heavy shells were bursting around the church tower, and when the smoke cleared away after the explosion of one big shell the statue was missing.

It was a great opportunity for the propagandists:

> Some of our newspapers said that the Germans had wantonly destroyed it, which I expect was believed by the people that read them at the time.[50]

But while the statue was still there, dangling below the horizontal, it was seen and interpreted by hundreds of thousands of men, who readily responded with significant moral metaphors and implicit allegorical myths. "The melodrama of it," says Carrington, "rose strongly in our hearts." [51] The most obvious "meaning" of the phenomenon was clear: it was an emblem of pathos, of the effect of war on the innocent, on women and children especially. For some, the Virgin was throwing the Child down into the battle, offering Him as a sacrifice which might end the slaughter. This was the interpretation of Paul Maze, a French liaison NCO, who half-posited "a miracle" in the Virgin's precarious maintenance of her position: "Still holding the infant Jesus in her outstretched arms," he says, "the statue of the Virgin Mary, in spite of many hits,

The Golden Virgin on the Basilica at Albert, 1916. (Imperial War Museum)

still held on top of the spire as if by a miracle. The precarious angle at which she now leaned forward gave her a despairing gesture, as though she were throwing the child into the battle." [52] Philip Gibbs interpreted the Virgin's gesture similarly, as a "peace-offering to this world at war." [53]

Others saw her action not as a sacrifice but as an act of mercy: she was reaching out to save her child, who—like a soldier—was about to fall. Thus S. S. Horsley in July, 1916: "Marched through Albert where we saw the famous church with the statue of Madonna and Child hanging from the top of the steeple, at an angle of about 40° as if the Madonna was leaning down to catch the child which had fallen." [54] Still others took her posture to signify the utmost grief over the cruelties being played out on the Somme. "The figure once stood triumphant on the cathedral tower," says Max Plowman; "now it is bowed as by the last extremity of grief." [55] And to some, her attitude seemed suicidal: she was "diving," apparently intent on destroying herself and her Child with her. [56] But regardless of the way one interpreted the Virgin's predicament, one's rhetoric tended to turn archaic and poetic when one thought of her. To Stephen Graham, what the Virgin is doing is "yearning":

"The leaning Virgin . . . hung out from the stricken tower of the mighty masonry of the Cathedral-church, and yearned o'er the city." The poeticism *o'er* is appropriate to the Virgin's high (if vague) portent. In the next sentence Graham lays aside that particular signal of the momentous and resumes with mere *over*, which marks the passage from metaphor back to mere cliché: "The miracle of her suspense in air over Albert was a never-ceasing wonder. . . ." [57]

Whatever myth one contrived for the leaning Virgin, one never forgot her or her almost "literary" entreaty that she be mythified. As late as 1949 Blunden is still not just remembering her but writing a poem of almost 100 lines, "When the Statue Fell," imagined as spoken to a child by her grandfather. The child has asked,

> "What was the strangest sight you ever saw?"

and the ancient responds by telling the story of Albert, its Basilica, the statue, its curious suspension, and its final fall, which he makes coincide with the end of the war. And in 1948, when Osbert Sitwell remembered Armistice Day, 1918, and its pitiful hopes for perpetual peace, he did so in imagery which bears the deep impress of the image of the golden Virgin, although she is not mentioned at all. His first image, remarkably, seems to fuse the leaning Virgin of one war with the inverted hanging Mussolini of another:

> After the Second World War, Winged Victory dangles from the sky like a gigantic draggled starling that has been hanged as a warning to other marauders: but in 1918, though we who had fought were even more disillusioned than our successors of the next conflict about a struggle in which it was plain that no great military leaders had been found, we were yet illusioned about the peace.

Having begun with a recall of the leaning Virgin as an ironic and broken Winged Victory, he goes on to remember, if subliminally, her bright gilding: "During the passage of more than four years, the worse the present had shown itself, the more golden the future . . . had become to our eyes." But now, remembering the joy on the first Armistice Day, his mind, he says, goes back to two scenes. In both gold is ironically intrusive: "First to the landscape of an early September morning, where the pale golden grasses held just the color of a harvest moon": but the field of golden grasses is covered with English and German dead. "It was a superb morning," he goes on,

> such a morning, I would have hazarded, as that on which men, crowned with the vast hemicycles of their gold helmets, clashed swords at Mycenae,

or outside the towers of Troy, only to be carried from the field to lie entombed in air and silence for millenniums under their stiff masks of virgin gold.[58]

Thirty years after Sitwell first looked up and wondered what to make of her, the golden Virgin persists, called up as a ghost in his phrase *virgin gold*. Perhaps he thought he had forgotten her. Her permanence is a measure of the significance which myth, with an urgency born of the most touching need, attached to her.

THE ROMANCE QUEST

A distinguished critic of our time has specified the following as characteristic of a certain kind of narrative. The protagonist, first of all, moves forward through successive stages involving "miracles and dangers" towards a crucial test. Magical numbers are important, and so is ritual. The landscape is "enchanted," full of "secret murmurings and whispers." The setting in which "perilous encounters" and testing take place is "fixed and isolated," distinct from the settings of the normal world. The hero and those he confronts are adept at "antithetical reasonings." There are only two social strata: one is privileged and aloof, while the other, more numerous, is "colorful but more usually comic or grotesque." Social arrangements are designed to culminate in "pompous ceremonies." Training is all-important: when not engaged in confrontations with the enemy, whether men, giants, ogres, or dragons, the hero devotes himself to "constant and tireless practice and proving." Finally, those engaged in these hazardous, stylized pursuits become "a circle of solidarity," "a community of the elect." The critic defining this kind of narrative is Erich Auerbach, and he is talking not about war memoirs, of course, but about medieval romance, of the sort written in France by Chrétien de Troyes in the twelfth century and in England by Sir Thomas Malory in the fifteenth.[59]

The experiences of a man going up the line to his destiny cannot help seeming to him like those of a hero of medieval romance if his imagination has been steeped in actual literary romances or their equivalent. For most who fought in the Great War, one highly popular equivalent was Victorian pseudo-medieval romance, like the versified redactions of Malory by Tennyson and the prose romances of William Morris. Morris's most popular romance was *The Well at the World's End*, published in 1896. There was hardly a literate man who fought between 1914 and 1918 who hadn't read it and been powerfully excited by it in his youth. For us it is rather boring, this protracted tale of 228,000 words about young Prince Ralph's adventures in search of the magic well at the end of the world, whose waters have the power to remove the scars of battle wounds. But for a generation to whom terms like *heroism* and *decency* and *nobility* conveyed meanings that were entirely secure, it was a heady read and an unforgettable source of images. The general familiarity with it and the ease with which it could be applied to the events of the war can be gauged from this: in May, 1915, an illustrated weekly headed an account of a trench skirmish won by the British with a caption in the stylish poetic-prose of the period, which here goes all the way and turns into blank verse: "How Three Encountered Fifty and Prevailed." [60] That caption could easily stand as one of the chapter titles in *The Well at the World's End*, like "How Ralph Justed with the Aliens." An audience to whom such "chapter headings" appealed in journalism was one implicitly learned in Morris's matter and style, or one which could easily come to value them. C. S. Lewis was only sixteen when the war began, but by 1917 he was nineteen and ready to go in. Just before he left he did a vast amount of reading, discovering books for the first, ecstatic time. "My great author at this period," he writes, "was William Morris. . . . In [his friend] Arthur's bookcase, I found *The Well at the World's End*. I looked— I read chapter headings—I dipped—and next day I went off into town to buy a copy of my own." [61]

There were many who arrived at Mametz Wood and Trones Wood and High Wood primed by previous adventuring in Morris's Wood Debateable and Wood Perilous. Both the literal and the literary are versions of what Frye calls the "demonic vegetable world" often associated with romance quests, "a sinister forest like the ones we meet in *Comus* or the opening of the *Inferno*, or a heath, which from Shakespeare to Hardy has been associated with tragic destiny, or a wilderness like that of Browning's *Childe Roland* or Eliot's *Waste Land*." [62] Morris's "end of the world" is a cliff overlooking a boundless sea, very unlike the world ending with the British front line. Yet in describing the landscape of the front Sassoon seems often to recall some of Morris's sinister settings as well as echoing Morris's title. "On wet days," he says, "the trees a mile away were like ash-grey smoke rising from the naked ridges, and it felt very much as if we were at the end of the world. And so we were: for that enemy world . . . had no relation to the landscape of life." [63] Again, "The end-of-the-world along the horizon had some obscure hold over my mind which drew my eyes to it almost eagerly, for I could still think of trench warfare as an adventure. The horizon was quiet just now, as if the dragons which lived there were dozing." [64] At the front, he finds another time, "we had arrived at the edge of the world." [65] And so literary an imagination as Blunden's was of course not behindhand in recalling and applying Morris. Thiepval in the winter of 1916 we today would call something like *sheer hell:* he designates it as a "filthy, limb-strewn, and most lonely world's-end. . . ." [66]

The prevailing ghostliness of the line was often registered in images deriving from such romances as Morris's. The infamous white chalk Butte of Warlencourt whose machine guns dominated the Somme lines for miles was like a terrible enormous living thing. Carrington says of it: "That ghastly hill . . . became fabulous. It shone white in the night and seemed to leer at you like an ogre in a fairy tale. It loomed up unexpectedly, peering into trenches where you thought yourself safe: it haunted your dreams." [67] (As well it might: many said that it was not a natural terrain feature at all but a Gallic burial tumulus, an antique mass grave.) Guy Chapman experienced the same sense of being secretly observed, secretly followed, the sense that Eliot dramatized in the final section of *The Waste Land* (lines 360–66, and note). "There is a secret magic about these waste lands [i.e., environs of the derelict villages on the edge of the battlefields]. While you wander through the corrupted overgrown orchard, there is always someone at your back. You turn. It is nothing but the creak of a branch. . . ." [68]

Hugh Quigley was thoroughly familiar with Victorian literary and aesthetic texts, which he recalled constantly at the front to help him "see." He does this with Ruskin's *Modern Painters*, as we have seen, and he knows *The Well at the World's End* as intimately. To him the ghastly canal at Ypres, clotted with corpses, is "like" the poison pool under the Dry Tree in Morris, around which lay the bodies of men with "dead leathery faces . . . drawn up in a grin, as though they had died in pain . . ." (Book III, Chap. 18). [69] One of Quigley's problems is how to remember. He seems to solve it by associating the thing to be remembered with an analog in a well-known literary text. This is what he does in recalling the bizarre look of the ruined Cloth Hall at Ypres in autumn, 1917. It was "so battered that not a single sculpture figure, or shadow of a figure, remained, except one gargoyle at the end, which leered down as jauntily as ever." He fixes this image by relating it to the Morris landscape, where similar figures leer: "When I come back, this incident will remain one of the treasured memories, something to recount time and again, as happening in a land of horror and dread whence few return, like that country Morris describes in the *Well at the World's End.*" [70]

But there was one English "romance" even better known than Morris's. This was Bunyan's *Pilgrim's Progress.* Everybody had been raised on it. When in the *Daily Express* on November 12, 1918, the columnist

"Orion" described his feeling upon hearing that peace had come, he wrote, "Like Christian, I felt a great burden slip from off my shoulders." The *Daily Express* was a "popular" paper, but "Orion" didn't have to say, "Like Christian in John Bunyan's *Pilgrim's Progress*." He knew he would be understood, not least by the troops, who had named one of the support trenches of the Hohenzollern Redoubt "Pilgrim's Progress." They would not fail to notice the similarity between a fully loaded soldier, marching to and from the line with haversack, ground-sheet, blanket, rifle, and ammunition, and the image of Christian at the outset of his adventures: "I saw a man clothed in rags . . . and a great burden upon his back." Recalling a terrible night march in the mud, Quigley writes: "The spirit takes note of nothing, perception dies, and, like Christian, we carry our own burden [here, rifle and ammunition], thinking only of it." [71] Christian's burden drops away when he beholds the Cross; Private Anthony French's when his equipment is blown off by the shell that wounds him in the thigh: "I had ceased to be a soldier. . . . Only my helmet remained. . . . I found myself without waterbottle, iron rations, gasmask. My watch had lost cover and glass. . . . Then an enormous burden of responsibility seemed to roll away as if this were the end of a pilgrim's progress. There was no pain. I felt at rest." [72] R. H. Tawney's burden falls away when, attacking the first day on the Somme, he realizes—it is his first action—that he is not going to be a coward after all:

> I hadn't gone ten yards before I felt a load fall from me. There's a sentence at the end of *The Pilgrim's Progress* which has always struck me as one of the most awful things imagined by man: "Then I saw that there was a way to Hell, even from the Gates of Heaven, as well as from the City of Destruction." To have gone so far and be rejected at last! Yet undoubtedly man walks between precipices, and no one knows the rottenness in him till he cracks, and then it's too late. I had been worried by the thought: "Suppose one should lose one's head and get other men cut up! Suppose one's legs should take fright and refuse to move!" Now I knew it was all right. I shouldn't be frightened and I shouldn't lose my head. Imagine the joy of that discovery! I felt quite happy and self-possessed. [73]

Even when Gunner William Pressey chooses the word *great* instead of *large* to describe the burdens carried by the French civilians retreating before the German advance in spring, 1918 ("Some had carts, others great bundles on their backs" [74]), we may suspect that *Pilgrim's Progress* is helping to determine his choice.

It is odd and wonderful that front-line experience should ape the pattern of the one book everybody knew. Or to put it perhaps more accurately, front-line experience seemed to become available for interpretation when it was seen how closely parts of it resembled the action of *Pilgrim's Progress*. Sassoon takes it for granted that his title *Sherston's Progress* will contribute significant shape to his episodic account of his passage through anxiety to arrive at his triumphant moment of relief as Rivers enters his hospital room. Like Christian, he has been looking for something and going somewhere, and at the end he knows fully what his goal has been all the while. And allusion to Bunyan can work sardonically as well, as it does in Henry Williamson's *The Patriot's Progress* (1930), whose hero, John Bullock, enters the war with enthusiasm and endures it stoically, only to end in bitterness, one leg missing, patronized by uncomprehending civilians.

The problem for the writer trying to describe elements of the Great War was its utter incredibility, and thus its incommunicability in its own terms. As Bernard Bergonzi has said, "The literary records of the Great War can be seen as a series of attempts to evolve a response that would have some degree of adequacy to the unparalleled situation in which the writers were involved." [75] Unprecedented meaning thus had to find precedent motifs and images. It is a case illustrating E. D. Hirsch's theory of the way new meanings get proposed:

> No one would ever invent or understand a new type of meaning unless he were capable of perceiving analogies and making novel subsumptions under previously known types. . . . By an imaginative leap the unknown is assimilated to the known, and something genuinely new is realized. [76]

The "new type of meaning" is that of the new industrialized mass trench warfare. The "previously known types" are the motifs and images of popular romance. The "something genuinely new" is the significant memories of the war we have been focusing on, where *significant* means, in fact, *artistic*. Because Dante has never really been domesticated in Protestant England, when an English sensibility looks for traditional images of waste and horror and loss and fear, it turns not to the *Inferno* but to *Pilgrim's Progress*.

It would be impossible to count the number of times "the Slough of Despond" is invoked as the only adequate designation for churned-up mud morasses pummeled by icy rain and heavy shells. It becomes one of the inevitable clichés of memory. So does "the Valley of the Shadow of Death," where, in Bunyan, "lay blood, bones, ashes, and mangled bodies of men, even of Pilgrims that had gone this way formerly." Major Pilditch invokes that valley to help him describe the indescribable:

> The bare poles and brick heaps of Souchez looked perfectly weird and unnatural as the sun came out and threw it all up into a livid pink-hued dis-

Near Zonnebeke, Ypres Salient. (Imperial War Museum)

tinctness. I knew I should never be able to describe its sinister appearance, but that I should never forget it. It reminded me of an old wood-cut in my grandfather's "Pilgrim's Progress," of the Valley of the Shadow of Death where Christian met Apollyon.[77]

When 2nd Lt. Alexander Gillespie was killed at Loos, on his body, we are told, was found a copy of *Pilgrim's Progress* with this passage marked: "Then I entered into the Valley of the Shadow of Death, and had no light for almost halfway through it. I thought I should have been killed there, and the sun rose, and I went through that which was behind with far more ease and quiet." [78] Apparently Gillespie had been using the passage as a sort of consolatory psalm, a version of Psalm 23:4 ("Yea, though I walk through the valley of the shadow of death, I will fear no evil") more appropriate for trench use because of its image of significant dawn.

Possessing so significant a first name, the artillery subaltern Christian Creswell Carver was in a special position to imagine himself re-enacting

Pilgrim's Progress. Writing his brother in March, 1917, he describes a mounted ammunition detail at night on the Somme:

> To our right and below us is the river stretching across a vista of broken stumps, running water and shell pools, to the skeleton gleaming white of another village on the far bank. If only an artist could paint the grim scene now while the hand of war and death is still hovering over it. In our steel helmets and chain visors we somehow recall *Pilgrim's Progress,* armored figures passing through the valley of the shadow. On—for Apollyon's talons are ever near.[79]

And of course he signs this letter "Christian."

The road which Good Will advises Christian to follow to get from the City of Destruction to the Celestial City is specifically "straight and narrow." So was the infamous miry twelve-kilometer road that led from Poperinghe to Ypres; and it was while moving up into the Salient on this road in darkness that soldiers seem most often to have recalled Christian's journey, in order to confer some shape and meaning on their suffering. "Hundreds of thousands of men must remember the road from Poperinghe to Ypres," says Henry Williamson. "Its straitness begins between two long lines of houses. . . ." [80] Carrington finds *Pilgrim's Progress* applicable in an almost uncanny way. "To find the way in the dark" up this road, he remembers, "was a task worthy of Bunyan's pilgrim." He then quotes Bunyan's account of Christian's progress through the Valley of the Shadow of Death:

> The pathway was here also exceeding narrow, and therefore good Christian was the more put to it; for when he sought in the dark to shun the ditch on the one hand, he was ready to tip over into the mire on the other. Thus he went on, for the pathway was here so dark that oft-times, when he lifted his foot to set forward, he knew not where, nor upon what, he should set it next. And ever and anon the flame and smoke would come out in . . . abundance, with sparks and hideous noises. . . . Thus he went on a great while; yet still the flames would be reaching towards him: Also he heard doleful voices and rushings to and fro, so that sometimes he thought he should be torn to pieces, or trodden down like mire in the streets.[81]

Such scenes of hazardous journeying constitute the essence of *Pilgrim's Progress,* whose title page itself specifies "His Dangerous Journey" as one of the three stages of Christian's experience, the other two being "The Manner of His Setting Out" and "Safe Arrival at the Desired Country." If the title of Sherriff's *Journey's End* alludes overtly to Othello's famous speech of acquiescence and surrender (V, ii, 263–85), it points implicitly as well to such a world of literary romance as Bunyan's, where combat-

as-journeying promises a meaning to be revealed at the end. Gordon
Swaine's poem of the Second War, "A Journey Through a War," picks
up the image of journeying, although by now the romance hero has attenuated to

> A figure through the pages of a fable,

and the allegorical meaning has clouded over:

> Obscure the moral and the fancy feeble.

It is the *Pilgrim's Progress* action of moving physically through some terrible topographical nightmare along a straight road that dominates Anthony French's memory, in 1972, of his whole war: "For me," he says,
"and probably for all who served and survive, there runs through this
panorama of memory, through the web of divergent, intersecting by-
ways, from the last sunset of rational existence to the morning of armistice and reckoning, one road along which the continuity of those days is
traced." [82]

Although the delicate, sensitive batman Alfred M. Hale was never
close to the line, he too interpreted his experience by calling *Pilgrim's
Progress* to his assistance. If he had no occasion to advert to the Valley of
the Shadow of Death, he did find a use for his memory of Christian's
anxious care of his "roll with a seal upon it," his all-important "certificate" which admits him, finally, to the Celestial City. Christian loses this
certificate once while asleep and spends an anxious time looking for it.
Hale's "certificate" takes two forms. The first is the slip of paper marked
C-2 attesting his status as a low-category man and constituting for a time
his defense against his own Slough of Despond—i.e., "overseas": "This
slip of paper I kept tight hold of and hugged to myself, as it were, until
one day . . . it was suggested to me by a sarcastic N.C.O. that I really
better keep it to myself, and not be so fond of trotting it out on all occasions." The second form Christian's "certificate" takes in Hale's agon is
that of the magical Yellow Paper admitting the bearer to the Celestial
City of demobilization and home.

The scene is the port at Dunkirk, March 4, 1919. Happy as a bride-
groom, Hale files with others marked for demobilization down "the long
cobbled roadway on our way at last for the sea and England." At the
quay the Celestial City awaits: "Then I saw our steamer waiting for us in
the mist and the rain." It is "the Big Ship"—the troops' mythological,
pseudo-Arthurian term for the once only imaginable boat plying between France and Demobilization. In *Pilgrim's Progress* the Prophets,
looking out over the gate of the City, are told: "These pilgrims are come

from the City of Destruction for the love that they bear to the King of this place." Then, Bunyan goes on, "The Pilgrims gave in unto them each man his certificate," whereupon the King "commanded to open the gate." Hale's welcome is less splendid but scarcely less gratifying: "The gangway was placed in position, and a very dirty looking individual in civilian garments stood on guard as we went one by one on board." Bunyan's account of Christian's salvation by certificate takes place in simple, passionate prose in which *then* is the principal connective, as in "Then the Pilgrims gave in unto them each man his certificate." Likewise Hale:

> Then it was I both saw and knew, as I had not done before, the value of that yellow paper. For now we were bidden to show it, and I held tight to it in the rain and displayed it to view as I went on board. . . . as did everybody else with but one exception. My paper held so tightly was all limp and sodden and torn with much handling, but I had it safe and sound right enough, and that was all I cared for.

As Bunyan's Christian enters the gates and looks back, he sees Ignorance, "a very brisk lad," soliciting admission. "Asked for his certificate," Ignorance "fumbled in his bosom for one, and found none." Asked by the Prophets "Have you none?" Ignorance is silent. What follows is the same thing that stuck with R. H. Tawney:

> The King . . . commanded the . . . Shining Ones . . . to go out and take Ignorance, and bind him hand and foot, and have him away. Then they took him up, and carried him through the air to the door that I saw in the side of the hill, and put him in there. Then I saw that there was a way to Hell, even from the gates of Heaven, as well as from the City of Destruction.

Hale's version of this is exquisite. Safe on the deck, ecstatic that his certificate has carried him to bliss everlasting, he looks back down to the quay one last time. "Not so the poor young devil pacing the quayside alone and forlorn. He had lost his paper; . . . I sincerely pitied him, . . . and have often wondered since what became of him." [83]

Bunyan's Celestial City is a fantasy of gold and jewels. It looks the way Dresden does when Vonnegut's Billy "Pilgrim," in *Slaughterhouse-Five*, sees it suddenly shining through the opened doors of a boxcar: "The doorways framed the loveliest city that most of the Americans had ever seen. The skyline was intricate and voluptuous and enchanted and absurd." Vonnegut's answer to the question "What did it look like?" suggests some of the continued capacity of *Pilgrim's Progress* to elicit the illusion of meaning from the wars of the twentieth century: "It looked like a Sunday school picture of Heaven to Billy Pilgrim." [84]

THE HONORABLE MISCARRIAGE OF *IN PARENTHESIS*

"My father sometimes read Bunyan aloud," recalls David Jones,[85] that odd, unassignable modern genius, half-English, half-Welsh, at once painter, poet, essayist, and engraver, a prodigy of folklore and liturgy and an adept at myth, ritual, and romance, the turgid allusionist of *In Parenthesis* and *The Anathémata.* He is one of those remarkable talents spawned in the late-Victorian atmosphere in whom aestheticism and religion met for mutual enrichment, and of whom T. S. Eliot, who in 1961 recommended *In Parenthesis* in an Introduction, is perhaps the most conspicuous example. Jones was born in 1895 at Brockley, Kent, of a Welsh father, who was a painter, and an English mother. His earliest bent was to drawing, which he began at the age of five. He drew impressively before he learned to read, around the age of nine. He was educated as a draughtsman and designer, from 1909 to 1914 at the Camberwell Art School, and, after the war, at the Westminster School of Art. In January, 1915, he enlisted in the Royal Welch Fusiliers and served at the front as a private from December, 1915, to March, 1918. He regarded himself as a "grotesquely incompetent [soldier], . . . a parade's despair," [86] but front-line existence made on him a permanent impact inseparable from the idea of medieval romance. As he says in the Preface to *In Parenthesis,*

> I think the day by day in the Waste Land, the sudden violences and long stillnesses, the sharp contours and unformed voids of that mysterious existence, profoundly affected the imaginations of those who suffered it. It was a place of enchantment. It is perhaps best described in Malory, book iv, chapter 15—that landscape spoke "with a grimly voice." [87]

Even physically Jones never got over the war. Like Hemingway selecting his table in the corner of a restaurant to secure his flanks and rear, Jones said around 1943, speaking of the kind of painting he liked to do: "I always work from a window of a house if it is at all possible. I like looking out on to the world from a reasonably sheltered position." [88]

In 1921 he became a Roman Catholic and learned of the artistic-religious community at Ditchling, Sussex, led by the sculptor, engraver, and typographer Eric Gill. The next year he joined this colony of devout craftsmen and for several years practiced engraving and watercolor there, developing his sacramental view of art and pursuing his interests in ritual and liturgy. He joined the Society of Wood Engravers in 1927, and thereafter worked as an illustrator for private presses, as well as for the firm of Faber and Faber. Whether engraving or painting, his favorite themes came from the Matter of Britain, the whole legendary narrative of King Arthur. Malory remained one of his favorite authors, as he had been Eric Gill's. Jones visited the Holy Land in 1934, and in 1937 published *In Parenthesis*, on which he had been working for ten years. His other long poem, dealing with (among other things) the Christianization of Britain, he titled *The Anathémata* and published in 1952. It was followed by a number of odd writings said to be elements of a continuing work in progress: *A, a, a, Domine Deus* (1955); *The Wall* (1955); *The Tribune's Visitation* (1958); *The Tutelar of the Place* (1961); *The Dream of Private Clitus* (1964); *The Hunt* (1965); *The Fatigue* (1965); and *The Sleeping Lord* (1967). These works, published as *The Sleeping Lord and Other Fragments* just before Jones's death in 1974, bespeak his lifelong obsession with the War: although many deal with Roman soldiers during the time of Christ, "the day-to-day routine and the language used to describe it," as David Blamires says, "are those of a British soldier of the First World War. Duckboards, bivvies, chitties, and the like mingle with the technical terms of Latin military vocabulary. . . ." [89] All these works are strenuously allusive—some think to the point of incoherence. All testify to Jones's serious, if perhaps Quixotic, desire to rescue and reinvigorate traditional pre-industrial religious and ethical connotations. His search is always for "valid signs," for an unimpoverished system of symbols capable of conveying even to a modern audience the rich complications of the Christian view of history. His method is that of association: he mines the "deposits" clustering around traditional meanings, anxious that not a one be lost. As he says, "If the painter makes visual forms, the content of which is chairs or chair-ishness, what are the chances that those who regard his painting will run to meet him with the notions 'seat,' 'throne,' 'session,' *'cathedra,'* 'Scone,' 'on-the-right-hand-of-the-Father,' in mind?" [90] In all his work, whether about trench life in the war or the military fatigues attending the Crucifixion, the experience of The Soldier is taken—as Conrad Aiken takes it too in *Ushant* (1952)—as representative of essential human experience. We are all on a hazardous advance through "a place of enchantment," towards, in Carrington's words, "No Man's Land and the unknown world beyond it." [91] The awful vulnerability of both nature and man—and their paradoxical privilege of glory—is what Jones learned at the front. "Wounded trees and wounded men," he says, "are very much an abiding image in my mind as a hang-over from the War." [92]

David Jones. (Mark Gerson, courtesy Faber & Faber)

In Parenthesis poses for itself the problem of re-attaching traditional meanings to the unprecedented actualities of the war. Jones expresses the problem in his Preface: "Some of us ask ourselves if Mr. X adjusting his box-respirator can be equated with what the poet envisaged, in 'I saw young Harry with his beaver on' " [93] (*Henry IV, Pt. I*, IV, i, 104). Jones believes such an equation can be made, and to assert it he associates the events of front-line fighting not only with Arthurian legend but with Welsh and English folklore, Old Testament history, Roman Catholic liturgy, Norse myth, Chaucer, *The Rime of the Ancient Mariner*, the poems of G. M. Hopkins, and even the works of Lewis Carroll. But by placing the suffering of ordinary modern British soldiers in such contexts as these, Jones produces a document which is curiously ambiguous and indecisive. For all the criticism of modern war which it implies, *In Parenthesis* at the same time can't keep its allusions from suggesting that the war, if ghastly, is firmly "in the tradition." It even implies that, once conceived to be in the tradition, the war can be understood. The tradition to which the poem points holds suffering to be close to sacrifice and individual effort to end in heroism; it contains, unfortunately, no precedent for an understanding of war as a shambles and its participants as victims. Actually, young Harry is not at all like Mr. X, but it is the ambition of *In Parenthesis* to obscure the distinction. The poem is a deeply conservative work which uses the past not, as it often pretends to do, to shame the present, but really to ennoble it. The effect of the poem, for all its horrors, is to rationalize and even to validate the war by implying that it somehow recovers many of the motifs and values of medieval chivalric romance. And yet, as Jones re-lives the experience of his actual characters, he is fully sympathetic with their daily painful predicament of isolation from home, from the past, and from values that could honestly be reported as heroic. The trouble is that the meddling intellect, taking the form this time of a sentimental Victorian literary Arthurianism after Tennyson and Morris, has romanticized the war. If we place *In Parenthesis* next to Masefield's *Gallipoli*, with its panoply of epigraphs from *The Song of Roland*, we can see its kinship with documents which are overtly patriotic and even propagandistic.

The poem is in seven parts, which bring the all-but-anonymous central character, Private John Ball, gradually closer to his climactic wounding in a wood on the Somme. John Ball, named after the priest who led the Peasants' Revolt in 1381, is the representative Briton. His name may have reached Jones through the agency of Morris's romance of 1888, *The Dream of John Ball*. Jones's John Ball is no leader of revolts: he is closer to Sad Sack, the constant butt of his corporal. He is in B Company of the Royal Welch Fusiliers, and when we first see him, at the opening of Part 1, he has just appeared late for final parade in a camp near an English port of embarcation:

Take that man's name, Sergeant Snell (1).[94]

It is December, 1915, and Ball's battalion is marching to the port. His platoon is commanded by twenty-year-old 2nd Lieutenant Piers Dorian Isambard Jenkins, whom Ball rather likes for his kindness, his flax-colored hair, and his well-cut uniform. Jones goes so far as to compare him to the noble foregound figure in Uccello's *Rout of San Romano*, which he has seen in the National Gallery. Economically Jones depicts the battalion's arrival at the quay, its boarding the ship and passage to France, its disembarcation there and train journey in cattle-cars up to the "new world" (9). Although we begin in an atmosphere of language very like that of a "realistic" post-Joycean novel, already Jones is beginning to make "equations," whose effect will be to relate military procedures to religious rituals and chivalric usages. Parade-grounds and refectories, he says, have this in common: they can be very quiet at times, even when crammed with men. The admonitions of the NCO's during the march—"Stop that talking. / Keep those chins in" (4)—are likened to a "liturgy." And the idiom of Shakespearian history plays is invoked to tell us that the transport left the dock: "They set toward France" (8).

Part 2 finds the battalion training behind the lines, listening to "lectures," propagating rumors, and generating anxieties. It moves still closer to the line by foot and bus-march, sees its first shell-damaged buildings, and gets its first whiff of gas. Ball hears his first shell arrive. It is aimed at a nearby battery, and this time damages only nature and a machine: "The sap of vegetables slobbered the spotless breech-block of No. 3 gun" (24). Ball's world is still vegetable and mineral, still innocent.

It begins to turn animal in Part 3, which takes Ball's unit into trenches for the first time, near Béthune. It performs the standard nightmare relief of a "quiet sector" in darkness, mud, and cold. The formulaic questions and answers exchanged between relievers and relieved are repeatedly conceived as "liturgy" and "ritual words" (28), and Jones explains the conventionality of this trench antiphon in a note:

> These coming from and these going to the front line used almost a liturgy, analogous to the seafaring "Who are you pray" employed by shipmasters hailing a passing boat. So used we to say: "Who are you," and the regiment would be named. And again we would say: "What's it like, mate," and the invariable reply, even in the more turbulent areas, would come: "Cushy, mate, cushy" (195).

Again, metaphors are doctrinal and chivalric:

> For John Ball there was in this night's parading, for all the fear in it, a kind of blessedness, here was borne away with yesterday's remoteness, an accumulated tedium, all they'd piled on since enlistment day: a whole unlovely order this night would transubstantiate, lend some grace to (27).

Mr. Jenkins's "little flock" is equated to an "armed bishopric" (31); the troglodytes Ball's platoon encounters peeping from their funk-holes are "Lazarus figures"; the neighboring corpses shrouded with chloride of lime were once "dung-making Holy Ghost temples"; and the whole theater of the relief is seen to be a "sepulchre" (43). When the relief is completed and everyone in his place, John Ball is ordered on sentry-duty, and in the light of a Very flare, he perceives a flooded shell hole in Arthurian—indeed, Tennysonian—terms: "corkscrew-picket-iron half submerged, as dark excalibur, by perverse incantation twisted" (50).

Part 3 ends with a "lyric" on rats testifying to Jones's admiration for the Eliot of

> I think we are in rats' alley
> Where the dead men lost their bones,

an admiration signaled earlier by Jones's choice of "Prickly Pear" to be the password for this first night.

You can hear the silence of it,

meditates David Jones–John Ball:

> you can hear the rat of no-man's-land
> rut-out intricacies,
> weasel-out his patient workings,
> scrut, scrut, sscrut,
> harrow out-earthly, trowel his cunning paw;
> redeem the time of our uncharity, to sap his own amphibious paradise (54).

Like men, the rats dig their own saps and trenches, go on their own carrying-parties, wear slimy "khaki," generally "dig in." After the sight and smell of his first corpses, Ball's world is now assuredly animal.

"A Day in the Trenches" is what someone else might have titled Part 4. Jones calls it "King Pellam's Launde," implying a similarity between the Waste Land of the front and the Waste Land of Malory's King Pellam, the maimed king of the grail story. Part 3 gave us trench life at night. Part 4 offers a typical day, from morning to evening stand-to. After rifle-cleaning, John Ball finds himself on sentry duty again, this time with a periscope. A lyric celebrates the woods of folklore and ballad, the bower of first love and the domain of the banished princes and dukes of romance. Merlin is mentioned, and so is Diana: the wood is being prepared to turn into the Sacred Wood Jones wants it to become when Ball's platoon is finally destroyed in it. Ball goes off on a pioneer fatigue, and on his way to draw picks and shovels with the section he notices—or is it Jones that notices?—the similarity of the trench scene to the modern urban, industrial squalor. Whoever is doing the noticing, it is a perceptive and prophetic moment:

> Slowly they made progress along the traverses, more easy to negotiate by light of day. Not night-bred fear, nor dark mystification nor lurking unseen snares any longer harassed them, but instead, a penetrating tedium, a boredom that leadened and oppressed, making the spirit quail and tire, took hold of them, as they went to their first fatigue. The untidied squalor of the loveless scene spread far horizontally, imaging unnamed discomfort, sordid and deprived as—

and here Jones conflates the vision of this horror with the standard urban backyard scene of the thirties—

> ill-kept hen-runs that back on sidings on wet weekdays where waste-land meets environs and punctured bins ooze canned-meats discarded, tyres to rot, derelict slow-weathered iron-ware disintegrates between factory-end and nettle-bed (75).

It is a way of implying the war's power to go on forever, however cunningly the postwar world of 1927 to 1937 may try to disguise itself as Paradise.

The next "lyric" is given to Ball's colleague Dai: he delivers what amounts to a history of wars and soldiers based on personal testimony through multiple incarnations. The stress is on Arthur's battles, but he includes some crucial moments of conflict in Christian history:

> I was in Michael's trench when bright Lucifer
> bulged his primal salient out.
> That caused it,
> that upset the joy-cart,
> and three parts waste (84).

Likewise,

> I was with Abel when his brother found him,
> under the green tree (79).

And he was, he says, a Roman soldier playing at the foot of the Cross the popular Great War gambling game of Crown and Anchor:

> I served Longinus that Dux bat-blind and bent;
> the Dandy Xth are my regiment;
> who diced
> Crown and Mud-hook
> under the Tree. . . (83).

The reader comes away from this persuaded that the state of the soldier is universal throughout history. But the problem is, if soldiering is universal, what's wrong with it? And if there is nothing in the special conditions of the Great War to alter cases drastically, what's so terrible about it? Why the shock? But Jones's commitment to his ritual-and-romance machinery impels him to keep hinting that this war is like others: he is careful to say of the NCO who issues the digging tools that he is "thickly greaved with mud" (89), and we get a sense that his condition is really no worse than that of the Roman soldiers wearing greaves who diced at the Crucifixion. Cushy, really.

"We shall be in it alright—it's in conjunction with the Frogs" (103)—so an estaminet acquaintance of Ball's prophesies near the beginning of Part 5, which covers several months. It is dominated first by rumors of the forthcoming summer Push on the Somme and then by the battalion's laborious marches southward as the prolonged though still distant artillery preparation grows louder and louder. Part 6 brings us closer still to the Somme catastrophe. It begins with a flurry of preparatory disjunct battle quotations from Malory. We are immediately in "the terrain of the bivouac" (135), a saucer valley directly behind the Somme jump-off line. The artillery drum fire goes on for days. The valley is a place not of three-man tents now but of "tabernacles" (144), and while the battalion waits to be committed, it enjoys something like a "proletarian holiday" (144): visits are paid, and skepticism and optimism collide in conversations like this:

> Fishy he don't put some back.
> He's legging it, mate—he's napoo (136).

But there are forebodings. Private Saunders, detached as a headquarters runner, returns with a vivid rumor couched in a medley of comic "Other-Ranks" and Malory:

> He said that there was a hell of a stink at Division . . . as to the ruling of this battle—and the G.S.O.2 who used to be with the 180th that long bloke and a man of great worship was in an awful pee. . . . this torf he forgot his name came out of ther Gen'ral's and say as how it was going to be a first clarst bollocks and murthering of Christen men and reckoned how he'd throw in his mit an' be no party to this so-called frontal attack never for no threat nor entreaty, for now, he says, blubbin' they reckon, is this noble fellowship wholly mischiefed (138).

The battalion's evolutions on the night of June 30 are scarcely promising. It moves forward in darkness during the height of the bombardment, but after considerable confusion moves back at daylight to find itself in oddly neat and orderly deserted German trenches. At nightfall, it is back in its saucer again. But during the night, it moves one last time, into assembly trenches.

Part 7 begins ominously with fragments of prayer and liturgy. Something terrible is about to happen. Ball's platoon is in a forward trench facing a wood, and there are casualties already. Finally they attack,

> and Mr. Jenkins takes them over
> and don't bunch on the left
> for Christ's sake (160),

and as they go over we see that they comprise all of Britain and all the
conditions and beliefs of uncomprehending, pathetically obedient mankind:

>   but we are rash levied
> from Islington and Hackney
> and the purlieus of Walworth
> flashers from Surbiton
> men of the stock of Abraham
> from Bromley-by-Bow
> Anglo-Welsh from Queens Ferry
> rosary-wallahs from Pembrey Dock. . . .

Everyone goes over, even

>   two lovers from Ebury Bridge,
>   Bates and Coldpepper
> that men called the Lily-white boys (160–61).

It is a disaster. Jones makes an effort to equate these men disemboweled
or torn apart by machine guns with dismembered antique gods in sacred
groves (with an eye to Sir J. G. Frazer, F.R.S, and T. S. Eliot, and "literature"), but now the poem doesn't work the way he wants it to, and
we focus only on innocent young Lieutenant Jenkins, who is shot almost
immediately, and on Sergeant Quilter, who gets it next, and on "the
severed head of '72 Morgan" (180). Despite Jones's well-intentioned urg-
ing, we refuse to see these victims as continuing the tradition of such
high-powered swordsman and cavalry heroes of romance, Renaissance
epic, and sacred history as Tristram, Lamorak de Galis, Alisand le Or-
phelin, Baumains, Balin and Balan, Jonathan, Absalom, Peredur, and
Taillefer. It is too much for "literature" to bear. We feel that Jones's
formula is wrong, all wrong, but it is exactly the vigor and pathos of his
own brilliant details that have taught us that.

What remains of the platoon struggles through German wire into the
wood, into the very center of Frye's "demonic vegetable world." There
is nasty fighting with grenades, and they take some prisoners and finally
dig in, surrounded by the dead of the battalion. Bates's friend Coldpep-
per has been killed, and

>   Bates without Coldpepper
> digs like a Bunyan muck-raker for his weight of woe (174).

At 9:35 that night they are ordered to attack through the woods and clear
it. Ball is machine-gunned through the legs. Like Christian dropping his
burden, he sloughs off rifle and ammunition as he drags himself toward

the rear, but not without a guilty reminiscence of months of earnest lectures about the inestimable value of his rifle. His reminiscence will remind us of what our fatigue and cynicism have made to seem the most comprehensive English poem about the Second World War, Henry Reed's "Naming of Parts." Ball remembers,

> It's the soldier's best friend if you care for the working parts and let us be 'aving those springs released smartly in Company billets on wet forenoons and clickerty-click and one up the spout and you men must really cultivate the habit of treating this weapon with the very greatest care and there should be a healthy rivalry among you—it should be a matter of very proper pride and
> Marry it man! Marry it!
> Cherish her, she's your very own (183).

With such clichés sounding in his ears, Ball, adding guilt to physical pain—where is young Harry with his beaver now?—abandons his rifle "under the oak" (186) and crawls back to wait for the stretcher-bearers.

There remains one final ritual. "The Queen of the Woods," a spirit combining aspects of the classical Diana and the Oak Spirit or Dryad of folklore, decorates the dead with appropriate wood-flowers, dealing with British and German alike:

> Some she gives white berries
> > some she gives brown
> Emil has a curious crown, it's
> > made of golden saxifrage.
> Fatty wears sweet-briar,
> He will reign with her for a thousand years (185).

In another writer that passage might be highly ironic, but here it's not, for Jones wants it to be true.

Jones has attempted in *In Parenthesis* to elevate the new Matter of Flanders and Picardy to the status of the old Matter of Britain. That it refuses to be so elevated, that it resists being subsumed into the heroic myth, is less Jones's fault than the war's. The war will not be understood in traditional terms: the machine gun alone makes it so special and unexampled that it simply can't be talked about as if it were one of the conventional wars of history. Or worse, of literary history. What keeps the poem from total success is Jones's excessively formal and doctrinal way of fleeing from the literal: the books and the words of Malory, Frazer, and Eliot are too insistently there, sometimes at the expense of their spirit. One result of this appliquéd literariness is rhetorical uncertainty and dramatic inconsistency. John Ball has never heard of the Greek

Premier Venizelos (140), and yet it may be thought that he is presumed to be learned in the mystical meanings of sacred woods according to the scholarly traditions of European folklore. As readers, we don't always know who's speaking, and to whom. The thirty-four pages of rather pedantic notes at the end bespeak the literary insecurity of the autodidact; they sometimes prop up the text where the author suspects the poetry has miscarried. Some of the poem is badly overwritten, just as the frontispiece drawing by Jones is too crowded with everything he can recall as relevant: a dead body, wire-pickets, rats, barbed wire, a tunic, a steel helmet, an ammunition belt, sandbags, blasted trees, mules, carrying parties, bully-beef tins, shattered houses, chicken-wire netting, and an entrenching tool. Too much. It is the visual equivalent of diction like *millesimal, brumous, pernitric, inutile*.

And yet for all these defects, *In Parenthesis* remains in many ways a masterpiece impervious to criticism. When it forgets momentarily its Romans and its Lancelots and its highminded medieval beadsmen and focuses on "that shit Major Lillywhite" (15) or on the visiting staff officer denominated as "that cissy from Brigade, the one wat powders" (40), the humanity of the poem seizes the reader. Jones's reading of physical details is accurate and evocative. Rifle bullets passing close overhead do crack just as Jones says: "Occasionally a rifle bullet raw snapt like tenuous hide whip by spiteful ostler handled" (42-43). And the poem is profoundly decent. When on his twenty-first birthday Mr. Jenkins receives both his promotion to full lieutenant and a nice parcel from Fortnum and Mason's, we are pleased. Details like these pull the poem in quite a different direction from that indicated by its insistent invocation of myth and ritual and romance. Details like these persuade us with all the power of art that the Western Front is not King Pellam's Land, that it will not be restored and made whole, ever, by the expiatory magic of the Grail. It is too human for that.

V

Oh What a Literary War

THE SENSE OF A NATIONAL LITERATURE

Looking back on a bad day spent worrying over a forthcoming night attack, Captain Oliver Lyttelton, educated at Eton and at Trinity College, Cambridge, says: "Well, that day dawdled away. Ovid and his mistress would not have addressed the gods that day: *O lente, lente currite noctis equi.*" [1] Remembering going up the line in a crowded third-class railway car, Private Stephen Graham, hardly educated at all, says:

> Huddled up in a dark corner of the carriage a-thinking of many such occasions in life when I have parted for the unknown, listening to the soldiers' tales, it recalled the mood of Clarence's dream when he was pacing on the hatches of the ship at night with the Duke of Gloucester, talking of the Wars of the Roses. [2]

"The mood of Clarence's dream" is that of guilt-ridden nightmare and "dismal terror":

As we pac'd along
Upon the giddy footing of the hatches,
Methought that Gloucester stumbled, and in falling
Struck me, that thought to slay him, overboard
Into the tumbling billows of the main. . . .
Methoughts I saw a thousand fearful wrecks,
A thousand men that fishes gnaw'd upon,
Wedges of gold, great anchors, heaps of pearls,
Inestimable stones, unvalued jewels,

> All scatt'red in the bottom of the sea;
> Some lay in dead men's skulls, and in the holes
> Where eyes did once inhabit there were crept,
> As 'twere in scorn of eyes, reflecting gems,
> That woo'd the slimy bottom of the deep
> And mock'd the dead bones that lay scatt'red by.
>   (*Richard III*, I, iv, 16–20; 24–33)

An allusion more proper to a sailor than a soldier, we may think, but still very pointed and effective, and certainly astonishing in view of the circumstances and the insecure idiom of the first half of Private Graham's sentence. Indeed, not the least interesting thing about his statement is the impression it gives of having been written by two different people: a virtual illiterate, who conducts the proceedings down to *recalled;* and a man of letters, who, because Shakespeare is now the focus, takes over at that point with impressive accuracy and economy. Equally striking is General Sir Ian Hamilton's literary behavior at Gallipoli. He has been expecting four crucial French divisions to arrive in the autumn of 1915. Told that they will be postponed a month, he is shattered. The terrible word *postponed* keeps sounding in his ear, and he thinks of Keats's "Ode to a Nightingale" and of

> Forlorn! the very word is like a bell.

Aware that this postponement will very likely mean the death of the Gallipoli expedition, he turns to his diary and writes: "Postponed! the word is like a knell." [3] It is a gesture which, together with Captain Lyttelton's and Private Graham's, suggests the unparalleled literariness of all ranks who fought the Great War.

Someone like Rupert Brooke we would expect to be full of literature, and we are not surprised when, writing Cathleen Nesbitt from the Aegean on March 12, 1915, he alludes to Olympus and Parnassus and Attica, as well as Shakespeare. Those things are in the common repertory of allusion. But what about this: "At any moment," he says, "we may be fetched along to kill the Paynim. *Or* we may stay here, the world forgetting, by the world forgot." [4] The allusion is to what is not the most popular poem by not the most popular English poet, *Eloisa to Abelard* (line 207), by Alexander Pope. The heavily facetious wits of the *Wipers Times* knew Pope's line too and expected their readers to recognize it. Reporting a cricket match in a rest area, a hearty editorial writer says (July 31, 1916): "This was a most interesting event, and the sight of some of our cheery Brigadiers (the war forgetting, by the war forgot from 2–7 p.m.) gambolling on the village green was most inspiring."

The American Civil War was the first, Theodor Ropp observes, "in which really large numbers of literate men fought as common soldiers." [5] By 1914, it was possible for soldiers to be not merely literate but vigorously literary, for the Great War occurred at a special historical moment when two "liberal" forces were powerfully coinciding in England. On the one hand, the belief in the educative powers of classical and English literature was still extremely strong. On the other, the appeal of popular education and "self-improvement" was at its peak, and such education was still conceived largely in humanistic terms. It was imagined that the study of literature at Workmen's Institutes and through such schemes as the National Home Reading Union would actively assist those of modest origins to rise in the class system. The volumes of the World's Classics and Everyman's Library were to be the "texts." The intersection of these two forces, the one "aristocratic," the other "democratic," established an atmosphere of public respect for literature unique in modern times. It was this respect for literature, unthinkable in today's milieu of very minor poets and journalist-novelists, that during the war was first nourishing and then ratifying the intellectual and artistic seriousness of James Joyce in Zürich, Ezra Pound in Kensington, and T. S. Eliot in his City bank. The literary earnestness of the readers of 1914–18 was these writers' stimulus and their opportunity. They could be assured of serious readers, like Private John Ball, who at one point takes from his haversack his India Paper edition of Sir Arthur Quiller-Couch's *Oxford Book of English Verse* and regales himself with bits of William Dunbar's *"Timor Mortis Conturbat Me."* [6] When the "trifling collection of verses" published by Lieutenant Edmund Blunden receives a favorable review in the *Times Literary Supplement*, Blunden's colonel, who has of course read the review in the trenches, removes him from line duty and gives him a nicer job back at battalion headquarters. [7]

There were few of any rank who had not been assured that the greatest of modern literatures was the English and who did not feel an appropriate pleasure in that assurance. If not everyone went so far as to agree with Samuel Johnson that "the chief glory of every people arises from its authors," [8] an astonishing number took literature seriously. While waiting in the saucer-shaped valley for their moment at the Somme, John Ball and some friends found a "sequestered place" on "a grassy slope" to sit and talk of "ordinary things." Ordinary things included "if you'd ever read the books of Mr. Wells" and "the poetry of Rupert Brooke"—interesting topics for a discussion among Englishmen, of the same status as "whether the French nation was nice or nasty." [9] Ball and his friends have no feeling that literature is not very near the center of normal experience, no sense that it belongs to intellectuals or aesthetes or teachers or critics. In 1914 there was virtually no cinema; there was no radio at all; and there was certainly no television. Except for sex and drinking, amusement was largely found in language formally arranged, either in books and periodicals or at the theater and music hall, or in one's own or one's friends' anecdotes, rumors, or clever structuring of words. It is hard for us to recover imaginatively such a world, but we must imagine it if we are to understand the way "literature" dominated the war from beginning to end, from Hardy's "What of the Faith and Fire Within Us?" to Kipling's "Their Name Liveth For Evermore."

It was not so in America, which has always done very well without a consciousness of a national literary canon. In the absence of a line of important "philosophic" poets running back to the fourteenth century, in a vacuum devoid of a Chaucer, a Spenser, a Shakespeare, a Milton, a Keats, a Wordsworth, a Tennyson, a Browning, an Arnold, and with no Malory or Bunyan either, American writing about the war tends to be spare and one-dimensional. The best-known American poem of the war, Alan Seeger's "I Have a Rendezvous with Death," operates without allusion, without the social instinct to invite a number of canonical poems into its vicinity for comparison or ironic contrast. It is unresonant and inadequate for irony compared with works like Blunden's "Vlamertinghe: Passing the Chateau, July, 1917," which opens with a line from Keats's "Ode on a Grecian Urn"; or Herbert Read's "The Happy Warrior," which brutally inverts Wordsworth's celebration of the honorable, well-conducted soldier; or Owen's "Exposure," which begins with a travesty-echo of the first line of Keats's "Ode to a Nightingale." It is unthinkable that any American poem issuing from the Great War would have as its title and its last two lines a tag from Horace familiar to every British schoolboy:

> Dulce et decorum est
> Pro patria mori.

Edgell Rickword's "Trench Poets" is what results from superadding to the stripped-down, ironic emotions of Hemingway a consciousness of an indispensable national literature:

> I knew a man, he was my chum,
> but he grew blacker every day,
> and would not brush the flies away,
> nor blanch however fierce the hum
> of passing shells; I used to read,
> to rouse him, random things from Donne—

like "Get with child a mandrake-root."
But you can tell he was far gone,
for he lay gaping, mackerel-eyed,
and stiff, and senseless as a post
even when that old poet cried
"I long to talk with some old lover's ghost."

I tried the Elegies one day,
but he, because he heard me say:
"What needst thou have more covering than a man?"
grinned nastily, and so I knew
the worms had got his brains at last.
There was one thing that I might do
to starve the worms; I racked my head
for healthy things and quoted *Maud*.
His grin got worse and I could see
he sneered at passion's purity.
He stank so badly, though we were great chums
I had to leave him; then rats ate his thumbs.

Rickword feels no need to append footnotes. John Ball could have explicated half the allusions by reaching into his haversack.

Indeed, the *Oxford Book of English Verse* presides over the Great War in a way that has never been sufficiently appreciated. I have in mind not merely Christopher Tietjens's obsession with "Virtue" (the first poem in the Herbert section), nor even the influence of the fifty-five-line personal effusion on light and darkness from *Paradise Lost* excerpted in the Milton section and including

                              Thus with the year
            Seasons return, but not to me returns
            Day, or the sweet approach of Ev'n or Morn. . . .
                              (III, 40–42)

This is echoed countless times, becoming a standard "turn" in recollections of the war. Thus C. E. Montague writes of the deterioration of a volunteer battalion as attrition necessitates larger and larger drafts of unwilling, untrained men: "Seasons returned, but not to that battalion returned the spirit of delight in which it had first learnt to soldier together. . . ." [10] Even more interesting, the *Oxford Book* was also the apparent source of one of the most popular of the front-line soldiers' sardonic songs, "The Bells of Hell." Imagined as sung either to the enemy or to one's own luckless comrades, it resembles a charm somehow protecting the singer:

> The bells of hell go ting-a-ling-a-ling
>   For you but not for me;
> And the little devils how they sing-a-ling-a-ling
>   For you but not for me.
> O Death, where is thy sting-a-ling-a-ling,
>   O Grave, thy victor-ee?
> The bells of hell go ting-a-ling-a-ling,
>   For you but not for me.

It would be hard to believe that the anonymous genius who contrived that song had not first encountered in his *Oxford Book*—like Ernest Parker, who quotes it in his memoir of the war [11]—this lyric of about 1600 which Quiller-Couch identifies as from a "Christ Church MS":

> Hey nonny no!
> Men are fools that wish to die!
> Is't not fine to dance and sing
> When the bells of death do ring? . . .

In the same way, Bret Harte's "What the Bullet Sang," one of the few American poems available in the *Oxford Book*, seems to lie behind both Sassoon's "The Kiss" and Owens's "Arms and the Boy," both of which, like Harte's poem, make much of the quasi-erotic desire of the bullet (and in Sassoon, the bayonet) to "kiss" or "nuzzle" the body of its adolescent target.

There was no *Oxford Book of American Verse* until 1950, but even if there had been one in 1918 Americans would have used it differently. Their way with literature was less assured, likely to be apologetic and self-conscious. An example is what E. E. Cummings makes of *Pilgrim's Progress* in *The Enormous Room* (1922), his fantastic recollection of his detention during the war in an absurd French military prison. Cummings's awareness of *Pilgrim's Progress* is verbal rather than substantive. His "Slough of Despond" is merely a stuffy ambulance section to which he has belonged, and his personal "Pilgrim's Progress"—the title of his third chapter—is not an active passage through trials to a ritual test: it is his passive transfer from a miserable place of detention to a happy one. These allusions evaporate away the meaning of *Pilgrim's Progress*, de-Christianize and de-mythify it; they use it as a framework for a sentimentality quite at odds with the import of Bunyan's work. Its terms and motifs remain as a mere "allusion"—ultimately to nothing—invoked as a schoolboy trick. The reason is that Cummings thinks *Pilgrim's Progress* funny and odd and assumes that his audience does too; indeed, he seems to expect praise for his cuteness in knowing about it at all. His refusal or affected inability to come to grips with traditional meanings can be seen blatantly in his encounter with a roadside crucifix, which he elaborately professes not to be able to identify or understand. All it looks like is "a little wooden man hanging all by himself . . .":

> The wooden body clumsy with pain burst into fragile legs with absurdly large feet and funny writhing toes; its little stiff arms made abrupt, cruel, equal angles with the road. About its stunted loins clung a ponderous and jocular fragment of drapery. . . .
>   Who was this wooden man? [12]

This is Positivism with a vengeance. From the stance he has chosen, Cummings would have to pretend not to be interested in the rumor of the Crucified Canadian or not to know what it was about.

For better or worse, the British intercourse with literature was not like that. It was instinctive and unapologetic—indeed, shameless. Consider a letter written home by Alexander Gillespie in May, 1915. Lurking behind Gillespie's words are the lively presences of Housman, Coleridge, Keats, Shelley, Hardy, Tennyson, and Wordsworth. Intact and generative are the traditional values associated with traditional symbols—white blossoms, stars, the moon, the nightingale, the heroes of the *Iliad*, pastoral flowers. What purports to be a letter is more like an unfledged poem:

> I wandered about among the ghostly cherry trees all in white, and watched the star-shells rising and falling to north and south. Presently a misty moon came up, and a nightingale began to sing. . . . It was strange to stand there and listen, for the song seemed to come all the more sweetly and clearly in the quiet intervals between the bursts of firing. There was something infinitely sweet and sad about it, as if the countryside were singing gently to itself, in the midst of all our noise and confusion and muddy work. . . . So I stood there, and thought of all the men and women who had listened to that song, just as for the first few weeks after Tom was killed I found myself thinking perpetually of all the men who had been killed in battle—Hector and Achilles and all the heroes of long ago, who were once so strong and active, and now are so quiet. Gradually the night wore on, until the day began to break, and I could see clearly the daisies and buttercups in the long grass about my feet. Then I gathered up my platoon together, and marched past the silent farms to our billets. [13]

To write like that you have to read all the time, and reading the national literature is what the British did a great deal of in the line.

THE USES OF THE CANON

The efficiency of the postal service made books as common at the front as parcels from Fortnum and Mason's, and the prevailing boredom of the static tactical situation, together with the universal commitment to the ideal of cultural self-improvement, assured that they were read as in no other war. No book was too outré: in fact, the gross inappropriateness of certain books was part of their value. At Ypres Geoffrey Keynes pored over Courtney and Smith's *Bibliography of Samuel Johnson* (Oxford, 1915), and in Macedonia R. W. Chapman edited Boswell's *Journal of a Tour to the Hebrides* and Johnson's *Journey to the Western Islands of Scotland* during idle moments at his artillery position:

> I had a hut made of sandbags, with a roof constructed of corrugated iron in layers, with large stones between . . . ; and here, in the long hot after-noons, . . . a temporary gunner, in a khaki shirt and shorts, might have been found collating the three editions of the *Tour to the Hebrides*, or re-read-ing *A Journey to the Western Islands* in the hope of finding a corruption in the text.[14]

Eighteenth-century writing was popular for the same reason that *Horizon* or the novels of Trollope were popular in the Second War. It offered an oasis of reasonableness and normality, a place one could crawl into for a few moments' respite from the sights, sounds, and smells of the twen-tieth century. Shelled at Passchendaele all through a night during which two of his best friends are killed, Alec Waugh is thankful that he has just received in the mail "a copy of Matthew Prior's poems." [15] Remember-ing his torments at the same time and place, Blunden says,

> My indebtedness to an eighteenth-century poet became enormous. At every spare moment I read in [Edward] Young's *Night Thoughts on Life, Death and Immortality*, and I felt the benefit of this grave and intellectual voice, speak-ing out of a profound eighteenth-century calm, often in metaphor which came home to one even in a pill-box. The mere amusement of discovering lines applicable to our crisis kept me from despair.

Thus, visiting a smoky dugout, Blunden enjoys recalling Young's

Dreadful post
Of observation! darker every hour.[16]
(VI, 32–33)

Just as it is entirely in character for Blunden to be reading a gentle, pious eighteenth-century poet, so it is typical of Robert Graves that in the trenches he could be found reading Samuel Butler's *Erewhon*, and for the sixth time.

"It is one of my aims," wrote an earnest young man, "to restore poetry to its true role of a *spoken* art. The music of words—the linking of sounds—the cadence of phrase—unity of action. Each poem should be *exact*, expressing in the *only* appropriate words the emotion experienced."

This manifesto issued not from a Kensington studio but from a trench, and the writer was the indefatigably literary Herbert Read. His ability to sustain the fragile artifact of literary theory undamaged within the world of shelling and raids and stink is almost theatrically dramatic: the next paragraph of the letter I have quoted begins, "I have been chosen for a death or glory job soon to come off." [17] To his correspondent he speaks of "the companionship of books," and says, "You must not grumble if they overwhelm my letters." [18] Overwhelm is the right word. During 1917 and 1918, Read corresponds about Wells's *The New Machiavelli*, Plato's *Republic*, *Don Quixote*, Bertrand Russell's *Principles of Social Reconstruction*, and the works of Henry James. He reads *Biographia Literaria*, Conrad's *Under Western Eyes*, Hazlitt's essays, Hardy's *Return of the Native*, W. T. Watts-Dunton's *Aylwin*, and Thoreau's *Walden*. He argues about the intellectual and artistic merits of Jules Romains, William Morris, Karl Marx, Robert Louis Stevenson, Gilbert Murray, William Archer, Edwin Muir, Meredith, Arnold, Masefield, Maupassant, Rossetti, Hawthorne, and the Brontës. He writes essays for the quarterly *Arts and Letters* and sees his first volume of poems published. Letters, corrected proofs, rejoinders, comments, and criticisms stream from his dugout, and at the same time he is commanding an infantry company, which means leading raids, supervising his subalterns, making sure his executive officer sends in the daily returns about strength and rations, writing the relatives of the dead and wounded, inspecting the living, and setting a good example of British phlegm. Read was of course already a distinguished critic who had amply demonstrated his capacity for disciplined work. But one did not have to be a professional critic to fall naturally into a front-line activity very much like literary criticism. Carrington once felt "a studious fit" and sent home for some Browning. "At first," he says, "I was mocked in the dugout as a highbrow for reading 'The Ring and the Book,' but saying nothing I waited until one of the scoffers idly picked it up. In ten minutes he was absorbed, and in three days we were fighting for turns to read it, and talking of nothing else at meals." [19]

In producing a multi-volume "Service" Kipling, the publishers badly misestimated the literary inclinations of those on the line. *Kim* was fairly popular, but most of the troops preferred anything of Conrad's, perhaps because, as Pilditch felt, his works offered characters caught in something like the troops' own predicament, people "who played their parts, half ignorant and yet half realizing the inexorable march of fate and their own insignificance before it." [20] It was the same instinct for dark and formal irony that turned the soldiers to Hardy. Sassoon speaks for the whole British Expeditionary Force when he says, "I didn't want to die—not before I'd finished *The Return of the Native* anyhow." [21] Hardy's most popular poem would seem to have been—it could be predicted—"The Man He Killed," published in 1909. Max Plowman has it in mind when he observes, "In this sunshine it seems impossible to believe that at any minute we in this trench, and they in that, may be blown to bits by shells fired from guns at invisible distances by hearty fellows who would be quite ready to stand you a drink if you met them face to face." [22] Everyone read Housman too, and there are moments in war memoirs when vignettes of rural irony seem the result of a conflation between Hardy and Housman. H. H. Cooper recalls one soldier's comment on a recently filled-in trench serving now as a mass grave: "Old madam will have a buckshee trench to grow celery in here when peace comes," he says, and with emphasis goes on to add the other element required for a fully ironic effect in the mode of Hardy and Housman: "*and* the lads of the village will be having their half pint in yonder estaminet." [23] H. G. Wells was popular, especially once the men discovered that *Mr. Britling Sees It Through* described the Army as "stupidly led." [24] The *Oxford Book of English Verse* was sufficient to satisfy the general appetite for poetry, although more serious readers—and writers—carried the latest volumes of verse. In the winter of 1917 Edward Thomas was reading Robert Frost's *Mountain Interval* at an artillery position near Arras and corresponding with Frost about it. [25]

With all this reading going on and with all this consciousness of the world of letters adjacent to the actual world—even louse-hunting was called "reading one's shirt"—it is to be expected that one's reports on experience will to an extraordinary degree lean on literature or recognize its presence and authority. A standard experience during the war was the company officer's discovery that his attitude toward his men, beginning in anxiety and formality, had turned into something close to devotion. Officers who before had been the wholly unwitting and inactive beneficiaries of the class system found now that the system involved reciprocal obligations. The men trusted their officer not just to safeguard their lives if he could but to deal with them decently when out of danger. They responded with wry admiration and affection, as to an odd twenty-year-old "father." Herbert Read's "My Company" is one of many poems registering the officer's emotion. Read realizes that someday this very English *comitatus* relation—suggestive even of that between older and younger boys at public school—will have an ending, either through death or demobilization, and he imagines the day when the "soul" of his company will be dissolved with something like authentic anguish:

> But, God! I know that I'll stand
> Someday in the loneliest wilderness,
> Someday my heart will cry
> For the soul that has been, but that now
> Is scattered with the winds,
> Deceased and devoid.
>
> I know that I'll wander with a cry:
> "O beautiful men, O men I loved,
> O whither are you gone, my company?"

Doing this in prose is harder, but Guy Chapman manages it with the assistance of the first stanza of Marvell's "The Definition of Love." Contemplating his men's impending demobilization, he writes:

> Looking back at those firm ranks as they marched into billets . . . , I found that this body of men had become so much a part of me that its disintegration would tear away something I cared for more dearly than I could have believed. I was it, and it was I.
>> My love is of a birth as rare
>> As 'tis for object strange and high:
>> It was begotten by Despair
>> Upon Impossibility.[26]

By merely quoting in the right context, Chapman endows *Despair*—originally a witty seventeenth-century erotic hyperbole, not fully to be believed—with some of the literal force which the war has given the word. In the same way, his context gives to *Impossibility* new meanings unthought of by readers who knew Marvell's poem as a celebration of "Platonic" devotion to a mistress. Chapman's feeling is distinctly English, seeming to require the transversion of an English poem to suggest it. His feeling may help explain why there was no serious mutiny in the British army, even under the most appalling conditions. And it is all done through the shrewd invocation of a known literary text. American, French, and German reminiscences of the war behave very differently.

One way of using canonical literature to help suggest the actuality of front-line experience was to literalize what before had been figurative. This is what Blunden does whimsically with Young's "post of observation," and it is what Chapman does more seriously with Marvell's word "Despair." When in Act IV, scene vi, Gloucester says to King Lear,

> O, let me kiss that hand!

the King, persuaded of the ubiquity of human moral corruption, says,

> Let me wipe it first; it smells of mortality.
>
> (132–33)

Lieutenant Bernard Pitt recalls this as a way of indicating the smell of corpses in the ruins of a French village, which is, he says, "merely a heap of bricks and stones, and it reeks to heaven of mortality." [27] In a similar way Oliver Lyttelton finds words adequate to the reality of Ypres in spring, 1916, only by remembering the title of James Thomson's poem of 1874, *The City of Dreadful Night*.[28] Captain James Crombie calls up the same words to suggest the desolation of Arras in March, 1917.[29]

Sometimes the assistance of literature is more subtle. Remembering a costly British attack, Carrington first thinks of the way a novelist with a leaning towards metaphorical cliché would render the spectacle of the battlefield afterwards: "It was then, turning back, that I knew what the novelists mean by a 'stricken field.' " Already prepared to think of the scene in literary terms, he searches for a simile and finds one: "Among the wire lay rows of khaki figures, as they had fallen to the machine-guns on the crest, thick as the sleepers in the Green Park on a summer Sunday evening." [30] His memory of the battlefield is fusing with his memory of *Paradise Lost* I, 302, where Milton describes the fallen legions of Satan spread out all over the "inflamed sea,"

> Thick as Autumnal Leaves that strow the Brooks
> In Vallombrosa . . . ,

and it would seem to be the long *e* of Milton's *Leaves* that suggests to Carrington his *sleepers*. What could be more English than to have Milton help one write a description of the aftermath of an aborted attack? And what could be more English than to call upon Milton to help one conceive of that dread scene in terms suggesting the antithetical world of pastoral?

Nothing, perhaps, except to call upon William Cowper to help one write an inverted, sardonic pastoral appropriate to the new kind of "harvest" being reaped in France. This is what Blunden does in his poem "Rural Economy." The straightfaced eighteenth-century title betrays not at all his ironic plan for the poem. The stanza form is that associated inseparably with Cowper's "The Castaway." Cowper's poem is about himself as one cast away from salvation like a seaman washed overboard. Blunden's is about soldiers in a bombarded wood as castaways from the human norm. While they are being cruelly shelled, not the least of their agonies is that from the edge of the woods they can see peaceful farming country some miles away and even the harvesting going forward there:

> In sight, life's farms sent forth their gear
> Here rakes and ploughs lay still,
> Yet, save some curious clods, all here
> Was raked and ploughed with a will.
> The sower was the ploughman too,
> And iron seeds broadcast he threw.

In the woods the war is plowing up the earth and reaping a human harvest, planting metal in flesh:

> What husbandry could outdo this?
> With flesh and blood he fed
> The planted iron that nought amiss
> Grew thick and swift and red,
> And in a night though ne'er so cold
> Those acres bristled a hundredfold.

This farmer performs miracles; the woods as well as the fields bear for him:

> Why, even the wood as well as field
> This ruseful farmer knew
> Could be reduced to plough and tilled,
> And if he'd planned, he'd do;
> The field and wood, all bone-fed loam,
> Shot up a roaring harvest home.

An English reader would find it hard to experience that stanza form without recalling at least bits of Cowper's poem which add dimension to Blunden's, like Cowper's apology for writing about so melancholy a subject at all:

> . . . tears by bards or heroes shed
> Alike immortalize the dead.

And Cowper's final explanation that he has written because

> . . . misery still delights to trace
> Its semblance in another's case

could be taken as expressing Blunden's concern and the reason for the artistic shape in which he has lodged it. By pointing the reader towards "The Castaway," he has traced the "semblance" between himself, William Cowper (the consciously dying castaway), and the men in the shell-plowed wood.

Of course Blunden is so literary that he can't contemplate the sight of a quartermaster who has just discovered that two jars of his service rum have been pinched without thinking of a text: "His 'eyes were wild,' " he says, remembering Keats's "La Belle Dame Sans Merci" and encouraging his reader to entertain an ironic comparison between the world of the thunderstruck quartermaster and that of Keats's entranced knight-at-arms.[31] The only thing they really have in common, the reader realizes, is a "cold hill's side." In the same way, Blunden can't admire a concert-party's cardboard train moving across the stage while the artistes sing sentimental songs from its windows without recalling Dryden's epigram on Milton: "The force of illusion could no further go." [32] Wordsworth's "Stepping Westward" adds resonance to a march back from the line on one occasion, and even the ruins of a village can be seen to be like a school of poetry: "The village of Brielen," he says, "had become the usual free-verse fandango of brick mounds and water-holes." [33]

But even the not very literate man had some poetry to fall back on. Speaking of the wounded left behind by the Germans after meeting a British night patrol, Herbert Yorke Theaker invokes Thomas Hood to help him do justice to an otherwise insupportable occasion: " 'Take them up tenderly, lift them with care' for these men are your reward for two hours hazardous work." [34] For many, however, the war was such that their main use of literature could not be artistic or ironic—only consolatory. Cowper's "God Moves in a Mysterious Way," says Vera Brittain, was a hymn very frequently sung at home. It became practically the hymn of the war, "sung at church services by congregations growing ever more anxious to be consoled and reassured." [35] Alexander Gillespie, the literary-minded man who marked the *Pilgrim's Progress* passage about moving safely through the Valley of the Shadow of Death, writes his father what he suspects will be his last letter. He knows his company is going to attack at Loos in the morning.

My dear Daddy,—
. . . Before long I think we shall be in the thick of it, for if we do attack, my company will be one of those in front, and I am likely to lead it. . . . It will be a great fight, and even when I think of you, I would not wish to be out of this. You remember Wordsworth's "Happy Warrior":
Who if he be called upon to face
Some awful moment to which Heaven has joined
Great issues, good or bad, for human kind,
Is happy as a lover, and attired
With sudden brightness like a man inspired.
Well, I could never be all that a happy warrior should be, but it will please you to know that I am very happy, and whatever happens, you will remember that. . . .
Always your loving
Bey.[36]

Did he know the poem by memory? It is not in the *Oxford Book*. Exactly like a character in a novel or play, Gillespie was killed next morning.

Gillespie's use of "The Happy Warrior" can be contrasted with Sassoon's. In one sense, *The Memoirs of George Sherston* can be considered a long series of adversary footnotes to Wordsworth's poem. Sassoon mentions it very often, and always ironically. At one point he writes a farewell letter like Gillespie's to Aunt Evelyn, and writes it in what he calls "the 'happy warrior' style": he is ashamed of it afterwards and ridicules it.[37] And perhaps he is rejecting something more than Wordsworth's poem. One of the "allegorical" paintings of George Frederic Watts admired by Sassoon's mother, and which she took him frequently to view at the National Gallery, was *The Happy Warrior*, depicting an armored youth being vaguely approached by a very sexless pre-Raphaelitish woman representing, doubtless, Death. In rejecting the Happy Warrior Sassoon is rejecting a whole Victorian moral and artistic style.

It is to be expected that a soldier like Anthony French, who has been brought up on *Pilgrim's Progress*, will know how to use related devotional texts for consolation. Attacking on the Somme in September, 1916, he moves forward at a high port with his friend Bert:

> The macabre surroundings strangely affected me. I thought of the pale horse in Revelations and of him that sat upon it. Clinging to calmness but desperate for sympathy I broke the silence and asked Bert what he thought of it all. He gave a wry smile. Then, "The Psalms. 'Though I walk through the Valley of the Shadow of Death'. . . ."
>
> "That's it," I said. To myself I repeated "I will fear no evil . . . no evil . . . I will fear no evil," and was soon comforted.[38]

Clearly, there are advantages in having a little literature, either sacred or profane. "Not being given to prayer," as he says, Cyril Falls got through a bad shelling by "repeating a school mnemonic for Latin adverbs, beginning: '*Ante, apud, ad, adversus.*' " [39]

PROBLEMS OF FACTUAL TESTIMONY

One of the cruxes of the war, of course, is the collision between events and the language available—or thought appropriate—to describe them. To put it more accurately, the collision was one between events and the public language used for over a century to celebrate the idea of progress. Logically there is no reason why the English language could not perfectly well render the actuality of trench warfare: it is rich in terms like *blood, terror, agony, madness, shit, cruelty, murder, sell-out, pain* and *hoax*, as well as phrases like *legs blown off, intestines gushing out over his hands, screaming all night, bleeding to death from the rectum,* and the like. Logically, one supposes, there's no reason why a language devised by man should be inadequate to describe any of man's works. The difficulty was in admitting that the war had been made by men and was being continued *ad infinitum* by them. The problem was less one of "language" than of gentility and optimism; it was less a problem of "linguistics" than of rhetoric. Louis Simpson speculates about the reason infantry soldiers so seldom render their experiences in language: "To a foot-soldier, war is almost entirely physical. That is why some men, when they think about war, fall silent. Language seems to falsify physical life and to betray those who have experienced it absolutely—the dead." [40] But that can't be right. The real reason is that soldiers have discovered that no one is very interested in the bad news they have to report. What listener wants to be torn and shaken when he doesn't have to be? We have made *unspeakable* mean indescribable: it really means *nasty.*

Whatever the cause, the presumed inadequacy of language itself to convey the facts about trench warfare is one of the motifs of all who wrote about the war. The painter Paul Nash found the look of the front "utterly indescribable," [41] and to H. H. Cooper "the smell rising from the bloated bodies was beyond description." [42] Even to a man destined to become a professional writer, the fact of the constant artillery thunder audible on the line seemed quite incommunicable. Robert Graves says in an interview:

> The funny thing was you went home on leave for six weeks, or six days, but the idea of being and staying at home was awful because you were with people who didn't understand what this was all about.
> [LESLIE] SMITH: Didn't you want to tell them?
> GRAVES: You couldn't: you can't communicate noise. Noise never stopped for one moment—ever. [43]

In August, 1915, Sergeant Ernest Nottingham tried to indicate what the noise was like by saying, "Ah! the exultation of the roar of the bombardment veritable! Hour on hour's ceaseless rolling reverberation." [44] But clearly the rhetoric of Byron confronting the ocean wouldn't do.

Nor would the hearty idiom of boys' adventure stories in which the young hero never failed to stand up and play the game serve to transmit the facts about modern mass man in the attack. This idiom was still being tried as late as 1918 by Edward G. D. Liveing in his book *Attack,* praised by Masefield in an introduction as "a simple and most vivid account of a modern battle." Liveing's account of the attack on

Gommecourt on July 1, 1916, proceeds through such clichés as "fleecy clouds" and "the calm before the storm" to observations like "battalions of infantry, with songs and jests, marched up . . . ready to give battle." "Men not too badly wounded," we are told, "were chatting gaily." We hear of "a very plucky young fellow" and are assured that "our boys rushed forward with splendid impetuosity." The more sensitive Alexander Aitken perceives both the impracticality of this sort of idiom and its source. Of the experience of being hit at the Somme, he says, "Even then no thought of death came, only some phrase like 'sledge-hammer blow,' from a serial read years before in a boys' magazine." [45]

A striking example of a man tied up in an unsuitable language is E. Norman Gladden, author of *Ypres, 1917: A Personal Account* (1967). Fifty years after the war he is still mired in clichés like Liveing's. The difference is that where Liveing's are those of the boys' serial, Gladden's are those of the weighty and judicious club man, fond of Elegant Variation and the idiom of solemn lunchtime calculation ("some alternative," "all reasonable odds"). Actually Gladden's powers of emotional recall are acute: his difficulties are all with style. What he wants to say is that before an important attack he felt scared but went to sleep anyway. He has been watching shell-bursts cutting the German wire "for the morrow":

> The morrow! My heart stood still at the thought. These last hours before a battle were always the most torturing. From the comparative safety of our fortified position the forthcoming experience assumed proportions of difficulty and horror that transcended the realms of possibility. If only some alternative were open. But I knew that I had no choice—and I could see no valid reason why I should again escape mutilation or death. All reasonable odds seemed to be against it. Fortunately I was very tired and decided that the best thing to be done was to get some sleep. Such is the resilience of youth that my invitation to Morpheus was completely accepted.[46]

(The troops, by the way, uninhibited by linguistic scruple and scornful of pretension, lowered the expression "the arms of Morpheus" to "the arms of Murphy"; from there it was a short way to "Murphyized," as in "No, I didn't see it: I was Murphyized.")

Some other problems of style attend one of Kipling's most honorable and decent works, the two-volume history *The Irish Guards in the Great War*, which he published in 1923 in part as a memorial to his dead son who had been in that unit. *Honorable* and *decent* do not go too far: Kipling performs the whole job without mentioning his son, who appears only in the list of dead, wounded, and missing at the end, together with hundreds of others. Again, the difficulty exemplified by Kipling's *Irish*

Guards is that of finding an appropriate rhetoric; and it implies the larger question of what style is suitable for history. More specifically, how are actual events deformed by the application to them of metaphor, rhetorical comparison, prose rhythm, assonance, alliteration, allusion, and sentence structures and connectives implying clear causality? Is there any way of compromising between the reader's expectations that written history ought to be interesting and meaningful and the cruel fact that much of what happens—all of what happens?—is inherently without "meaning"? Kipling's admirably professional description of the advantageous German positions at the beginning of the British attack on the Somme illustrates the problems:

> Here the enemy had sat for two years, looking down upon France and daily strengthening himself. His trebled and quadrupled lines of defense, worked for him by his prisoners, ran below and along the flanks and on the tops of ranges of five-hundred-foot downs. Some of these were studded with close woods, deadlier even than the fortified villages between them; some cut with narrowing valleys that drew machine-gun fire as chimneys draw draughts; some opening into broad, seemingly smooth slopes, whose every haunch and hollow covered sunk forts, carefully placed mine-fields, machine-gun pits, gigantic quarries, enlarged in the chalk, connecting with systems of catacomb-like dug-outs and subterranean works at all depths, in which brigades could lie till the fitting moment. Belt upon belt of fifty-yard-deep wire protected these points, either directly or at such angles as should herd and hold up attacking infantry to the fire of veiled guns. Nothing in the entire system had been neglected or unforeseen, except knowledge of the nature of the men who, in due time, should wear their red way through every yard of it.[47]

A version of this designed to do nothing but describe would go like this: "The Germans had been strengthening their defenses for two years. It was a defense in depth, exploiting the smooth slopes ascending to their high ground. They fortified forests and villages and dug deep underground shelters in the chalky soil. They protected their positions with copious wire, often arranged to force attackers into the fire of machine guns." But would anyone want to read that? Would it satisfy those who expected the performance of the Irish Guards in the Great War to be interesting? More important, what fun would it be for anyone to write it? It is more challenging for the professional writer and gratifying to the reader to confront effects like the almost poetic symmetrical, interlocking assonance and alliteration with which Kipling orchestrates his climax:

> men who . . . should wear their red way through every yard of it.

Strictly speaking, it would seem impossible to write an account of anything without some "literature" leaking in. Probably only a complete illiterate who very seldom heard narrative of any kind could give an "accurate" account of a personal experience. Stephen Spender says that he knew such a one, named Ned, while with the National Fire Service in the Second War:

> Because of his illiteracy he was the only man in the station who told the truth about his fire-fighting experiences. The others had almost completely substituted descriptions which they read in the newspapers or heard on the wireless for their own impressions. "Cor mate, at the docks it was a bleeding inferno," or "Just then Jerry let hell loose on us," were the formulae into which experiences such as wading through streams of molten sugar, or being stung by a storm of sparks from burning pepper, or inundated with boiling tea at the dock fires, had been reduced. But Ned had read no accounts of his experiences and so he could describe them vividly.[48]

Actually, to narrate anything Ned would have had to learn somewhere the principles of sequence and unity and transition and causality, but let that pass. Spender's illustration is instructive.

Charles Carrington is anything but illiterate. Indeed, he is so consciously literary that in the interests of an accurate documentary narrative he has tried to refine out of *A Subaltern's War* everything that might appear too artful. His memoirs of the war, he says in his Preface, "were not intended to be literary studies." He does not claim for them, he insists, "any literary merit: they are simply records of everything I could recall, every action, word or emotion, in a vivid personal experience which I felt [in 1920] to be beginning to fade from my memory." But observe what happens when he sets out "simply" to "record" experience which has made a deep impact: literature comes rushing in and takes over, just as Impressionist painting floods in to dominate Hugh Quigley's *Passchendaele and the Somme*, determining what he chooses to notice. Carrington thus describes the variety of night activity on the line:

> Some skilled officers and men crawled, boy-scout fashion, to listen and observe near the enemies' lines; some wrestled like the slaves that built the Pyramids at dragging baulks of timber, coils of wire and engineering tools by main force to the line; some with torn and bleeding hands struggled Laocoön-like to twist and strain new strands of barbed wire into the entanglement a few yards in front of the posts.[49]

A man dying with a bullet through his head becomes, as Carrington thinks about the melodrama of his last moments, almost admittedly a character in a novel: "A gurgling and a moaning came from his lips, now high and liquid, now low and dry—a 'death-rattle' fit for the most blood-thirsty of novelists." And he goes on: "Old Mills [wounded in the stom-ach], tough, bronzed, ginger-moustached and forty-one years old, lay be-side this text 'that taught the rustic moralist to die.' " [50] One can hardly get more English-literary than to compel into one's account of "what happened" a line, even if slightly misquoted, from Gray's *Elegy*. (The poem must have offered a terrible temptation, occupying as it did a prominent place virtually in the center of the *Oxford Book*.)

The point is this: finding the war "indescribable" in any but the avail-able language of traditional literature, those who recalled it had to do so in known literary terms. Joyce, Eliot, Lawrence, Pound, Yeats were not present at the front to induct them into new idioms which might have done the job better. Inhibited by scruples of decency and believing in the historical continuity of styles, writers about the war had to appeal to the sympathy of readers by invoking the familiar and suggesting its re-semblance to what many of them suspected was an unprecedented and (in their terms) an all-but-incommunicable reality. Very often, the new reality had no resemblance whatever to the familiar, and the absence of a plausible style placed some writers in what they thought was an impossi-ble position. Speaking of the sights and sounds on the Somme in Sep-tember, 1916, Alexander Aitken writes: "The road here and the ground to either side were strewn with bodies, some motionless, some not. Cries and groans, prayers, imprecations, reached me. I leave it to the sensitive imagination; I once wrote it all down, only to discover that horror, truth-fully described, weakens to the merely clinical." [51] But what was needed was exactly the clinical—or even obscene—language the literary Aitken regards as "weak." It would take still another war, and an even worse one, before such language would force itself up from below and propose itself for use. It was a matter of leaving, finally, the nineteenth century behind.

THE PROGRESS OF EUPHEMISM

Actually, the war was much worse than any description of it possible in the twenties or thirties could suggest. Or, of course, while it was going on. Lloyd George knew this at the time. "The thing is horrible," he said, "and beyond human nature to bear, and I feel I can't go on any longer with the bloody business." He was convinced that if the war could once be described in accurate language, people would insist that it be stopped. "But of course they don't—and can't know. The correspondents don't write and the censorship wouldn't pass the truth." [52] If censorship was one inhibition on the truth, another was the British tendency towards heroic grandiosity about all their wars. Americans and British have always been divided on the matter of the style thought suitable for war. It is significant that what the Americans call "The Unknown Soldier" the British elevate to "The Unknown Warrior."

The American historian Barbara Tuchman has studied this British habit of "raising" the idiom of warfare, and in speaking of British accounts of their military performance in Burma in the Second War, she says with some acidity,

> No nation has ever produced a military history of such verbal nobility as the British. Retreat or advance, win or lose, blunder or bravery, murderous folly or unyielding resolution, all emerge alike clothed in dignity and touched with glory. . . . Everyone is splendid: soldiers are staunch, commanders cool, the fighting magnificent. Whatever the fiasco, aplomb is unbroken. Mistakes, failures, stupidities, or other causes of disaster mysteriously vanish. Disasters are recorded with care and pride and become transmuted into things of beauty. . . . Other nations attempt but never quite achieve the same self-esteem.[53]

Writing in the *Daily Mirror* on November 22, 1916, W. Beach Thomas managed to assert that the dead British soldier even lies on the battlefield in a special way bespeaking his moral superiority: "Even as he lies on the field he looks more quietly faithful, more simply steadfast than others." He looks especially modest and gentlemanly too, "as if he had taken care while he died that there should be no parade in his bearing, no heroics in his posture." Notice the date of this: nine days after the four-month-long Somme attack had finally petered out. The number of British soldiers killed and wounded since the first of July had reached 420,000.

The war began for the British in a context of jargon and verbal delicacy, and it proceeded in an atmosphere of euphemism as rigorous and impenetrable as language and literature skillfully used could make it. The crucial cable sent home by His Britannic Majesty's Consul in Sarajevo on June 28, 1914, was couched in these terms: "Heir apparent and his consort assassinated this morning by means of an explosive nature." The Cunard liner on which Matthew Arnold had traveled to the United States in 1883 was named the *Servia*, and no one thought an international breach of taste had occurred. But once the war began that designation for a friendly country wouldn't do—it was too suggestive of *servility*. Sometime between August, 1914, and April, 1915, the name of the country was quietly "raised" by the newspapers to *Serbia*, and *Serbia* it has remained.[54] Once the war broke out, shepherd dogs which had been

"German" became "Alsatian," and when the young Evelyn Waugh returned to school in September, 1914, he found several boys who "turned up with names suddenly anglicized—one, unfortunately named Kaiser, appeared as 'Kingsley.' . . ." [55] (These were the days when it was possible for the young Graham Greene to witness the stoning of a dachshund in the High Street of Birkhamsted.[56])

No one was to know too much. Until 1916, the parents of soldiers executed for "acts prejudicial to military discipline" were given the news straight, but after agitation by Sylvia Pankhurst, they were informed by telegram that their soldier had "died of wounds." The writers of the daily official communiqué became experts in their work. Just before the Somme attack, Aitken says, he "began to take notice of the communiqués which tersely described, or in most cases avoided describing, the doings of each night on the Western Front. In particular, I noticed a limited selection of euphemistic adjectives, such as 'sharp,' 'brisk'; e.g., 'the enemy was ejected after brisk fighting'; 'there was sharp retaliation,' and the like." He learned later that *sharp* or *brisk* had quite precise meanings: they meant that about fifty per cent of a company had been killed or wounded in a raid. Aitken adds: "A future historian, if he leaned at all on these carefully sieved accounts, would be quite misled." [57] The communiqué issued on the evening of July 1, 1916, shows what can be done by equivocating and implying and not saying too much:

> Attack launched north of River Somme this morning at 7:30 a.m., in conjunction with French.
> British troops have broken into German forward system of defenses on front of 16 miles.
> Fighting is continuing.
> French attack on our immediate right proceeds equally satisfactorily.

Even those who drew the maps for newspapers became skilled in equivocation. It was their habit to flatten out the Ypres Salient to make it appear a harmless little protuberance, not at all a vast enclosure of some eighteen square miles in which the British were exposed to shelling from three—and sometimes four—sides. One of the highest achievements in saying the thing that is not was the poster announcing the terms of the Military Service Act of 1916:

EVERY UNMARRIED MAN
of
MILITARY AGE . . .
CAN CHOOSE
ONE OF TWO COURSES:

(1) He can ENLIST AT ONCE and join the Colors without delay;
(2) He can ATTEST AT ONCE UNDER THE GROUP SYSTEM
and be called up in due course with his Group.

And then the climax, anticipating the world of Heller's *Catch-22*:

If he does neither, a third course awaits him:
HE WILL BE DEEMED TO HAVE ENLISTED. . . .

If the authorities relied on euphemism to keep the truth from others—
the French mutinies of 1917 became acts of "collective indiscipline"—the
troops relied on it to soften the truth for themselves. The whole concept
of the "Blighty wound" is an example, where *Blighty*, connoting home,
comfort, and escape, is felt to remove a large part of the terror of *wound*.
Even Frank Richards, no friend at all of equivocation, can't bring himself
to say that a man has been wounded in the genitals: he is "hit low
down." Of course there was a whole set of euphemisms for getting
killed. "Going West" was thought to be too la-di-da; more popular were
"to be knocked out," "going out of it," "going under." And the troops
became masters of that use of the passive voice common among the cul-
turally insecure as a form of gentility ("A very odd sight was seen
here" [58]); only they used it to avoid designating themselves as agents of
nasty or shameful acts. For example: "We were given small bags to
collect what remains could be found of the bodies, but only small por-
tions were recovered." [59] E. Norman Gladden is clearly a little ashamed
of his conduct at Messines in 1917 when under German shelling his
group disperses in not quite a military manner and throws away its grenades to make flight easier. He comes to grips with the problem this
way: "We broke into scattered groups and most of the Mills bombs were
dumped on the way." [60] Similarly, General Jack relies on passive con-
structions to avoid designating malefactors or the blameworthy too
openly. Acts of thievery can be made to seem performed virtually without human agency: "The West Yorks have by no means wasted the day.
A 'dump' of equipment salvaged from the battlefields . . . catches their
eyes and, in the absence of any referee, some New Army leather equipment in our possession is exchanged for the lighter Regular Army web,
while other suitable stores are also selected. . . ." [61]

A man who has killed thirty-eight of his own countrymen needs all the
help he can get from the passive voice to remove his will and agency the
greatest possible distance from the fate of his victims. Lt. Col. Graham
Seton Hutchison tells how, during the frantic retreat of March, 1918,
he stopped a rout when he encountered a group of forty men preparing
to surrender to the swarming Germans. Hutchison explains:

Such an action as this will in a short time spread like dry rot through an army and it is one of those dire military necessities which calls for immediate and prompt action. If there does not exist on the spot a leader of sufficient courage and initiative to check it by a word, it must be necessary to check it by shooting. This was done. Of a party of forty men who held up their hands, thirty-eight were shot down with the result that this never occurred again.[62]

(William Moore's wry comment is worth having: "Had he shot two and inspired thirty-eight, there would have been greater logic in his action. To shoot thirty-eight to inspire two does not make sense." [63]) No wonder, thinking about it later, Hutchison craftily deploys the passive voice.

Active voice also became a "casualty" (speaking of euphemisms) in the *Official History*, which will often invoke the passive to throw a scene into merciful soft focus, as in the cinema. "The extended lines [attacking at Beaumont Hamel] started in excellent order, but gradually melted away. There was no wavering or attempting to come back, the men fell in their ranks—mostly before the first hundred yards of No Man's Land had been crossed." [64] A triumphant use of passive voice to veil and soften the outlines of failure is by the anonymous author of the 392-page apologia, *Sir Douglas Haig's Big Push: The Battle of the Somme*, undated but doubtless belonging to 1917. When this author comes to the point where he has to explain why the forward movement finally ended, he writes: "Winter, which interrupts Victory and gives Defeat a respite, was now at hand, and during the rest of the year no further forward movement was attempted by us." [65]

It would be going too far to trace the impulse behind all modern official euphemism to the Great War alone. And yet there is a sense in which public euphemism as the special rhetorical sound of life in the latter third of the twentieth century can be said to originate in the years 1914–18. It was perhaps the first time in history that official policy produced events so shocking, bizarre, and stomach-turning that the events had to be tidied up for presentation to a highly literate mass population. Perhaps one would not be mistaken in finding in the official euphemisms of the Great War the ultimate progenitors of such later favorites as "combat," "combat-fatigue," "mental illness," "nuclear device," "Department of Defense" (formerly "War Department"), "Free World," "homicide," and "fatality." Americans have not too long ago heard an invasion conducted in their name termed an "incursion," bombing called "protective reaction strikes," and concentration camps called "pacification centers." Carelessly aimed or wildly dropped bombs are now "incontinent ordnance." There's little question that Great War com-

muniqué-writers would admire that one and be proud to acknowledge it
as their progeny.

VERNACULAR USAGES AND FORM-RHETORIC

In an atmosphere so heavily charged with a consciousness of language
and literature, even the unliterary man rose on occasion to genuine, if
minor, literary triumphs. Like calling "cold feet" *frigiped*, for example, or
the cemetery *the rest camp;* or speaking of going over the top (a great
image all on its own, of course) as *jumping the bags.* The names of French
and Belgian towns offered constant opportunities for something very like
wit. Doingt became *Doing It;* Etaples was either *Eatables* or *Eat-apples.*
Auchonvillers was *Ocean Villas,* Foncquevillers was *Funky Villas* (appropriate for the staff), Biefvillers *Beef Villas.* Since Poperinghe was shortened to *Pop,* Albert had to be shortened to *Bert.* Ypres was *Ée-priss, Eeps,*
or *Wipers.* The popular "estaminet" was metamorphosed first into *ésta-
mint* and finally into the eloquent, evocative *just-a-minute.* The modern
designation *plonc* for cheap wine of any color, usually red now, derives
from the troops' way with the term *vin blanc.* Certain ad hoc place names
at the front attained undoubted literary quality: Idiot Crossroads, for
example, as well as Dead Dog Farm, Jerk House, Vampire Point, Shell-
trap Barn, and of course the familiar Hellfire Corner. Even the unremitting profanity and obscenity were managed so as to achieve literary effects. The intensifier of all work was *fucking*, pronounced *fuckin'*, and one
exhibited one's quasi-poetic talents by treating it with the greatest possible originality as a movable "internal" modifier and placing it well inside
the word to be modified. As in "I can't stand no more of that Mac-
bloody-fuckin'-Conochie." Perhaps *hell* was overworked, although occasionally, when used in some surprising context, it could be telling:
"During the night," one soldier writes in his diary, "we dug in shell
holes. A night of great suspense. Verily Verily Hell." [66]
  J. B. Priestley has remembered the tendency of the troops to use *old* to
indicate half-affectionate familiarity, as in the trench song about the
whereabouts of the battalion:

> It's hanging on the old barbed wire.
> I've seen 'em, I've seen 'em
> Hanging on the old barbed wire.

"To this day [1962]," says Priestley, "I cannot listen to it unmoved. There
is a flash of pure genius, entirely English, in that 'old,' for it means that
even that devilish enemy, that death-trap, the wire, has somehow been

accepted, recognized and acknowledged almost with affection, by the deep rueful charity of this verse. I have looked through whole anthologies that said less to me." [67] The constant meeting and passing of marching units on the roads generated a large repertory of road humor based on musical and literary patterns of orderly insult. Passing a well-known battalion, a marcher would shout out "Bugger!" His colleagues would take this as the first two notes of the *Colonel Bogey March*, and after a suitable pause would sing out the rest of the melody: "—the Worcestershires!" Some standard texts of antiphonal road humor involving the affectionate *old* shaped themselves into virtual poetic couplets, as Anthony French testifies: "We approached a signaller . . . repairing a broken line, and a voice cried: 'Some say "Good old Signals!" ' to which a second voice replied: 'Others say "——— old Signals!" ' The verb was irrelevant and its execution biologically absurd, but the couplet was invariable whoever might be 'good old this' or 'good old that.' It was always the curious adjective 'old' that preserved the affection." [68] Even the enemy could be called "old Jerry."

The literary instinct of the troops showed itself also in the ad hoc epitaphs which they would scrawl on boards and erect over bodies, or parts of them, exposed by shellfire and clumsily re-buried. One, instinct with paradox, anticipates the formal epitaph later written for the Tomb of the Unknown Warrior in Westminster Abbey:

> In loving memory
> of an
> *Unknown* British soldier.

Another represents the triumph of literature over all-but-unimaginable stinks and horrors:

> Sleep on, Beloved Brother;
> Take thy Gentle Rest.[69]

Perhaps the most impressive of these trench epitaphs is the one written by someone fully familiar with the *Greek Anthology* and set over the mass grave of the Devonshires that was made of a trench near Mansel Copse, on the Somme, which they had seized, defended, and been killed in. It read:

> The Devonshires held this trench.
> The Devonshires hold it still.[70]

For economical eloquence, this recalls Simonides' epitaph on the Spartan dead at Thermopylae:

> Stranger, tell the Spartans how we die:
> Obedient to their laws, here we lie,[71]

and anticipates some of Kipling's later "Epitaphs of the War," described by him as "naked cribs of the Greek Anthology." [72]

Unashamed sentiment like that in these epitaphs was one of the prevailing emotional styles of the front. Another was the antithetical style of utter *sang-froid*, or what we would have to call the style of British Phlegm. The trick here is to affect to be entirely unflappable; one speaks as if the war were entirely normal and matter-of-fact. Thus Clive Watts writes his sister, "It was most interesting being in the trenches this morning and seeing the effect of the shelling." [73] And S. S. Horsley writes home: "Got a new type of gas-goggles with rubber eye-pieces, very comfortable and useful for motoring after the war." [74] P. H. Pilditch traced this style to the "stoical reticence" learned by young officers at the public schools and spoke of it as "a sort of euphemism": "everything is toned down. . . . Nothing is 'horrible.' That word is never used in public. Things are 'darned unpleasant,' 'Rather nasty,' or, if very bad, simply 'damnable.' " [75] But the effect is less euphemistic than ironic and comic, as when A. Surfleet writes in his diary, "I don't think I'll ever get used to those '5.9's,' " [76] or when General Jack says, "On my usual afternoon walk today a shrapnel shell scattered a shower of bullets around me in an unpleasant manner." [77] Originating as an officers' style, this one was quickly embraced by Other Ranks as well. Private Frank Richards's *Old Soldiers Never Die* is a triumph in this style: Graves ascribes its success exactly to "its humorous restraint in describing unparalleled horrors." [78] And Private R. W. Mitchell was a master of it, writing in his diary: "June 21 [1916]. Move to trenches Hebuterne. Strafing and a certain dampness." [79] If terror could be masked by British Phlegm, so could joy and relief. Clive Watts writes his sister on November 11, 1918: "Well, today we ought to feel very jolly, we have heard that hostilities cease at 11 AM today." [80] Although Hemingway constructed his personal tonality in part by imitating the Cable-ese of the foreign journalist, he also learned by imitating the Great War style of British Phlegm.

The refusal of the men to say anything in their letters home indicates how pervasive the style of British Phlegm became. The censorship of the letters on the line doubtless imposed inhibitions, but this hardly accounts for the unique style of almost unvarying formulaic understatement in which the letters of Other Ranks were written. The substance and style of the genre "Other Ranks' Letter Home" became one of the inviolable literary usages of the war, and because they were governed by conventions so strict, such letters, as Stuart Cloete says, "formed a kind of literature of their own." [81]

Any officer who had censored thousands of them could make up a model, as Graves does:

> This comes leaving me in the pink which I hope it finds you. We are having a bit of rain at present. I expect you'll have read in the papers of this latest do. I lost a few good pals but happened to be lucky myself. Fags are always welcome, also socks. [82]

Herbert Read's model letter retains the "in the pink" convention but also essays a little cliché humor:

> Dear old pal—Just a line hoping as how you are in the pink of condition as this leaves me at present. Well, old pal, we are out of the line just now in a ruined village. The beer is rotten. With good luck we shall be over the top in a week or two, which means a gold [wound] stripe in Blighty or a landowner in France. Well, they say it's all for little Belgium, so cheer up, says I: but wait till I gets hold of little Belgium.
>
> From your old pal,
Bill. [83]

The trick was to fill the page by saying nothing and to offer the maximum number of clichés. *Bearing the brunt* and *keep smiling* were as popular as *in the pink*. For some writers the form of the letter home was so rigid that no variation was allowed to violate it. R. E. Vernède reports on "a rather nice boy in my platoon who writes a family letter daily always beginning

> Dear Mum and Dad, and dear loving sisters Rosie, Letty, and our Gladys,—
> I am very pleased to write you another welcome letter as this leaves me. Dear Mum and Dad and loving sisters, I hope you keeps the home fires burning. Not arf. The boys are in the pink. Not arf. Dear Loving sisters Rosie, Letty, and our Gladys, keep merry and bright. Not arf.

It goes on like that for three pages," says Vernède, "absolutely fixed; and if he has to say anything definite, like acknowledging a parcel, he has to put in a separate letter—not to interfere with the sacred order of things." [84]

The main motive determining these conventions was a decent solicitude for the feelings of the recipient. What possible good could result from telling the truth? In R. H. Mottram's *Spanish Farm Trilogy* (1927) Lieutenant Geoffrey Skene tries several times to write his uncle what is actually going on. He tears up his first version, fearing the censor will stop it. He tears up the second, too, fearing that it will give his uncle "the horrors." And he tears up the third because he finds that it too frivolously juxtaposes the emptying of latrines and the digging of graves. Skene finally resorts to the Other Ranks formula and writes this:

Dear Uncle,—
   Thanks for the woolly cap and the cigarettes and chocolate. They were most welcome and it is jolly good of you. The weather is not too bad for the time of year. Give my kind regards to everyone at the office and tell them we are winning fast, and I shall soon be home.
                                           Ever yours,
                                           Geoffrey.[85]

Clearly, any historian would err badly who relied on letters for factual testimony about the war. Clearly, Harvey Swados labors under misapprehensions when, introducing Remarque's *All Quiet on the Western Front*, he asserts that Remarque's fiction is as real and intimate "as a series of personal letters from a doomed brother or friend at the front." [86] Ironically, the reticence which originated in the writers' sympathy for the feelings of their addressees was destined in the long run simply to widen the chasm of incomprehension which opened between them.

   If a man was too tired to transcribe the clichés of the conventional phlegmatic letter, he could always turn to the famous Field Service Post Card, which allowed him simply to cross out the phrases which "did not apply." It was officially called "Form A. 2042" but was known to its users as the "Whizz Bang" or "Quick Firer"; millions and millions were sent home from the front.

   Jorge Luis Borges in his short story "The Aleph," that brilliant comic testament to the powers of memory and imagination—and at the same time lament for their limitations—gets it profoundly right as usual. Staring at The Aleph (a thing like and not like a crystal ball), an observer sees all places and events and things simultaneously and from every possible point of view. Borges's "examples" of what one can see in The Aleph include such emblems of the multitudinous as "every grain of sand" in "convex equatorial deserts"; "the multitudes of America"; "interminable eyes"; "all the mirrors in the planet"; "all the ants on earth"; and "a terraqueous globe between two mirrors which multiplied it without end." And Borges, who sometimes gives the impression of knowing everything, singles out an image which brings the essence of the Great War—that is, its multitudinousness—within his vision: in The Aleph, he

NOTHING is to be written on this side except the date and
signature of the sender. Sentences not required may be erased.
If anything else is added the post card will be destroyed.

I am quite well.

I have been admitted into hospital
{sick   }   and am going on well.
{wounded}   and hope to be discharged soon.

I am being sent down to the base.

I have received your
            {letter dated _____
            {telegram ”   _____
            {parcel ” _____

Letters follows at first opportunity.

I have received no letter from you
                        {lately
                        {for a long time.

Signature}
    only. }
Date _____

says, "I saw the survivors of a battle sending out post cards." [87] The
Field Service Post Card was most commonly used just this way: it was
sent—with everything crossed out except "I am quite well"—immediately after a battle which relatives might suspect their soldiers had been
in. Such were the hazards of occupying newly blown mine-craters that,
according to George Coppard, "Before starting a twelve-hour shift in a
crater, each man had to complete a field postcard for his next of kin,
leaving the terse message 'I am quite well' undeleted." [88]

The implicit optimism of the post card is worth noting—the way it offers no provision for transmitting news like "I have lost my left leg" or "I have been admitted into hospital wounded and do not expect to recover." Because it provided no way of saying "I am going up the line again," its users had to improvise. Wilfred Owen had an understanding with his mother that when he used a double line to cross out "I am being sent down to the base," he meant he was at the front again.[89] Close to brilliant is the way the post card allows one to admit to no state of health between being "quite" well, on the one hand, and, on the other, being so sick that one is in hospital. The egregious *quite* seems to have struck users of the card as so embarrassing that they conveniently forgot it as the years passed. Recalling the war after fifteen years, Llewellyn Wyn Griffith remembers the words as "I am well"; so does Stuart Cloete; and P. W. Turner and R. H. Haigh remember them as "I am well and unwounded."[90] One paid for the convenience of using the post card by adopting its cheerful view of things, by pretending to be in a world where belated mail and a rapidly healing wound are the worst that can happen, and where there is only one thinkable direction one can go—to the rear.

The Field Service Post Card has the honor of being the first widespread exemplar of that kind of document which uniquely characterizes the modern world: the "Form." It is the progenitor of all modern forms on which you fill in things or cross out things or check off things, from police traffic summonses to "questionnaires" and income-tax blanks. When the Field Service Post Card was devised, the novelty of its brassy self-sufficiency, as well as its implications about the uniform identity of human creatures, amused the sophisticated and the gentle alike, and they delighted to parody it, as Owen does in a letter to Sassoon:

> My dear Siegfried,
>   Here are a few poems to tempt you to a letter. I begin to think your correspondence must be intercepted somewhere. So I will state merely
>
> I have had no letter from you $\begin{cases} \text{lately} \\ \text{for a long time.} \end{cases}$[91]

Although he was only fifteen years old when the war ended, the Quick Firer made an indelible impression on Evelyn Waugh. In 1950, arranging for a luncheon and an afternoon's outing with a friend's son, he sent him, as the friend says, "a postcard modelled on those printed forms that make it easy for soldiers to write home":

I $\begin{cases} \text{can} \\ \text{cannot} \end{cases}$ lunch on Saturday June 17th

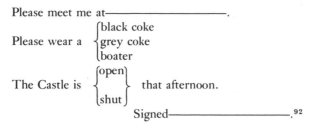

Please meet me at————————.

Please wear a $\begin{cases} \text{black coke} \\ \text{grey coke} \\ \text{boater} \end{cases}$

The Castle is $\begin{cases} \text{open} \\ \text{shut} \end{cases}$ that afternoon.

Signed————————.[92]

One even suspects that it is the Field Service Post Card to which Waugh is paying distant homage in the semi-facetious post cards that he—and Edmund Wilson as well—had printed to be sent to importunate strangers presenting inconvenient invitations or requests. Waugh's read, "Mr. Evelyn Waugh deeply regrets that he is unable to do what is so kindly proposed"; Wilson's listed a number of things which "Mr. Edmund Wilson does not do," like granting interviews, reading manuscripts, delivering commencement addresses, receiving honorary degrees, etc., each with a small box to be checked.

Infinite replication and utter uniformity—those are the ideas attached to the Field Service Post Card, the first wartime printing of which, in November, 1914, was one million copies. As the first widely known example of dehumanized, automated communication, the post card popularized a mode of rhetoric indispensable to the conduct of later wars fought by great faceless conscripted armies. (Something like it made a memorable appearance during the Second War. Such official printed cards, where one struck out what was not applicable, were the only means by which ordinary people were allowed to communicate between occupied and unoccupied France during 1942.) The perversion of a fully flexible humane rhetoric betokened by the post card has seemed so typical not just of the conduct of later wars but of something like their "cause" that satirists have made it one of their commonest targets. In Heller's *Catch-22* Colonel Cathcart writes form-letters of condolence implicitly requiring the recipient to strike out "what does not apply":

> Dear Mrs, Mr, Miss, or Mr and Mrs Daneeka:
>   Words cannot express the deep personal grief I experienced when your husband, son, father or brother was killed, wounded, or reported missing in action.[93]

This of course takes the implications of the Field Service Post Card one step further: it obliges the recipient to understand not only that the dead man possessed no individuality but that the reader of the "form," the hapless addressee, is not to have any either. The association of "forms" with euphemisms or with palpable lies, and the identity of both with modern experience—that is, the experience of war and such of its attendants as rationing and perpetual disability pensions—is likewise implied in one of Tyrone Slothrop's overheated, melodramatic meditations in Pynchon's *Gravity's Rainbow*. Slothrop is thinking about the true character of the Second War. It was never, he concludes in his half-paranoid way, about "politics" at all. "The politics was all theater," he finds, "all just to keep the people distracted. . . . Secretly, it was being dictated instead by the needs of technology . . . by a conspiracy of human beings and techniques, by something that needed the energy burst of war, crying, 'Money be damned, the very life of [insert name of Nation] is at stake.' " A similar rhetoric presides over Major Theodore Bloat's obituary letter to the *Times* about Slothrop's friend Oliver Tantivy:

> True charm . . . humble-mindedness . . . strength of character . . . fundamental Christian cleanness and goodness . . . we all loved Oliver . . . his courage, kindness of heart and unfailing good humor were an inspiration to all of us . . . died bravely in battle leading a gallant attempt to rescue members of his unit who were pinned down by German artillery. . . .[94]

Both these performances of Pynchon's catch nicely the essential point made by the Field Service Post Card: that we are all interchangeable parts, and that so long as we embrace that fact, our world will be fairly satisfactory and even cheerful—only belated mail and slight wounds need be feared. Another recent work in the tradition is "Lederer's Legacy," a short story by James Aitken about an exhausted senior clerk in Vietnam. His uniquely "modern" job consists in shifting around the clichés that make up the "citations" accompanying soldiers' medals. His repertory includes *tenacious devotion to duty, intrepid actions, a total lack of regard for his personal safety, inspired leadership, gallant display of heroic action, disregarding the danger involved, during the initial exchange of fire, overrunning the enemy positions*, and—most reminiscent of the Field Service Post Card—*displaying the utmost of personal bravery/courage/intrepidity*.[95] As Colonel Cathcart puts it, "Words cannot express. . . ."

SURVIVALS

Nobody alive during the war, whether a combatant or not, ever got over its special diction and system of metaphor, its whole jargon of techniques and tactics and strategy. (One reason we can use a term like *tactics* so readily, literally or in metaphors, is that the Great War taught it to us.) And often what impressed itself so deeply was something more than lan-

guage. Not a few works written during the war, and written about matters far distant from the war, carry more of the war about them than is
always recognized. Strachey's *Eminent Victorians*, published in 1918, is an
example. Strachey was working on it all during the long years of heart-
breaking frontal assaults on the German line, and the repeated and costly
aborting of these attacks seems to have augmented his appreciation of the
values of innuendo and the oblique, of working up on a position by
shrewd maneuver. In working up on the Victorians, the "direct
method," as he says in his Preface, won't do: "It is not by the direct
method of a scrupulous narration that the explorer of the past can hope
to depict [the Victorian Age]. If he is wise, he will adopt a subtler strategy. He will attack his subject in unexpected places; he will fall upon the
flank, or the rear. . . ." [96] (We may be reminded of I. A. Richards
blowing up the Hindenburg Line in 1926.) In the same way we can observe how Strachey's impatience with Haig's Scottish rigidity and faith
in a Haig-like God underlies some of the acerbity with which he treats
the last of his Eminent Victorians, General Gordon, slaughtered at
Khartoum for his stubbornness and megalomania.

Data entering the consciousness during the war emerge long afterward
as metaphor. Blunden conceives of memory itself as very like a trench
system, with main stems and lesser branches, as well as "saps" for reaching minor "positions": "Let the smoke of the German breakfast fires, yes,
and the savor of their coffee, rise in these pages [of *Undertones of War*],
and be kindly mused upon in our neighboring saps of retrogression." [97]
In 1948 Sassoon, searching for a figure to indicate the relation of his wartime poems to their context of civilian complacency, chooses to see them
as Very lights, "rockets sent up to illuminate the darkness." [98] Stephen
Spender reports on his father's "fairly extensive vocabulary of military
metaphor." He remembers: "Whenever one of us asked him a favor, he
would hold his head down with a butting gesture, and, looking up from
under his shaggy sandy eyebrows, say: 'You are trying to get round my
flank.' " [99] One result of the persistence of Great War rhetoric is that the
contours of the Second War tend to merge with those of the First. In
1940 eighteen-year-old Colin Perry was observing the London Blitz from
the perspective of Great War memoirs, of which he was a devotee. As he
says, "I like reading books about the last war in this one." [100] The result
was a conflation of wars through language, or, as one might put it, the
perception of one Great War running from 1914 to 1945. On one occasion Perry spent a weekend at St. Albans after a long period in
bombed London. His respite over, he had to re-enter the world of the
Blitz. Or, in his words for returning to London, "We went up the line

again at about 3 on the Sunday afternoon." Another time Perry tells us
that the town of Mitcham was severely bombed: indeed, it "caught a
packet." [101] A character in *Journey's End* couldn't put it more precisely.

Even if now attenuated and largely metaphorical, the diction of war
resides everywhere just below the surface of modern experience. American football has its two-platoon system, and medical research aspires to
breakthroughs. One says, "We were bombarded with forms" or "We've
had a barrage of complaints today" without, of course, any sharp awareness that one is recalling war and yet with a sense that such figures are
somehow most appropriate to the modern situation. The word *crummy*,
felt to apply to numerous phenomena of the modern industrial environment ("This is a crummy town"), is a Great War word, originally meaning itchy because lousy. And there is the word *lousy* itself, without
which, it sometimes seems, modern life could hardly be conducted.
Before the war *keepsake* was the English word for a small thing kept for
remembrance; after the British had lived in France for so long, the word
became *souvenir*, which has now virtually ousted *keepsake*. Popular discussion of economics relies heavily on terms like *sector* ("The public vs. the
private sector"), and the conduct of labor politics would be a very different thing without the military jargon (like *rank and file*) which it appropriated from the war.

Of the *trench* words like *trenchmouth*, *trenchfoot*, *trenchknife*, and *trenchcoat*, only *trench fever* has not survived. *Trenchcoat* (originally *Burberry*) has
recently spawned the adjective *trenchy* to describe a style in women's
fashions: "This year it's the trenchy look." In the *Sunday Times Magazine*
for August 8, 1971, a caption celebrating the vogue of the trenchy announces that "Fashion Goes Over the Top." That phrase is now also attached to charity drives without much residue of its original meaning:
"Help our fund drive over the top!" The most torpid television talk programs have sensed the need for stirring titles drawn from the war, like
"Firing Line" for William H. Buckley's interview program and the rather
complex pun "Behind the Lines" for a 1970 series of programs analyzing
the press.

The phrase No Man's Land has haunted the imagination for sixty
years, although its original associations with fixed positions and static
warfare are eroding. One of the photographs in Jacques Nobécourt's
Hitler's Last Gamble: The Battle of the Bulge (1967) shows three German
soldiers in snow-capes kneeling in a snow-covered open field with a pine
forest in the distant background. The caption reads, "A German patrol
in no-man's land," but there's really no parallel at all since the picture is
taken in the open in broad daylight with no damage visible anywhere—

even the pine trees are unscarred—and with no identifiable direction in which danger is conceived to lie. As it detaches itself further and further from its original associations, the phrase No Man's Land becomes available for jokes, like the title of Tony Parker's book of popular sociology, *In No Man's Land: Some Unmarried Mothers* (1972).

And a *French 75* is now a drink consisting of three parts champagne and one of cognac.

dramatic actions, then military training is very largely training in melodrama.

"Our own death is indeed, unimaginable," Freud said in 1915, "and whenever we make the attempt to imagine it we can perceive that we really survive as spectators." [2] It is thus the very hazard of military situations that turns them theatrical. And it is their utter unthinkableness: it is impossible for a participant to believe that he is taking part in such murderous proceedings in his own character. The whole thing is too grossly farcical, perverse, cruel, and absurd to be credited as a form of "real life." Seeing warfare as theater provides a psychic escape for the participant: with a sufficient sense of theater, he can perform his duties without implicating his "real" self and without impairing his innermost conviction that the world is still a rational place. Just before the attack on Loos, Major Pilditch testifies to "a queer new feeling these last few days, intensified last night. A sort of feeling of unreality, as if I were acting on a stage. . . ." [3]

Carrington testifies to the division of the psyche into something like actor, on the one hand, and spectator, on the other, especially during moments of heightened anxiety when one is "beside oneself." Before an attack, he says, "we checked and explained to one another the details of our plans. By this time I was beside myself and noticed how one ego calmly talked tactics while the other knew that all this energy was moonshine. One half of me was convinced that all this was real; the other knew it was illusion." The actor acts, the spectator observes and records: "I sent word to Sinker to take the morning rum issue round the platoons, and was then left to dissemble before my orderlies." Similarly at Passchendaele, where Carrington and his colleagues found themselves undergoing a bad shelling in an exposed position:

> We had nothing to do but to sit and listen for the roar of the 5.9's, lasting for five seconds each, perhaps twice a minute. One would be talking aimlessly of some unimportant thing when the warning would begin. The speaker's voice would check for an infinitesimal fraction of a second; then he would finish his sentence with a studied normality marvellously true to life. [4]

Countless wounded soldiers recall dividing into actor and audience at moments of the highest emergency. Stuart Cloete's "Jim Hilton," wounded in the shoulder, makes his way to the rear thus:

> The curious thing was that he was not here; he was somewhere else. On a high place, . . . looking down at this solitary figure picking its way between the shell holes. He thought: that's young Captain Jim Hilton, that little figure. I wonder if he'll make it. . . . He was an observer, not a participant. It

```
┌─────────────────────────────────────────────────┐
│                                                 │
│                    VI                           │
│               Theater of War                    │
│                                                 │
└─────────────────────────────────────────────────┘
```

VI
Theater of War

BEING BESIDE ONESELF

Peculiar to military language is the use of terms with significant unintended meanings which to the outsider may easily seem ironic. *Mess* is an example; so is *fatigue*. The conscript ordered to *Fall out!* at the end of a thirty-mile training march will find it hard to avoid an ironic response. In *Canto LXXVIII*, musing on modern wars, Pound lights on the cliché

> theatre of war

and thinking about it, immediately adds:

> "theatre" is good.

The most obvious reason why "theater" and modern war seem so compatible is that modern wars are fought by conscripted armies, whose members know they are only temporarily playing their ill-learned parts. Describing a Canadian brigadier, one military historian resorts instinctively to the idiom of the program-note associated with the theater, and the amateur theater at that: "He was in real life a bond and insurance broker in Vancouver." [1] If "real life" is "real," then military life must be pretense. The wearing of "costumes" not chosen by their wearers augments the sense of the theatrical. So does the availability of a number of generically rigid stage character-types, almost like those of Comedy of Humors: the hapless Private, the vainglorious Corporal, the sadistic Sergeant, the adolescent, snobbish Lieutenant, the fire-eating Major, the dotty Colonel. If killing and avoiding being killed are the ultimate melo-

was always like that in war though he had not realized it before. You were never you. The I part of you was somewhere else.[5]

And even when the emergency is past the theatrical remains, as Sassoon indicates. Playing the part of wounded young officer for various visitors to the hospital, he alters the role to make it effective with different audiences, just as an actor will often play a matinee broader than an evening performance. If the audience is "Some Senior Officer under whom I'd served," Sassoon is "modest, politely subordinate, . . . quite ready to go out again." His key line here is: "Awfully nice of you to come and see me, sir." Confronted by some "middle-aged or elderly Male Civilian," he notes in himself a "tendency . . . to assume haggard facial aspect of one who had 'been through hell.' " The line to be delivered now is: "Oh, yes, I'll be out there again by the autumn." To a "Charming Sister of Brother Officer," he is "jocular, talkative, debonair, and diffidently heroic," saying, "Jolly decent of you to blow in and see me." And to a "Hunting Friend," he makes "jokes about the Germans, as if throwing bombs at them was a tolerable substitute for fox-hunting." Now he must say: "By Jingo, that must have been a nailing good gallop!" The "real" Sassoon, he perceives, is the one that surfaces when the audiences have all gone, the one "mainly disposed toward a self-pitying estrangement from everyone except the troops at the front line." [6] And he hams it up even more flagrantly, he knows, when at Craiglockhart Hospital he mimes the character of a heroic pacifist before an appreciative audience consisting entirely of himself. "In what . . . I will call my soul (Grand Soul Theater: performances nightly), protagonistic performances were keeping the drama alive. (I might almost say that there was a bit of 'ham' acting going on at times.)" [7]

Curiously, theater and war were beginning to overlap each other's precincts even before the war began. A popular West End play during the season of 1911–12 was *An Englishman's Home,* by Major Guy du Maurier, an Army officer in South Africa. The play was one of the popular prewar threat-of-invasion works designed to convey a *frisson* by persuading the middle-class audience that the Empire was in jeopardy of being carved up by Continental powers as a result of democracy, sentimentality, and the deteriorating physique of the lower orders.[8] The play depicted the siege of an Englishman's house by invading soldiers. At the climax, the British father and householder shoots an enemy officer from a window, and is then taken out and executed. It was a long-running success, and its melodramatic effectiveness was sufficiently well known to be recalled as all-but-proverbial by Stanley Casson in 1915:

> As a cheerful party of us were enjoying a dinner-party at a [London] restau-
> rant, an orderly knocked at the door and handed me a pink army telegraphic
> form. My orders to leave the following day for France with a draft. The
> manner of my going was as dramatic as *The Englishman's Home*, though less
> futile. My friends chaffed me on the way it had been staged and accused me
> of having given the party a nice dramatic finish.[9]

The devices of the theater were frequently invoked at home to stimulate
civilian morale and to publicize the War Loan. There was not merely the
set of stage trenches in Kensington Gardens; there was the "ruined village" erected out of flats in Trafalgar Square. Told about it by his sister,
Clive Watts comments from the front: "I should like to be in Trafalgar
Square to see the ruined village, it must be rather good. I don't suppose
however they can reproduce the smell with anything like accuracy. Then
there is the dead horse question, without several specimens of which no
ruined village can be regarded as complete."[10] When the United States
entered the war, it followed suit and erected a wooden battleship, just
like a set for a Navy musical, which filled one side of Union Square in
New York City.

There was vivid irony in the practice of officers hurling themselves
into theatergoing when on leave and then returning to the theater of war
in France. Liddell Hart thinks it anomalous that he lusted after theater
when on leave. "I . . . felt an intense desire to see plays again," he
remembers, "more intense than ever before or since. . . . I went twice a
day to the theater."[11] P. H. Pilditch found himself unable to stay away
from the Oxford Theater, where he saw his own experience refracted in
The Better 'Ole: "The play was fully of the dirty, noisy scenes I know so
well and hate so heartily."[12] And yet he went, propelled perhaps by a
sense that because all the life he knew had turned theatrical, the theater
was the most seemly place to spend a leave. If he ever saw the *Wipers
Times*, he would have been quite accustomed to having the war conceived
in theatrical terms. Its pages specialized in facetious theatrical advertisements very close in idiom to the real ones in London:

OVER THE TOP
A SCREAMING FARCE.

. . .

A Stirring Drama,
ENTITLED:
M I N E D
A Most Uplifting Performance.

And there were ads for variety turns:

MISS MINNIE WERFER
Always Meets with a Thunderous Reception.

. . .

THE ATKINS FAMILY.
Celebrated "A" Frame Equilibrists and Duckboard Dodgers.

These were announced as appearing at the Cloth Hall, Ypres, "Best Ventilated Hall in the Town." If the *Wipers Times* thus brought the music hall to the war, Sassoon in "Blighters" felt obliged to bring the war to the music hall. It is notable that melodrama and music hall are the modes in which the British have twice memorably recalled the war on the stage, once in *Journey's End* in 1928, once in *Oh What a Lovely War* in 1963.

A temporary army consisting of strangers forcibly accumulated will offer constant opportunities for theatrical artifice—that is, fraud, illusion, and misrepresentation—in its members' relations with each other. Blunden recalls the temptations such an army puts in the way of those anxious to assume characters not their own. Arriving near the line at Béthune, he meets again an unlovely person he had first met at a camp at home: "He was noted for hairy raggedness and the desire to borrow a little money." Although Blunden, with characteristic decency, calls him merely "a doubtful blessing," it is clear that he was bad news in every way: the first thing he tells the new arrivals is that the line is "hell."

> In the huts at Shoreham, months before, he had been wont to quote soulfully the wild-west verses of Robert Service. . . , cantering rhetoric about huskies and hoboes on icy trails; at length he had said, with the modest yet authoritative tone suitable to such a disclosure, "I AM—Robert Service." Some believed. He never retreated from the claim; we heard it again in France; and the poor fellow was at last killed at Richebourg . . . in a hell more sardonic and sunnily devilish than ten thousand Robert Services could evolve, or wolves and grizzlies inhabit.[13]

(Actually, the real Robert Service lived until 1958.) This is a case simply more bizarre and pitiful than most. The temptation to melodramatic self-casting was everywhere.

Casson speaks of the satisfaction with which some soldiers fleeing from the German gas attack near Ypres in April, 1915, delivered the line, "We are the last of the Buffs,"[14] just as groups wandering rearward during the Mons retreat conventionally said of themselves, "We're the sole survivors of the Blankshire Regiment." They were told by an officer—who might be thought on this occasion to be playing the role of a dramatic critic—"All right. Keep straight on for a couple of miles or more and you will find three or four hundred other sole survivors of your regiment bivouacking in a field." [15]

There is hardly a book about the war that at some point does not avail itself of a theatrical figure. Blunden's fifteenth and sixteenth chapters in *Undertones of War* are titled "Theatre of War" and "A German Performance," and so different a book as George Adam's propagandistic, optimistic emission of 1915 is titled *Behind the Scenes at the Front*. Its antithesis, Barrie Pitt's sophisticated and embittered modern history, *1918*, is sub-titled *The Last Act*. And speaking of the last act, there is one unforgettable vignette which if true is fine, and which if aprocryphal is even better. It is Herbert Essame's memory of the German machine-gunner signaling the closing of a long run on November 11, 1918: "On the Fourth Army front, at two minutes to eleven, a machine gun, about 200 yards from the leading British troops, fired off a complete belt without a pause. A single machine-gunner was then seen to stand up beside his weapon, take off his helmet, bow, and turning about walk slowly to the rear." [16] It is with considerable acuteness that Mary McCarthy, discoursing on the character Archie Rice in John Osborne's *The Entertainer*, notices that "the actor and the soldier have the same mythology." What they require alike is "timing, co-ordination, a cool nerve, resourcefulness." Furthermore, "before a performance every actor experiences a slight case of battle-nerves, and actors, like soldiers, are superstitious." [17]

THE BRITISH THEATRICAL FLAIR

If soldiers in general are like actors, British soldiers are more like actors than others, as a reading of German, French, and American memoirs of the war will confirm. The machine-gun actor taking his final bow may have been German, but the eye that noticed and the hand that recorded were British. Instead of reaching toward the cool metaphor of stage plays, Remarque, in *All Quiet*, and Jünger, in *Storm of Steel*, invoke overheated figures of nightmares and call upon the whole frenzied machinery of Gothic romance. Chapter 4 of *All Quiet* enacts a mad and quite un-British Gothic fantasia as a group of badly disorganized German troops is shelled in a civilian cemetery. Graves are torn asunder, coffins are hurled in the air, old cadavers are flung out—and the narrator and his chums preserve themselves by crawling into the coffins and covering themselves with the stinking cerements. This will remind us less of *Hamlet* than of, say, *The Monk*. In a similar way Jünger, seeking an imagery adequate to his feelings as the German bombardment reaches a climax before the great attack in March, 1918, conjures up all the traditional violence of Gothic and Renaissance Europe that a Romantic imagination can conceive and packs it into one sentence. After asserting that "the tremendous force of destruction that bent over the field of battle was concentrated in our brains," he writes: "So may men of the Renaissance have been locked in their passions, so may a Cellini have raged or werewolves have howled and hunted through the night on the track of blood." [18] As imagery this is disastrous: it sounds worse than the war. The British way is more phlegmatic and ironic, more conscious that if the war is not real, it must be not real in a more understandable, social way.

I think there are two main reasons for the British tendency to fuse memories of the war with the imagery of theater. One is the vividness of the sense of role enjoined by the British class system. The other is the British awareness of possessing Shakespeare as a major national asset.

Before the war, as Orwell noticed, the British exhibited the most conspicuous class system in Europe. "In 1910," he says, "every human being in these islands could be 'placed' in an instant by his clothes, manners and accent." [19] Even army regiments were ranked in a strict hierarchy of class, from the Guards at the top to the Territorials at the bottom. In an attack, the antiquity of a regiment determined its position on the line: the "senior" regiment expected the position of honor, on the right, just as in the dramaturgy of a formal dinner-party. After the war problems of caste greatly complicated the work of the Imperial War Graves Commission, which very early discovered that many officers' families assumed that their people would be decently segregated from Other Ranks in the cemeteries. It was with a sense of promulgating a bold innovation that the Commission finally concluded that, since the dead could now be regarded as practically "equal," the cemeteries in France should recognize no traditional social distinctions.[20] Christopher Isherwood remembers what it was like to live as a child in a prewar "middle-class" establishment:

> As long as he was trotting around after [the maids] or doing play jobs which Cook invented for him in the kitchen, he was like a stagehand behind the scenes in a theatrical production, he was part of the show. But when the curtain finally went up and some of the maids put on starched aprons and became actresses who served lunch in the dining room, then Christopher was excluded. He had to sit still at the table and be waited on. He was just a member of the audience.[21]

In contrast to the American scene, British life is pervaded by the sense of theater allied to this instinct for class distinctions. It is the British who praise one another by saying, "Good show!" During the war, it was the British, rather than the French, the Americans, the Italians, the Portuguese, the Russians, or the Germans, who referred to trench raids as "shows" or "stunts," perhaps because it is in British families that charades and amateur theatricals are a convention. And it is English playwrights—or at least Anglo-Irish ones—like Wilde and Shaw who compose plays proclaiming at every point that they are plays. It is the British, not the Americans, who understand acting to be a mode of artificial representing—i.e., *mimesis*—rather than a mode of natural "expressing." It is England that enjoys using such words as *posh* and *toff* and until very recently insisted on a distinction between the Public and the Saloon bars.

And it all dies very slowly indeed. J. B. Priestley observes that when representatives of labor and management face each other in a television debate, "I feel I am watching character actors playing—and often overplaying—their allotted parts." The date is 1972:

> The trade union leaders, producing their flattest accents, contrive to be both mumbling and pontifical, and seem to love phrases like "primary function of the Executive" and similar committee jargon.
>   At the same time the representatives of the management, perhaps carefully chosen, offer us caricatures of upper-class types, and insist upon saying "Naturalleh" and "Couldn't agray with yoh moah!" and might almost be great-nephews of Wodehouse characters.[22]

On the other hand, to indicate the special presence of Shakespeare in English writing about the war all one has to do is point to Frederic Manning's novel *Her Privates We* (1930) and try to imagine it as a French, German, or American performance. From its title (*Hamlet*, II, ii, 233) to its Shakespearian chapter-headings and its constant awareness of Henry V at Agincourt, it is permeated with a consciousness of Shakespeare not just as a literary but specifically as a theatrical resource. We have seen David Jones's attempt to deal with the war as if it had some relation to the battle where one saw "young Harry with his beaver on." Like Manning's, Jones's sense of Shakespeare is not just literary; in his understanding of the supreme difficulty of rendering the war in mere language, it is theatrical too: "No one, I suppose," he says, "however much not given to association, could see infantry in tin-hats, with ground-sheets over their shoulders, with sharpened pine-stakes in their hands, and not recall 'or may we cram/Within this wooden O . . .' "[23] (*Henry V*, Prologue, 12–13). When lonely and scared on the Somme, J. S. Wane used to recite speeches from *Henry V*,[24] and W. L. P. Dunn likewise found that play a very present help in time of trouble. Walking forward on July 1, 1916, watching his friends fall and writhe as the machine-gun fire went through them, "the thought that was uppermost in my mind," he remembers, "was the phrase 'For England,' which I seemed to be repeating continually. This is the truth and not put in for heroics. To be perfectly truthful, I was scared stiff." [25] As perhaps Shakespeare's Henry really was when he urged his men to "Cry 'God for Harry, England, and St. George!' " (*Henry V*, III, i, 34).

ON SET

Most people were terrified, and for everyone the dramaturgic provided a dimension within which the unspeakable could to a degree be familiarized and interpreted. After all, just as a play must have an ending, so might the war; just as an actor gets up unhurt after the curtain falls on his apparent murder, so might the soldier. And just as a play has a structure, so might a war conceived as analogous to a play have a structure— and with it, a meaning. It is meaning that Sassoon is searching for as he divides his experience into "Acts"—three of them, we will notice—of a play notable for its structure of preparation, climax, and release. First come the four weeks at the army school at Flixécourt: "Those four weeks had kept their hold on my mind, and they now seemed like the First Act of a play—a light-hearted First Act. . . ." And he goes on: "Life at the Army School, with its superb physical health, had been like a prelude to some really conclusive sacrifice of high-spirited youth." After this preparation, climax: "Act II had carried me along to the fateful First of July." Finally, release: "Act III had sent me home to think things over." [26] The model play Sassoon appears to have in mind would seem to be a melodrama. Others as well equated their experiences with conventional elements of this kind of play. To Richard Blaker, a sniper leaning against his shield "eavesdropping" on the enemy is like "a theatrical eavesdropper in some corridor scene." [27] As J. R. Ackerley watches his brother prepare to lead a suicidal raid in which he is going to be killed, only theatrical clichés seem appropriate. "Here was a situation," says Ackerley, "which would have appealed to the actor in him, drama indeed, the lime-lit moment, himself in the leading role, all eyes on him." As Ackerley sees him off on his mission, he notices that "it was impossible to speak the most commonplace word or make the most ordinary gesture without its at once acquiring the heavy over-emphasis of melodrama." [28]

For others, the idea of an attack was closer to comedy, or to the "screaming farce" of the *Wipers Times*. Carrington speaks of the luckless

General Sir Robert Fanshawe, "Fanny" to his subordinates. The minute he achieved a command, "Fanny" submitted a plan for a vainglorious attack, which was, says Carrington, "duly pigeonholed by the staff as 'Fanny's First Play,' since they knew their man." [29] (Shaw's 1911 play of that name is identifiably English, being a play about a play as well as about theatrical criticism and the varieties of theatrical genres.) The unreality of pantomime and the impression it makes on the child are what Guy Chapman thinks of when he recalls his first fire-trench,[30] and "Uncle Harry," whose memoirs are in the Imperial War Museum, invokes the image of an immense papier-maché pantomime costume head to do justice to the size and oddity of a German head seen for a moment raised above the top of its trench. Unreality and futility are the characteristics assigned to pantomime: R. Rolleston West recalls that firing at German planes with small-arms was "a pantomime." [31]

The frequent "concerts" offered behind the line by official concert-parties kept the idea of revue and music hall lively as an ironic analogy. Blunden exploits it in his poem "Concert Party: Busseboum," where, after the men have delighted in the dancing, jokes, and music of the traveling troupe, they emerge from the show at cold sunset to find in the distance another "show" in progress, with its own music, jokes, and dancing:

> We heard another matinée,
> We heard the maniac blast
>
> Of barrage south by Saint Eloi,
> And the red lights flaming there
> Called madness: Come, my bonny boy,
> And dance to the latest air.
>
> To this new concert, white we stood;
> Cold certainty held our breath;
> While men in the tunnels below Larch Wood
> Were kicking men to death.

A raid is going on. To Blunden, the word *raid* "may be defined as the one in the whole vocabulary of the war which most instantly caused a sinking feeling in the stomach. . . ." [32] To Sassoon as well the idea of a raid is so shocking that in defense he betakes himself to the imagery of the concert-party, speaking of raids as "entertainments" and of the men who have blackened their faces with burnt cork as a troupe of "minstrels" about to put on a "concert." [33] Guy Chapman sensed something about one highly organized attack that resembled music hall, especially

the way it was "advertised" to the enemy well before it began.[34] The
name of Heath Robinson, cartoonist and illustrator, and designer of fantastic "Rube Goldberg" comedy sets for the Alhambra and Empire music
halls, was often pressed into service. Looking down on the trench lines
from the air, Cecil Lewis calls the scene a "spectacle," which, as a theatrical term, has about it some of the exasperation of Goneril's "An interlude!" (*King Lear*, V, iii, 90). The trench system, says Lewis, "had all
the elements of grotesque comedy—a prodigious and complex effort,
cunningly contrived, and carried out with deadly seriousness, in order to
achieve just nothing at all. It was Heath Robinson raised to the *n*th
power—a fantastic caricature of common sense." [35] Just what one would
see at the Empire. And as Philip Gibbs remembers, "Any allusion to 'the
Empire' left [the troops] stone-cold unless they confused it with the Empire Music Hall, when their hearts warmed to the name." [36] For the artistically more sophisticated, the idea of music-hall "concerts" could con-
note something like Commedia del Arte, as it does to the aesthete Hugh
Quigley. As Passchendaele degenerated into the ridiculous, all pretence
at serious purpose disappeared: "Columbine degenerated to a frowsy
Fleming dispensing miserable coffee. Harlequin a naked wretch with not
a spangle on him, not a single sequin to clothe his frail humanity. We
were just Puncinellos without the lustre, fools doing foolish
things. . . ." [37]

Those who actually fought the war tended to leave the inviting analogy to Greek or any other kind of tragedy to the journalists. On the
spot, it looked less like a tragedy than like a melodrama, a farce, or a
music-hall turn. Or even a school-pageant played by enthusiastic but
unprepared boy actors impersonating soldiers, trying to look like Tennysonian heroes. The mental equipment of the public school boys who
went "straight from the cricket-field and the prefect's study to the
trenches," says Douglas Goldring, was as useful "for withstanding the
shock of experience . . . as the imitation suit of armor, the dummy lance
and shield of the actor in a pageant." [38]

But the theater requires audiences as well as players. The audiences
were supplied sometimes by the enemy, sometimes by the British. Attacking into German rifle fire near Ypres, Casson "had precisely the feeling that comes over one when the curtain goes up at amateur theatricals.
Here were we, the performers, until so recently, idly sitting in the
wings. There was the audience waiting to give us the reception we de-
served." [39] When the British withdrew most secretly from the Gallipoli
peninsula, they played to the Turks. "Small parties carrying towels as if
for bathing parade," says Aitken, "had gone down open saps and come

back up covered ones, circling on their steps like a stage army." [40] On this occasion the whole theatrical *ruse de guerre* was a smashing success, and the Turks saw through the illusion only after the ships had taken all the actors away.

Evening stand-to was perhaps the most theatrical time: the lights went down as if by rheostat, and the British audience stared intently into the darkness. One could almost hear the pit-orchestra tuning up and the last thumps behind the curtain. It was exciting, as David Jones remembers:

> A long way off a machine-gunner seemed as one tuning an instrument, who strikes the same note quickly, several times, and now a lower one, singly; while scene-shifters thud and scrape behind expectant curtaining; and impatient shuffling of the feet—in the stalls they take out watches with a nervous hand, they can hardly bear it. [41]

Excessive curiosity about what was taking place on the stage was sometimes fatal to the more impulsive members of the audience. As Kipling imagines it in one of his *Epitaphs of the War* called "The Beginner,"

> On the first hour of my first day
> In the front trench I fell.
> (Children in boxes at a play
> Stand up to watch it well.)

Touring the Somme cemeteries two years after the war, Stephen Graham observed many placarded "full" and "closed": "A notice says 'Cemetery closed' as one might read outside a theater at night—'Pit full,' 'Gallery full,' 'Stalls full.' " [42]

To all these theatrically minded perceivers, the trench setting became something very close to a "trench set." To Wyndham Lewis such ubiquitous details as German skulls, "festoons of mud-caked wire," and "trees like gibbets" constituted props quite naturally attending "those titanic casts of dying and shell-shocked actors, who charged this stage with a romantic electricity." [43] And the actual dummy items manufactured for purposes of camouflage and deception were close to literal theatrical props. In 1916 S. S. Horsley visited "the camouflage factory" in Amiens. He found it "a wonderful workshop where imitation trees, guns, figures, sandbags, etc., were made of plaster and paper. Most realistic. . . . Amongst other subjects was a dead horse, and a dead German, to be used as an observation post." [44]

It is amazing how much technical knowledge of stage engineering some of these soldiers display as a matter of course. In describing the view from the forward trench they speak of cycloramas, backcloths, and drop-scenes, just as if they were stage managers weary with expertise.

And they delight to relate changes in the landscape to the kind of effect associated with that ever-popular staple of Georgian theater, the transformation scene. Such a scene was achieved by very gradually changing the lighting on a painted scrim—a sheet of gauze or a fine shrimp-net. What before had appeared solid—the side of a brick building, say— could be made gradually transparent by simultaneously reducing the light on the front of the scrim and increasing the light behind it. When the "transformation" was complete, the now invisible scrim was raised quickly to the flies. The effect was the equivalent of "dissolve" or "mix" in cinema; but where cinematic dissolve is largely a mere transitional device, not designed to be appreciated for its own sake, "transformation" was so remarkable, so breathtaking and miraculous, that it became an effect treasured for itself, a distinct constituent of "the magic of the theater." No serious production could satisfy without one.

At the front, "transformation scenes" were to be witnessed at morning and evening stand-to. Richard Aldington's short story "At All Costs" focuses on a character, Hanley, who looks out across No Man's Land after a long night:

> Hanley got on the fire-step in a shell-smashed abandoned bay, and watched with his glasses slung round his neck. The artillery had died down to a couple of batteries, when the first perceptible lightening of the air came. . . . Very gradually, very slowly, the darkness dissipated, as if thin imperceptible veils were being rolled up in a transformation scene.[45]

It was the same at night. Gladden recalls: "As the dusk began to fall the landscape towards the town gradually took on a new aspect. It was like a transformation scene on a vast stage." [46] It was just as familiar and just as marvelous as what one used to see on an evening out in London.

THE CARICATURE SCENES OF ROBERT GRAVES

Of all memoirs of the war, the "stagiest" is Robert Graves's *Good-bye to All That,* published first in 1929 but extensively rewritten for its reissue in 1957. Like James Boswell, who wrote in his journal (October 12, 1780), "I told Erskine I was to write Dr. Johnson's life in scenes," Graves might have said in 1929 that it was "in scenes" that he was going to write of the front-line war. And working up his memories into a mode of theater, Graves eschewed tragedy and melodrama in favor of farce and comedy, as if anticipating Friedrich Dürrenmatt's observation of 1954 that "comedy alone is suitable for us," because "tragedy presupposes guilt, despair, moderation, lucidity, vision, a sense of responsibility," none of which we have got:

> In the Punch-and-Judy show of our century . . . there are no more guilty
> and also, no responsible men. It is always, "We couldn't help it" and "We
> didn't really want that to happen." And indeed, things happen without any-
> one in particular being responsible for them. Everything is dragged along
> and everyone gets caught somewhere in the sweep of events. We are all col-
> lectively guilty, collectively bogged down in the sins of our fathers and of
> our forefathers. . . . That is our misfortune, but not our guilt. . . . Com-
> edy alone is suitable for us.[47]

And in Graves's view, not just comedy: something close to Comedy of
Humors, a mode to which he is invited by the palpable character conventions of the army, with its system of ranks, its externalization of per-
sonality, its impatience with ambiguity or subtlety, and its arcana of
conventional "duties" with their invariable attendant gestures and
"lines." "Graves," says Randall Jarrell, "is the true heir of Ben Jonson." [48] Luxuriating in character types, Graves has said few things more
revealing about his art than this: "There is a fat boy in every school (even
if he is not really very fat), and a funny-man in every barrack-room (even
if he is not really very funny). . . ." [49]

In considering *Good-bye to All That*, it is well to clear up immediately
the question of its relation to "fact." J. M. Cohen is not the only critic to
err badly by speaking of the book as "harshly actual" and by saying, "It
is the work of a man who is not trying to create an effect." Rather than
calling it "a direct and factual autobiography," Cohen would have done
better to apply to it the term he attaches to Graves's Claudius novels.
They are, he says, "comedies of evil." [50] Those who mistake *Good-bye to
All That* for a documentary autobiography (Cohen praises its "accurate
documentation") should find instructive Graves's essay "P.S. to 'Good-
bye to All That,' " published two years after the book appeared.
Confessing that he wrote the book to make "a lump of money" (which he
did—he was able to set himself up in Majorca on the royalties), he enu-
merates the obligatory "ingredients" of a popular memoir:

> I have more or less deliberately mixed in all the ingredients that I know are
> mixed into other popular books. For instance, while I was writing, I re-
> minded myself that people like reading about food and drink, so I searched
> my memory for the meals that have had significance in my life and put them
> down. And they like reading about murders, so I was careful not to leave
> out any of the six or seven that I could tell about. Ghosts, of course. There
> must, in every book of this sort, be at least one ghost story with a possible
> explanation, and one without any explanation, except that it was a ghost. I
> put in three or four ghosts that I remembered.
>   And kings. . . . People also like reading about other people's mothers.

. . . And they like hearing about T. E. Lawrence, because he is supposed to be a mystery man. . . . And, of course, the Prince of Wales.

People like reading about poets. I put in a lot of poets. . . . Then, of course, Prime Ministers. . . . A little foreign travel is usually needed; I hadn't done much of this, but I made the most of what I had. Sport is essential. . . . Other subjects of interest that could not be neglected were school episodes, love affairs (regular and irregular), wounds, weddings, religious doubts, methods of bringing up children, severe illnesses, suicides. But the best bet of all is battles, and I had been in two quite good ones—the first conveniently enough a failure, though set off by extreme heroism, the second a success, though a little clouded by irresolution.

So it was easy to write a book that would interest everybody. . . . And it was already roughly organized in my mind in the form of a number of short stories, which is the way that people find it easiest to be interested in the things that interest them. They like what they call "situations."

Furthermore, "the most painful chapters have to be the jokiest." [51] Add "the best bet of all is battles" to "the most painful chapters have to be the jokiest" and divide by the idea of "situations" and you have the formula for Graves's kind of farce. The more closely we attend to Graves's theory and practice, the more we can appreciate the generic terminology used by "Odo Stevens," in Anthony Powell's *Temporary Kings*. Stevens was one who "hovered about on the outskirts of the literary world, writing an occasional article, reviewing an occasional book. . . . [He] had never repeated the success of *Sad Majors*, a work distinguished, in its way, among examples of what its author called 'that dicey art-form, the war reminiscence.' " [52]

"Anything processed by memory is fiction," as the novelist Wright Morris has perceived.[53] In *Leviathan*, Thomas Hobbes puts it this way: "Imagination and memory are but one thing, which for divers considerations hath divers names." [54] And in *An Apology for Poetry*, Sir Philip Sidney apprehends the "poetic"—that is, fictional—element not just in all "history" but specifically in history touching on wars and battles:

Even historiographers (although their lips sound of things done, and verity be written in their foreheads) have been glad to borrow both fashion and perchance weight of poets. . . . Herodotus . . . and all the rest that followed him either stole or usurped of poetry their passionate describing of passions, the many particularities of battles, which no man could affirm, or . . . long orations put in the mouths of great kings and captains, which it is certain they never pronounced.[55]

We expect a memoir dealing with a great historical event to "dramatize" things. We have seen Sassoon's memoir doing just that. But with Graves we have to expect it more than with others, for he is "first and last," as Jarrell sees, "a poet: in between he is a Graves." [56] A poet, we remember Aristotle saying, is one who has mastered the art of telling lies successfully, that is, dramatically, interestingly. And what is a Graves? A Graves is a tongue-in-cheek neurasthenic farceur whose material is "facts." Hear him on what happens to the wives of brilliant mathematicians:

> Mathematic genius is . . . notoriously short-lasting—it reaches a peak at the age of about twenty-three and then declines—and is as a rule colored by persistent emotional adolescence. Since advanced mathematicians are too easily enticed into the grey political underworld of nuclear physics, a remarkably high percentage of mental breakdowns among their wives is everywhere noted.[57]

Asked by a television interviewer whether his view that homosexuality is caused by the excessive drinking of milk is "based on intuition or on what we would call scientific observation," Graves replies: "On objective reasoning." [58] His "objective reasoning" here is as gratuitously outrageous as the anthropological scholarship of *The White Goddess*, the literary scholarship of his translation (with Omar Ali Shah) of *The Rubaiyat of Omar Khayyam*, or the preposterous etymological arguments with which he peppers his essays.

But to put it so solemnly is to risk falling into Graves's trap. It is to ignore the delightful impetuosity, the mastery, the throw-away fun of it all. Graves is a joker, a manic illusionist, whether gaily constructing flamboyant fictional anthropology, re-writing ancient "history," flourishing erroneous or irrelevant etymology, over-emphasizing the importance of "Welsh verse theory," or transforming the White Goddess from a psychological metaphor into a virtual anthropological "fact." And the more doubtful his assertions grow, the more likely he is to modify them with adverbs like *clearly* or *obviously*. Being "a Graves" is a way of being scandalously "Celtish" (at school "I always claimed to be Irish," he says in *Good-bye to All That*" [39] [59]). It is a way—perhaps the only way left—of rebelling against the positivistic pretensions of non-Celts and satirizing the preposterous scientism of the twentieth century. His enemies are always the same: solemnity, certainty, complacency, pomposity, cruelty. And it was the Great War that brought them to his attention.

Actually, any man with some experience and a bent toward the literal can easily catch Graves out in his fictions and exaggerations. The unsophisticated George Coppard explodes one of the melodramatic facilities in *Good-bye to All That* with simple common sense. Graves asserts—it is a popular cynical vignette—that machine-gun crews often fired off several belts without pause to heat the water in the cooling-jacket for making tea (109–10). Amusing but highly unlikely—Coppard quietly notes that no one wants tea laced with machine oil.⁶⁰ Another of Graves's machine-gun anecdotes collapses as "fact" upon inquiry. At one point he says,

> There was a daily exchange of courtesies between our machine-guns and the Germans' at stand-to; by removing cartridges from the ammunition belt one could rap out the rhythm of the familiar prostitutes' call: "MEET me DOWN in PICC-a-DILL-Y," to which the Germans would reply, though in slower tempo, because our guns were faster than theirs: "YES, with-OUT my DRAWERS ON!" (170–71)

Very nice. But the fact is that if you remove cartridges from the belt the gun stops working when the empty space encounters the firing mechanism. (These stories are like the popular legend that in a firing squad one man is given a rifle secretly loaded with a blank so that no member of the squad can be certain that he has fired one of the fatal bullets. But attractive as this is as melodrama, there's something wrong with it: the rifle containing the blank is the only one that will not recoil when fired, with the result that every man on the squad will end by knowing anyway. The story won't do.)

But we are in no danger of being misled as long as we perceive that *Good-bye to All That* is no more "a direct and factual autobiography" than Sassoon's memoirs. It is rather a satire, built out of anecdotes heavily influenced by the techniques of stage comedy. What Thomas Paine says of Burke's *Reflections on the Revolution in France* applies exactly: Burke, says Paine, makes "the whole machinery bend to produce a stage effect." ⁶¹ No one has ever denied the brilliance of *Good-bye to All That,* and no one has ever been bored by it. Its brilliance and compelling energy reside in its structural invention and in its perpetual resourcefulness in imposing the patterns of farce and comedy onto the blank horrors or meaningless vacancies of experience. If it really were a documentary transcription of the actual, it would be worth very little, and would surely not be, as it is, infinitely re-readable. It is valuable just because it is not true in that way. Graves calls on paradox to suggest the way it is true:

> The memoirs of a man who went through some of the worst experiences of trench warfare are not truthful if they do not contain a high proportion of falsities. High-explosive barrages will make a temporary liar or visionary of anyone; the old trench-mind is at work in all over-estimation of casualties, "unnecessary" dwelling on horrors, mixing of dates and confusion between trench rumors and scenes actually witnessed.⁶²

In recovering "the old [theatrical] trench-mind" for the purposes of writing the book, Graves has performed a triumph of personal show business.

He was in an especially rebellious mood when he dashed off the book in eight weeks during May, June, and July of 1929 and sent the manuscript to Jonathan Cape. His marriage with Nancy Nicolson had just come apart, he owed money, he had quarreled with most of his friends, his view of English society had become grossly contemptuous, and he was still ridden by his wartime neurasthenia, which manifested itself in frequent bursts of tears and bouts of twitching. His task as he wrote was to make money by interesting an audience he despised and proposed never to see again the minute he was finished. Relief at having done with them all is the emotion that finally works itself loose from the black humor which dominates most of the book.

The first nine chapters detail his prewar life. He was, he says, a perceptive, satiric, skeptical infant, from the outset an accurate appraiser of knaves and fools, including Swinburne, "an inveterate pram-stopper and patter and kisser" (1). His Scotch-Irish father was a school inspector, but also a composer, collector, and anthologist of Anglo-Irish songs. In addition, he was a popular dramatist, one of whose plays ran for two hundred performances. His first wife, who was Irish, died after bearing five children, and he then married a German woman who bore him five more, including Robert, born in 1895. The family lived at Wimbledon in ample, literate middle-class style while Robert attended a succession of preparatory schools and spent summers roaming through the romantic castles near Munich belonging to relatives of his mother's. At fourteen he entered Charterhouse School, which he despised. He was humiliated and bullied, and saved himself only by taking up boxing. He mitigated his loneliness by falling in love with a younger boy, "exceptionally intelligent and fine-spirited. Call him Dick" (48). (The name *Dick* was becoming conventional for this sort of thing. Sassoon's *Memoirs of a Fox-Hunting Man*, with its "Dick Tiltwood," had appeared a year before Graves wrote this.) Graves's devotion to Dick and his friendship with one of the masters, the mountaineer George Mallory, were about all he enjoyed at Charterhouse. Before he could go on to Oxford, the war began, and he enlisted immediately. He was nineteen.

In this nine-chapter prologue Graves practices and perfects the form of the short theatrical anecdote or sketch which he will proceed to impose upon the forthcoming matter offered by the war. His wry anecdotes take the shape of virtual playlets, or, as he is fond of calling them, especially when he is one of the players, "caricature scenes." They are "theatrical" because they present character types entirely externally, the way an au-

Robert Graves. (Black Star)

dience would see them. The audience is not vouchsafed what they are or
what they think and feel or what they were last Thursday, but only visible or audible signs of what they do and say, how they dress and stand
or sit or move or gesture. Their remarks are not paraphrased or rendered
in indirect discourse: they are presented in dialogue. Many of these
playlets have all the black-and-white immediacy of cartoons with cap-
tions, and indeed Graves's skill at writing pithy "lines" will suggest the
dynamics of the standard two-line caption under a cartoon in *Punch*. It is
a model that is always before him. In 1955, ridiculing Yeats's shrewd ir-
rationalism, he dramatizes Yeats's reliance on his wife as a medium
whose maunderings can be turned into salable poems:

UNDERGRADUATE: Have you written any poems, recently, Sir?
YEATS: No, my wife has been feeling poorly and disinclined.[63]

One can see it as on a stage and hear the burst of laughter at the end.
    Whatever material they embody, the effect of Graves's "caricature
scenes" is farcical, and they rely on a number of techniques associated

with comic writing for the theater. Some depend upon astonishing coincidences. Some deploy the device of climactic multiple endings—the audience thinks the joke is over and is then given an additional one or sometimes two even funnier lines. Some expose the disparity between the expected and the actual. Some offer bizarre characters borrowed from what would seem to be a freak show. Some, like sketches in music hall, present comic encounters between representatives of disparate social classes. Some involve the main character's not knowing some crucial fact. And some, more melodramatic, depict rescues or salvations in the nick of time. All operate by offering the audience a succession of little ironies and surprises. By the time we have reached the fifth paragraph of *Good-bye to All That*, we are convinced that we are in the hands of a master showman who is not going to let us down. "My best comic turn," says the author, "is a double-jointed pelvis. I can sit on a table and rap like the Fox sisters with it." Indeed, so extraordinary is this puppet-master that, as he says proudly, "I do not carry a watch because I always magnetize the main-spring" (3).

Graves had been in the Officers Training Corps at Charterhouse, and when he presented himself at the regimental depot of the Royal Welch Fusiliers at Wrexham, he was commissioned after a few weeks' training. His account of his early days in the Army is full of caricature scenes. One of the funniest, the grave judicial inquiry into the "nuisance" deposited by Private Davies on the barrack square, Graves introduced into the book only in 1957. In 1929 he said, "I have an accurate record of the trial, but my publishers advise me not to give it here." [64] It is a perfect Jonsonian comic scene, each man in his humor, and it is ready to be staged by a cast of six:

> SERGEANT-MAJOR (*off stage*): Now, then, you 99 Davies, "F" Company, cap off, as you were, cap off, as you were, cap off! That's better. Escort and prisoner, right *turn!* Quick *march!* Right wheel! (*On stage*) Left wheel! Mark time! Escort and prisoner, *halt!* Left *turn!*
>
> COLONEL: Read the charge, Sergeant-Major.
>
> SERGEANT-MAJOR: No. 99 Pte. W. Davies, "F" Company, at Wrexham on 20th August: improper conduct. Committing a nuisance on the barrack square. Witness: Sergeant Timmins, Corporal Jones.
>
> COLONEL: Sergeant Timmins, your evidence.
>
> SERGEANT TIMMINS: Sir, on the said date about two p.m., I was hacting Horderly Sar'nt. Corporal Jones reported the nuisance to me. I hinspected it. It was the prisoner's, Sir.
>
> COLONEL: Corporal Jones! Your evidence.
>
> CORPORAL JONES: Sir, on the said date I was crossing the barrack square, when I saw prisoner in a sitting posture. He was committing excreta, Sir. I took his name and reported to the orderly-sergeant, Sir.

COLONEL: Well, Private Davies, what have you to say for yourself?
99 DAVIES (*in a nervous sing-song*): Sir, I came over queer all of a sudden, Sir.
  I haad the diarrhoeas terrible baad. I haad to do it, Sir.
COLONEL: But, my good man, the latrine was only a few yards away.
99 DAVIES: Colonel, Sir, you caan't stop nature!
SERGEANT-MAJOR: Don't answer an officer like that! (*Pause*)
SERGEANT TIMMINS (*coughs*): Sir?
COLONEL: Yes, Sergeant Timmins?
SERGEANT TIMMINS: Sir, I had occasion to hexamine the nuisance, Sir, *and it
  was done with a heffort, Sir!*
COLONEL: Do you take my punishment, Private Davies?
99 DAVIES: Yes, Colonel, Sir.
COLONEL: You have done a very dirty act, and disgraced the regiment and
  your comrades. I shall make an example of you. Ten days' detention.
SERGEANT-MAJOR: Escort and prisoner, left *turn!* Quick *march!* Left wheel!
  (*Off stage*): Escort and prisoner, *halt!* Cap on! March him off to the Guard
  Room. Get ready the next case! (77–78).

Despite such moments, Graves was proud to be in so self-respecting a
regiment as the Royal Welch Fusiliers, a mark of whose distinction was
the "flash," a fan-like cluster of five black ribbons attached to the back of
the tunic collar. The Army Council had some doubts about permitting
the regiment this irregular privilege, but the Royal Welch resisted all attempts to take it away. Graves's pride in it is enacted in this little bit of
theater, warmly sentimental this time, set in Buckingham Palace:

> Once, in 1917, when an officer of my company went to be decorated with
> the Military Cross at Buckingham Palace, King George, as Colonel-in-Chief
> of the Regiment, showed a personal interest in the flash. . . . The King
> gave him the order "About turn!," for a look at the flash, and the "About
> turn!" again. "Good," he said, "You're still wearing it, I see," and then, in a
> stage whisper: "Don't ever let anyone take it from you!" (85).

That is typical of Graves's theatrical method: the scene is a conventional,
almost ritual confrontation between character types representative of
widely disparate classes who are presented externally by their physical
presence and their dialogue. We feel that the King would not be playing
the scene properly if his whisper were anything but a *stage-whisper:* after
all, the audience wants to hear what he's saying.

Posted to France as a replacement officer in the spring of 1915, Graves
disgustedly finds himself assigned to the sad and battered Welsh Regiment, consisting largely of poorly trained scourings and leavings. His
platoon includes a man named Burford who is sixty-three years old, and
another, Bumford, aged fifteen. These two draw together with a theatrical symmetry which might be predicted from the similarity of their

names: "Old Burford, who is so old that he refuses to sleep with the other men of the platoon, has found a private doss in an out-building among some farm tools. . . . Young Bumford is the only man he'll talk to" (107–8). We are expected to credit this entirely traditional symmetrical arrangement with the same willing suspension of disbelief which enables us to enjoy the following traditional turn. Two men appear before the adjutant and report that they've just shot their company sergeant major.

> The Adjutant said: "Good heavens, how did that happen?"
> "It was an accident, Sir."
> "What do you mean, you damn fools? Did you mistake him for a spy?"
> "No, Sir, we mistook him for our platoon sergeant" (109).

Punch again.

After some months in and out of the line near Béthune, Graves finally joins the Second Battalion of his own regiment near Laventie, and his pride in it suffers a sad blow. He is horrified to find the senior regular officers bullies who forbid the temporary subalterns, or "warts," whiskey in the mess and ignore them socially for a period of six months except to rag and insult them whenever possible. He is humiliated by the colonel, the second-in-command, and the adjutant just as he had been humiliated by the "Bloods" at Charterhouse. But he finds one man to respect, Captain Thomas, his company commander. It is he who must direct the company's part in a preposterous attack, which begins as farce and ends as Grand Guignol.

The operation order Thomas brings from battalion headquarters is ridiculously optimistic, and as he reads it off, Graves and his fellow officers—including a subaltern called "The Actor"—can't help laughing.

> "What's up?" asked Thomas irritably.
> The Actor giggled: "Who in God's name is responsible for this little effort?"
> "Don't know," Thomas said. "Probably Paul the Pimp, or someone like that." (Paul the Pimp was a captain on the Divisional Staff, young, inexperienced, and much disliked. He "wore red tabs upon his chest, And even on his undervest.")

Thomas reveals that their attack is to be only a diversion to distract the enemy while the real attack takes place well to the right.

> "Personally, I don't give a damn either way. We'll get killed whatever happens."
> We all laughed (145).

The attack is to be preceded by a forty-minute discharge of gas from cylinders in the trenches. For security reasons the gas is euphemized as "the accessory." When it is discovered that the management of the gas is in the hands of a gas company officered by chemistry dons from London University, morale hits a comic rock-bottom. "Of course they'll bungle it," says Thomas. "How could they do anything else?" (146). Not only is the gas bungled: everything goes wrong. The storeman stumbles and spills all the rum in the trench just before the company goes over; the new type of grenade won't work in the dampness; the colonel departs for the rear with a slight cut on his hand; a crucial German machine gun is left undestroyed; the German artillery has the whole exercise taped. The gas is supposed to be blown across by favorable winds. When the great moment proves entirely calm, the gas company sends back the message "Dead calm. Impossible discharge accessory," only to be ordered by the staff, who like characters in farce are entirely obsessed, mechanical, and unbending: "Accessory to be discharged at all costs" (152). The gas, finally discharged after the discovery that most of the wrenches for releasing it won't fit, drifts out and then settles back into the British trenches. Men are going over and rapidly coming back, and we hear comically contradictory crowd noises: " 'Come on!' 'Get back, you bastards!' 'Gas turning on us!' 'Keep your heads, you men!' 'Back like hell, boys!' 'Whose orders?' 'What's happening?' 'Gas!' 'Back!' 'Come on!' 'Gas!' 'Back!' " (154–55). A "bloody balls-up" is what the troops called it. Historians call it the Battle of Loos.

(A word about the rhetoric of "Impossible discharge accessory." That message falls into the category of Cablegram Humor, a staple of Victorian and Georgian comedy. Graves loves it. Compare his 1957 version of Wordsworth's "The Solitary Reaper" in Cable-ese: "SOLITARY HIGHLAND LASS REAPING BINDING GRAIN STOP MELANCHOLY SONG OVERFLOWS PROFOUND VALE." [65])

As the attack proceeds, farce gradually modulates to something more serious but no less theatrical. One platoon officer, attacking the untouched German machine gun in short rushes, "jumped up from his shell-hole, waved and signalled 'Forward!' "

Nobody stirred.
He shouted: "You bloody cowards, are you leaving me to go on alone?"
His platoon-sergeant, groaning with a broken shoulder, gasped: "Not cowards, Sir. Willing enough. But they're all f——ing dead." The . . . machine-gun, traversing, had caught them as they rose to the whistle (156).

At the end of the attack Graves and the Actor were the only officers left in the company.

After this, "a black depression held me," Graves says (170). And his worsening condition finds its correlative in the collapse of his ideal image of Dick, at home. The news reaches him that sixteen-year-old Dick has made "a certain proposal" to a Canadian corporal stationed near Charterhouse and has been arrested and bound over for psychiatric treatment (171). "This news," says Graves, "nearly finished me. I decided that Dick had been driven out of his mind by the War. . . . with so much slaughter about, it would be easy to think of him as dead" (171). (The real Dick, by the way, was finally "cured" by Dr. W. H. R. Rivers, Sassoon's and Owen's alienist at Craiglockhart.) This whole matter of Dick and his metamorphosis from what Graves calls a "pseudo-homosexual" into a real one lies at the heart of *Good-bye to All That*. Its importance was clearer in the first edition, where Graves says,

> In English preparatory and public schools romance is necessarily homosexual. The opposite sex is despised and hated, treated as something obscene. Many boys never recover from this perversion. I only recovered by a shock at the age of twenty-one. For every one born homo-sexual there are at least ten permanent pseudo-homo-sexuals made by the public school system. And nine of these ten are as honorably chaste and sentimental as I was.[66]

In 1957 Graves deleted one sentence: "I only recovered by a shock at the age of twenty-one." The shock was his discovery that he had been deceived by pleasant appearances: a relation he had thought beneficially sentimental now revealed itself to have been instinct with disaster. It was like the summer of 1914. It makes a telling parallel with Graves's discovery—"Never such innocence again"—that the Second Battalion of the Royal Welch Fusiliers, a few company-grade officers and men excepted, is a collection of bullies, knaves, cowards, and fools.

He is delighted to find himself transferred to the more humane First Battalion in November, 1915. There he meets Sassoon, as well as Sassoon's "Dick," Lieutenant David Thomas. The three become inseparable friends while the battalion begins its long rehearsals for the breakout and open warfare it assumes will follow the Somme attack in the spring. Life in billets offers opportunities for numerous caricature scenes. One takes place in the theater-like setting of a disused French schoolroom, where the officers of the battalion are addressed by their furious colonel. He has noticed slackness, he says, and as he designates an instance of it, he falls naturally into the Graves mode of theatrical anecdote, complete with a consciousness of social distinctions and the "lines" appropriate to different social players:

I have here principally to tell you of a very disagreeable occurrence. As I left my Orderly Room this morning, I came upon a group of soldiers. . . . One of these soldiers was in conversation with a lance-corporal. You may not believe me, but it is a fact that he addressed the corporal by his Christian name: *he called him Jack!* And the corporal made no protest. . . . Naturally, I put the corporal under arrest. . . . I reduced him to the ranks, and awarded the man Field Punishment for using insubordinate language to an N.C.O (179).

Listening to this as a member of the "audience," Graves is aware of the "part" he himself is playing in this absurd costume drama:

Myself in faultless khaki with highly polished buttons and belt, revolver at hip, whistle on cord, delicate moustache on upper lip, and stern endeavor a-glint in either eye, pretending to be a Regular Army captain. . . .

But "in real life" he is something quite different, "crushed into that inky desk-bench like an overgrown school-boy" (179–80).

Back in the line again in March, 1916, the battalion has three officers killed in one night, including David Thomas. "My breaking-point was near now," Graves recognizes, and he speculates on the way his nervous collapse, when it comes, will look to spectators. His view of it is typically externalized, the tell-tale gestures visualized as if beheld by someone watching a character on stage: "It would be a general nervous collapse, with tears and twitchings and dirtied trousers; I had seen cases like that" (198). His transfer back to the hated Second Battalion is hardly a happy omen, and in early July, 1916, he finds himself in incredible circumstances near High Wood on the Somme. On July 20, his luck runs out: a German shell goes off close behind him, and a shell fragment hits him in the back, going right through his lung. He is in such bad shape at the dressing-station that his colonel, assuming he's dying, kindly writes his parents, informing them that he has gone. As a result his name appears in the official casualty list: he has "Died of Wounds."

A few days later Graves manages to write home and assure his parents that he is going to recover. There is some discrepancy about dates here: for symbolic and artistic reasons, Graves wants the report of his death to coincide with his twenty-first birthday (July 24), although his father remembers the date as earlier. "One can sympathize with Graves," says George Stade, "who as a poet and scholar has always preferred poetic resonance to the dull monotone of facts; and to die on a twenty-first birthday is to illustrate a kind of poetic justice.[67]

Back in hospital in London, Graves is delighted by the combined com-

edy and melodrama of a clipping from the Court Circular of the *Times:*
"Captain Robert Graves, Royal Welch Fusiliers, officially reported died
of wounds, wishes to inform his friends that he is recovering from his
wounds at Queen Alexandra's Hospital, Highgate, N" (227). Almost immediately, he quotes another funny document, the infamous propaganda
pamphlet containing a letter by "a Little Mother" reprehending any
thought of a negotiated peace and celebrating the sacrifice of British
mothers who have "given" their sons. It is sentimental, bloodthirsty,
complacent, cruel, fatuous, and self-congratulatory, all at once, and ("of
course," Graves would say) it is accompanied by a train of earnest illiterate testimonials from third-rate newspapers, non-combatant soldiers, and
bereaved mothers, one of whom says: "I have lost my two dear boys, but
since I was shown the 'Little Mother's' beautiful letter a resignation too
perfect to describe has calmed all my aching sorrow, and I would now
gladly give my sons twice over" (231).

It is at this point in *Good-bye to All That* that we may become aware of
how rich the book is in fatuous, erroneous, or preposterous written
"texts" and documents, the normal materials of serious "history" but
here exposed in all their farcical ineptitude and error. Almost all of
them—even Sassoon's "A Soldier's Declaration"—have in common some
dissociation from actuality or some fatal error in assumption or conclusion. Their variety is striking, and there are so many that Graves felt he
could cut one entirely from the 1957 edition, the priceless letter at the
end of Chapter II [68] from an amateur gentlewoman poet, instinctively
praising Graves's very worst poem and at the same time slyly begging a
loan with a long, rambling, self-celebrating paranoid tale of having been
cheated of an inheritance. There is the "question and answer history
book" of his boyhood, which begins

> *Question:* Why were the Britons so called?
> *Answer:* Because they painted themselves blue (17).

There are the propaganda news clippings about the priests of Antwerp,
hung upside down as human clappers in their own church bells (67–68).
There is the laughable Loos attack order (144–47), and the optimistic orders, all based on false premises, written on field message forms
(216, 242). There is the colonel's letter deposing not merely that Graves is
dead but that he was "very gallant" (219). There is the erroneous casualty list (218–20) and the Letter of the Little Mother. There is the farcical mistransmission in Morse code that sends a battalion destined for
York to Cork instead (270). There is an autograph collector's disoriented
letter to Thomas Hardy, beginning

Dear Mr. Hardy,
   I am interested to know why the devil you don't reply to my request . . .
(306).

There are the lunatic examination-papers written by three of Graves's students of "English Literature" at the University of Cairo (334–37). And in the new Epilogue, written in 1957, there is the news that one reason Graves was suspected of being a German spy while harboring in South Devon during the Second War is that someone made a silly, lying document out of a vegetable marrow in his garden by surreptitiously scratching "HEIL HITLER!" on it (345). The point of all these is not just humankind's immense liability to error, folly, and psychosis. It is also the dubiousness of a rational—or at least a clear-sighted—historiography. The documents on which a work of "history" might be based are so wrong or so loathsome or so silly or so downright mad that no one could immerse himself in them for long, Graves implies, without coming badly unhinged.

The Letter from the Little Mother is the classic case in point and crucial to the whole unraveling, satiric effect of *Good-bye to All That*. One of Graves's readers, 'A Soldier Who Has Served All Over the World," perceived as much and wrote Graves:

> You are a discredit to the Service, disloyal to your comrades and typical of that miserable breed which tries to gain notoriety by belittling others. Your language is just "water-closet," and evidently your regiment resented such an undesirable member. The only good page is that quoting the beautiful letter of The Little Mother, but even there you betray the degenerate mind by interleaving it between obscenities.[69]

A pity that letter wasn't available to be included in *Good-bye to All That*. It is the kind of letter we can imagine Ben Jonson receiving many of.

By November, 1916, Graves is well enough to put on his uniform again—the entry and exit holes in the tunic neatly mended—and rejoin the Depot Battalion for re-posting. He is soon back with the Second Battalion on the Somme, where he is secretly delighted to find that all his enemies, the regular officers, have been killed or wounded: it makes the battalion a nicer place and fulfills the angry prophecy Graves had uttered when he first joined and had been bullied in the Officers' Mess at Laventie: "You damned snobs! I'll survive you all. There'll come a time when there won't be one of you left in the Battalion to remember this Mess at Laventie" (129). But the weakness in Graves's lung is beginning to tell, and he is returned to England with "bronchitis." He finds himself first in the hospital at Somerville College, Oxford, and then recuperating ano-

malously and comically at Queen Victoria's Osborne, on the Isle of
Wight.

It is while at Osborne that he sees Sassoon's Declaration. He is appalled at the risk of court-martial Sassoon is taking and distressed by
Sassoon's political and rhetorical naïveté: "Nobody would follow his example, either in England or in Germany." The public temper had already found its spokesman in the Little Mother: "The War would inevi-
tably go on until one side or the other cracked" (261). Graves gets out of
Osborne, rigs Sassoon's medical board, and testifies before it. He bursts
into tears "three times" and is told, "Young man, you ought to be before
this board yourself" (263). His dramaturgy is successful, and Sassoon is
sent to Craiglockhart for "cure." Graves tells us that there Sassoon first
met Wilfred Owen, "an idealistic homosexual with a religious background." At least that is what he wanted to tell us in the American edition (the Anchor Books paperback) of the 1957 re-issue.[70] The phrase
was omitted from the British edition at the request of Harold Owen, and
it was subsequently canceled in the American edition. It does not now
appear in any edition of *Good-bye to All That*.[71] Just as Graves always
knew they would, respectability and disingenuousness have won. Just as
Graves learned during the war, written documents remain a delusive
guide to reality.

He is now classified B-1, "fit for garrison service abroad," but despite
his hope to be sent to Egypt or Palestine, he spends the rest of the war
training troops in England and Ireland. In January, 1918, he married the
feminist Nancy Nicolson: "Nancy had read the marriage-service [another
funny document] for the first time that morning, and been so disgusted
that she all but refused to go through with the wedding . . . .":

> Another caricature scene to look back on: myself striding up the red carpet,
> wearing field-boots, spurs and sword; Nancy meeting me in a blue-check
> silk wedding dress, utterly furious; packed benches on either side of the
> church, full of relatives; aunts using handkerchiefs; the choir boys out of
> tune; Nancy savagely muttering the responses, myself shouting them in a
> parade-ground voice (272).

The news of the Armistice, he says, brought him no pleasure; rather, it
"sent me out walking alone . . . , cursing and sobbing and thinking of
the dead" (278).

Demobilized, he instantly catches Spanish influenza and almost dies of
it. He recovers in Wales, where for almost a year he tries to shake off the
war:

I was still mentally and nervously organized for War. Shells used to come bursting on my bed at midnight, even though Nancy shared it with me; strangers in daytime would assume the faces of friends who had been killed. When strong enough to climb the hill behind Harlech . . . , I could not help seeing it as a prospective battlefield. I would find myself working out tactical problems, planning . . . where to place a Lewis gun if I were trying to rush Dolwreiddiog Farm from the brow of the hill, and what would be the best cover for my rifle-grenade section. . . .

Some legacies of the war ran even deeper, and one, perhaps, has had literary consequences: "I still had the Army habit of commandeering anything of uncertain ownership that I found lying about; also a difficulty in telling the truth . . ." (287). His experience of the Army had ratified his fierce insistence on his independence, and he swore "on the very day of my demobilization never to be under anyone's orders for the rest of my life. Somehow I must live by writing" (288). In October, 1919, he entered Oxford to study English Literature, living five miles out, at Boar's Hill, where he knew Blunden, Masefield, and Robert Nichols. There he and Nancy briefly ran a small general store while he wrote poems as well as his academic thesis, brilliantly titled—the war had certainly handed him the first three words—*The Illogical Element in English Poetry.*

"The Illogical Element in the Experience of Robert Graves" might be the title of the episode that closes *Good-bye to All That.* He takes up the position of Professor of English Literature at the ridiculous Royal Egyptian University, Cairo. The student essays are so funny and hopeless that as an honest man he can't go on. After saying that "Egypt gave me plenty of caricature scenes to look back on" (341), he approaches the end of the book in a final flurry of anecdotes and vignettes, most of them farcical, and concludes with a brief paragraph summarizing his life from 1926 to 1929, which he says has been "dramatic," with "new characters [appearing] on the stage" (343). All that is left is disgust and exile.

Compared with both Blunden and Sassoon, Graves is very little interested in "nature" or scenery: human creatures are his focus, and his book is built, as theirs are not, very largely out of dialogue. And compared with Sassoon, who is remarkably gentle with his characters and extraordinarily "nice" to them, Graves, who had, as Sassoon once told him, "a first-rate nose for anything nasty," [72] sees his as largely a collection of knaves and fools. Almost literally: one can go through *Good-bye to All That* making two lists, one of knaves, one of fools, and the two lists will comprise ninety per cent of the characters. As a memoirist, Graves seems most interested not in accurate recall but in recovering moments when he most clearly perceives the knavery of knaves and the foolishness of fools. For him as for D. H. Lawrence, knavery and folly are the style of the war, and one of the very worst things about it is that it creates a theater perfectly appropriate for knavery and folly. It brings out all the terrible people.

If Graves, the scourge of knaves and fools, is the heir of Ben Jonson, it can be seen that Joseph Heller is the heir of Graves. And the very theatricality of *Catch-22* is a part of what Heller has learned from *Good-bye to All That*. *Catch-22* resembles less a "novel" than a series of blackout skits, to such a degree that it was an easy matter for Heller to transform the work into a "dramatization" in 1971. Another legatee of Graves is Evelyn Waugh, whose *Sword of Honor* trilogy does to the Second War what Graves did to the First. Waugh's book is made up of the same farcical high-jinks, the same kind of ironic reversals, all taking place in the Graves atmosphere of balls-up and confusion. Indeed, both Graves and Waugh include characters who deliver the line, "Thank God we've got a Navy." [73] If Loos is the characteristic absurd disaster to Graves, Crete is Waugh's version. Waugh's sense of theater is as conspicuous as Graves's, although it tends to invoke more pretentious genres than farce. During the rout on Crete, a small sports car drives up: "Sprawled in the back, upheld by a kneeling orderly, as though in gruesome parody of a death scene from grand opera, lay a dusty and bloody New Zealand officer." [74] Both Graves and Waugh have written fiction-memoirs, although Graves's is a fiction disguised as a memoir while Waugh's is a memoir disguised as a fiction. To derive Waugh's trilogy, one would superadd the farce in *Good-bye to All That* to the moral predicament of Ford's Tietjens in *Parade's End:* this would posit Guy Crouchback, Waugh's victim-hero, as well as establish a world where the broad joke of Apthorpe's thunder-box co-exists harmoniously with messy and meaningless violent death. And both Waugh and Heller would be as ready as Graves to agree with the proposition that comedy alone is suitable for us.

RECALLING THE WAR AS THEATER

The sense of proscenium theater attaching to personal experience in the Great War seems appropriate for a war settled for years into fixed positions and fought by people adhering very largely to "fixed" traditional images. "The slow pace at which the soldier of those days had to move," says Blunden, "and the long months spent in the same areas, helped to engrave the picture on the mind." [75] Just as appropriate will seem the more kinetic sense of cinema attaching to the experience of later wars, wars characterized by a new geographical remoteness, mobility, rapidity, complex technology, and ever-increasing incredibility. "This was like a movie," thinks Hennessey, landing on a remote South Pacific beach in Mailer's *The Naked and the Dead*.[76] To Americans, the Great War in France was as remote as, say, the Second War in the Solomon Islands: "Over There" (meaning *way* Over There) is characteristically an American, not a British, song of the war. And the most notable piece of Great War writing which makes extensive use of cinematic rather than traditional stage parallels is American. It is John Dos Passos's *Three Soldiers* (1921), and in it we find the heroic fantasies of Fusselli—he hasn't encountered the reality of the war yet—assuming the conventions of cinematic melodrama, as if unwinding from "long movie reels of heroism." [77] His dreams "of glory" fall naturally into the shape of the visual clichés of such cinematic costume melodramas as Griffith's *Birth of a Nation* (1915). Even the conventional speeded-up marching of the troops seems a special index of "Americanness." Fusselli is having "a dream of what it would be like over there":

> He was in a place like the Exposition ground, full of old men and women in peasant costumes, like in the song "When It's Apple Blossom Time in Normandy." Men in spiked helmets who looked like firemen kept charging through, like the Ku Klux Klan in the movies, jumping from their horses and setting fire to buildings with strange outlandish gestures, spitting babies on their long swords. Those were the Huns. Then there were flags blowing very hard in the wind, and the sound of a band. The Yanks were coming. Everything was lost in a scene from a movie in which khaki-clad regiments marched fast, fast across the scene.[78]

That takes us some distance towards Pynchon's identification, in *Gravity's Rainbow*, of war with film: "On D-Day," says a German film-maker, "when I heard General Eisenhower on the radio announcing the invasion of Normandy, I thought it was really Clark Gable, have you ever noticed? the voices are *identical*. . . ." [79] "Yes, it *is* a movie," Pynchon announces, arrived almost at the final disclosure that modern life itself—equivalent, as his book has shown, to modern war—is a film too.[80] If not literally, then in this way: those who fought in the Second War couldn't help noticing the extra dimension of drama added to their experiences by their memories of the films about the Great War. Pinned down on the Anzio beachhead in 1944, Raleigh Trevelyan sees that what he is doing has a relation if not to life at least to art: "We were jammed head to toe, completely immobile, with volleys of tracer like whiplashes a matter of inches overhead. It was a complete *All Quiet on the Western Front* film set once more." [81]

And if one's perception of the Second War naturally takes the form of one's response to cinema, one's perception of the Vietnam War equates that experience with the films of the Second War—and with those films as seen on late-night television. The hero of Wayne Karlin's short story "Medical Evacuation" has trouble conceiving of himself as anything but a player in some obsolete Second War film:

> Standing by for medevacs, he always had the feeling he was being centered in a movie camera. Flight suits and sun glasses, casually late show and focused in sharply by the magnifying clearness of the Danang air. . . . . A long shot: the crews running self-consciously to the helicopters. . . . Switch to the interior of the helicopter, the close-up lens zooming in for a really fine shot of his sweaty grease-stained face framed over his machine gun by the open port. Then a full shot of the interior. . . . [82]

Having witnessed mankind in a state of war all his life, William Burroughs is capable of the very modern (because post–Great War) perception that the human race has really been "on set" from the very beginning, being "shot" when not "shooting." [83]

The intercourse between the idea of the Great War and the idea of the dramaturgic has been happily notated by Anthony Burgess in his novel *The Wanting Seed* (1962), set in England in the twenty-first century. The continuation of present tendencies has resulted in a vast increase in population and a corresponding scarcity of food. The adversaries are fecundity, on the one hand, and starvation on the other. The British government attacks the problem in many ways: encouraging homosexuality; punishing families who exceed the one-birth-per-family rule; extending the living space by attaching steel "Annex Islands" to the land mass of the United Kingdom; importing "canned man" (in the form of bully-beef) from China; and staging fake Great War trench battles to trim the population of such anti-social elements as female "cretinous over-producers" and male "corner boys and . . . criminals" (216). [84]

Burgess's victim-hero is Tristram Foxe, a schoolmaster. He is married to Beatrice-Joanna Foxe, who secretly loves Tristram's brother Derek, a high government official posing as a homosexual both to hold his job and to facilitate access to Beatrice-Joanna. Impregnated by Derek, she flees to the Midlands to conceal her traitorous fecundity. There she has twins. Tristram is deprived of his job and sets off on foot to find his wife. He fails to locate her, but during his picaresque wanderings he himself becomes a pariah and ends up, like other pariahs, as a conscript in a vague "war." It is vague because no one knows or will say who the

enemy is. But because there is an army there must be a war, as one
soldier tries to make clear to Tristram:

> ". . . it stands to reason you've got to have a war. Not because anybody
> wants it, of course, but because there's an army. . . . Armies is for wars
> and wars is for armies. That's only plain common sense."
> "War's finished," said Tristram. "War's outlawed. There hasn't been any
> war for years and years and years."
> "All the more reason why there's got to be a war," said the [soldier], "if
> we've been such a long time without one."
> "But," said Tristram, agitated, "you've no conception what war was like.
> I've read books about the old wars. They were terrible, terrible. . . . We
> can't have all that again. I've seen photographs," he shivered. "Films,
> too . . ." (156).

The forthcoming "war," designed actually as part of the final solution
of the population problem, seems distinctly anomalous and archaic. For
one thing, "The British Army . . . was pure infantry with minimal support of specialist corps." For another, the men in it behave curiously,
"up-thumbing at the [newsreel] cameras with a partially dentate leer, the
best of British luck and pluck" (174). Tristram finds himself a sergeant-
instructor with "the new armies." When he begins to wonder in front of
the men what the war is about, he is taken off instructional duties and
assigned as a platoon sergeant in a rifle company. His suspicions increase
when he finds his platoon, commanded by a credulous young subaltern,
Mr. Dollimore, engaged in such exercises as naming the parts of Great
War rifles. "The organization, nomenclature, procedure, armament of
this new British Army all seemed to have come out of old books, old
films" (180). Curious.

But Mr. Dollimore doesn't notice that anything is amiss, so full is his
mind of Great War clichés, tags from Rupert Brooke, and hearty verbal
formulas devised for use on the Western Front a century and a half earlier. Tristram, obsessed by not knowing who the enemy is supposed to
be, is struck by the drama of it all, the palpable theatricality of the proceedings, the way all the theater of the Great War—that classic—is invoked to validate this war by making it "look right." In the hold of a
troop transport, moving toward the first encounter with the enemy, the
troops are singing Great War songs ("We were up at Loos / When you
were on the booze") while Tristram broods in his bunk:

> It seemed to him that he had been suddenly transported to a time and place
> he had never visited before, a world out of books and films, ineffably an-
> cient. Kitchener, napoo, Bottomley, heavies, archies, zeppelins, Bing
> Boys—. . . this was really a film, really a story, and they had all been

caught in it. The whole thing was fictitious, they were all characters in somebody's dream (190).

He decides that whatever may be the sinister farce shaping up around him, he must somehow save himself and survive. He will continue "playing his part" until he can escape. But it all seems so real that even Tristram's emotions get engaged. Just before disembarcation, when Mr. Dollimore says to the platoon, "This old country we love so well. We'll do our best for her, won't we, chaps?" and the men lower their eyes in embarrassment, "Tristram suddenly felt a great love for them" (193), as if instructed in the proper style by· something like Wilfred Owen's "Greater Love."

As the battalion forms up on the quay, all the details are right, all perfectly "Great War":

> Lamps were few, as if some modified blackout regulation prevailed. RTO men loped round with clipboards. MPs strolled in pairs. A red-tabbed Major with a false patrician accent haw-hawed, slapping his flank with a leather-bound baton. Mr. Dollimore was summoned, with other subalterns, to a brief conference near some sheds. Inland the crumps of heavies and squeals of shells, sheet lightning, all part of the war film (194).

The battalion marches to the "base camp" down a road on both sides of which "the flashes showed pruned trees like stage cut-outs . . ." (194). A corporal begins to suspect out loud that they have landed in Ireland to fight "the Micks" (195). Tristram, on the other hand, knows for a certainty that the war is being "staged" when he perceives that the sound of artillery up ahead emanates from an amplifier playing a cracked record: "Dada *rump*, dada *rump*, dada *rump*. . . . *too* regular" (195). It is, all too clearly now, a "gramophony war" (196).

The next morning the platoon is told that they will be going up the line that evening, "a 'show' of some sort being imminent":

> Mr. Dollimore shone with joy at the prospect. "Blow out, you bugles, over the rich dead!" he quoted, tactless, to his platoon.
> "You seem to have the death-urge pretty strongly," said Tristram, cleaning his pistol.
> "Eh? Eh?" Mr. Dollimore recalled himself from his index of first lines. "We shall survive," he said. "The Boche will get what's coming to him."
> "The Boche?"
> "The enemy. Another name for the enemy. During my officers' training course," said Mr. Dollimore, "we had films every evening. It was always the Boche. No, I'm telling a lie. Sometimes it was Fritz. And Jerry, sometimes."

"I see. And you also had war poetry?"
"On Saturday mornings . . ." (197).

An ancient airplane, "—strings, struts, an open cockpit and waving goggled aeronaut—lurched over the camp and away again. 'One of ours,' Mr. Dollimore told his platoon. 'The gallant R.F.C.' " (197).

At evening, during a flamboyant sunset—this is, after all, exactly like "that prototypical war" (198)—the men draw ammunition and begin their final march up to the trenches. "They marched through a hamlet, a contrived Gothic mess of ruins," and to a shattered country house which proves to be the entrance to a trench. " 'Queer sort of 'ouse this is,' grumbled a man. . . . It was a mere shell, like something from a film-set" (199). Tristram's platoon arrives at its assigned section of trench, and Tristram is ordered by Mr. Dollimore to get the men under cover in the dugouts and to post sentries. While "the record-players banged away at their simulacra of passionate war," Tristram's fury finally boils over:

"What," said Tristram, "is the point of posting sentries? There's no enemy over there. The whole thing is a fake. . . . Don't you see? This is the new way, the modern way, of dealing with excess population. The noises are fakes. The flashes are fakes. Where's our artillery? Did you see any artillery behind the lines? Of course you didn't. Have you seen any shells or shrapnel?" (200).

Horrified by this spreading of alarm and despondency, Mr. Dollimore takes out his pistol and fires at Tristram. He misses, and Tristram decides he must be more careful. He now "knew that there was no way back, behind the lines. If there was any way at all, it lay ahead, over the top and the best of luck" (202). But not, Tristram realizes, in joining the attack, set for 2200 hours. Mr. Dollimore gives the order to fix bayonets and to load all rifles, " 'not forgetting one up the spout.' Mr. Dollimore stood upright, looking important. 'England,' he suddenly said, and his nose filled with tears" (203). With ten minutes to go the bombardment stops, and "in this unfamiliar hushed dark the enemy could more clearly be heard, coughing, whispering, in the light tones of small-boned Orientals" (203–4). At a few minutes before the attack Mr. Dollimore is so excited and ennobled that he begins conflating Rupert Brooke with *Henry V*:

Pistol trembling, eyes never leaving his wrist-watch, [he] was ready to lead his thirty over (some corner of a foreign field) in brave assault, owing God a death (that is for ever England) (204).

With five minutes to go Tristram has decided that if Mr. Dollimore is
going to lead the platoon over, he himself will push them over and then
shrewdly stay behind.

> 2155. "O God of battles," Mr. Dollimore was whispering, "steel my sol-
> dier's heart." . . . 2158. The bayonets trembled. Somebody started hic-
> coughs and kept saying, "Pardon." 2159. "Ah," said Mr. Dollimore, and he
> watched his second-hand. . . . "We're coming up now to, we're coming up
> now to—"
>   2200. Whistles skirled shrilled deadly silver all along the line, and the
> phonographic bombardment clamored out at once dementedly (204).

Mr. Dollimore takes them over:

> Rifles bitterly cracked and spattered. . . . Bullets dismally whinged. There
> were deep bloody curses, there were screams. Tristram, his head above the
> parapet, saw etched black cut-out bodies facing each other in hand-to-hand,
> clumsy, falling, firing, jabbing, in some old film about soldiers. He distinctly
> observed Mr. Dollimore falling back . . . and then crashing with his mouth
> open. . . . It was slaughter, it was mutual massacre, it was impossible to
> miss (204–5).

And then Tristram has a clear view of the enemy and realizes that they
are not Orientals: they are women.

Three minutes after the battle has begun it is all over. Sanitary squads
move over the battlefield finishing off the wounded with pistol shots.
Tristram plays dead in the trench and they pass him by. When everything is quiet, he scoops up the pool of the platoon's money left behind
in a dugout as "a prize for the survivors—an ancient tradition" (207) and
makes his way directly across the battlefield. A mile behind the enemy
trench-line he comes to a wire-mesh fence with a gate in it. It marks the
perimeter of the large "set" which is the battlefield. He discovers where
he is: in Southwestern Ireland, near Dingle. He manages to escape to
Killarney, to Rosslare, to Fishguard, to Brighton, and finally to London,
where he sardonically reports himself to the Army as "the sole survivor"
and threatens to expose the whole murderous plot. A major explains that
that won't do: Tristram is compromised, having taken part in an "E.S.":

> "What's an E.S.?"
> "Extermination Session. That's what the new battles are called. . . ."
> "It was . . . murder," said Tristram violently. "Those poor, wretched,
> defenseless—"
> "Oh, come, they weren't exactly defenseless, were they? Beware of cli-
> chés, Foxe. . . . They were well trained and well armed and they died

gloriously, believing they were dying in a great cause. And, you know, they really were" (216).

Besides, the major hints, the dead have been useful in another way. They have been turned into bully-beef (219); it is the classic Great War rumor about the German Corpse Rendering Factory all over again. After all this it is not much of a comfort to find Tristram reunited with a contrite Beatrice-Joanna in a half-parodic happy ending.

What is impressive about *The Wanting Seed* is not, of course, its plausibility. It is hardly more plausible than *Good-bye to All That*. It is not far-fetched to compare the two, for the contours of Graves's caricature scenes are dimly visible underneath many of Burgess's. What does make *The Wanting Seed* an impressive work is its power of imagination and understanding. Assisted by extensive, indeed obsessed, reading in Great War memoirs and animated by a profound appreciation of the way the Great War establishes the prototype for modern insensate organized violence, Burgess has sensed both the alliance between war and theater and the necessity of our approach to the Great War through the channels of myth and cliché. Which is to say that he has located War as a major source of modern myth. The book is permeated by an understanding of the modern world as the inevitable construct of a whole series of wars. Housewives buy "rations"; the Population Police ("Poppol"), whose duty is to enforce the laws against illegal pregnancies, wear Great War pips and crowns to denote rank; people lodge in "billets"; a standard greeting is "Keep Smiling!": the whole of the new world is very like that of the old Army. In the same way, the battlefield "set" manages to suggest some theatrical things very much of our time, like the false railway station constructed at the Treblinka concentration camp in the 1940's, with its "synthetic ticket window, painted first- and second-class waiting rooms, clock fixed at 3 P.M." [85]

A similar emblem of the modern world's instinct for dissimulation (or self-deception) assisted by technology is the phonograph record or "recorded message," and this fascinates Burgess as much as "sets" do. One of Tristram's conversations with Lt. Dollimore is interrupted when "a scratchy record hissed from the loudspeakers, a synthetic bugle blared its angelus" (181). Just before departing for "overseas," Tristram's battalion is addressed by its colonel, but not in person: he delivers his replicated valedictory exhortation on a record—unfortunately cracked—played over a loudspeaker:

> "You will be fighting an evil and unscrupulous enemy in the defense of a noble cause. I know you will cover yourselves in glo in glo in glo craaaaaark

come back alive and therefore I say godspeed and all the luck in the world go with you." A pity, thought Tristram, drinking tea-substitute in the sergeants' mess, a pity that a cracked record should make that perhaps sincere man sound so cynical. A sergeant from Swansea . . . got up from the table singing in a fine tenor, "Cover yourselves in glow, in glow, in glo-o-ow" (187).

It is the cracked artillery record that convinces Tristram that the battlefield is a stage, and later, as he meditates on what has happened to him, he associates the colonel's cracked farewell message with the cracked artillery record: "Cover yourselves in glow, in glow, dada *rump*" (212). The image of electronic fraud implicating listeners in great public or mortal issues points towards *Gravity's Rainbow*, where the glove compartment of a California automobile turns out to contain a "library" of sound-effect tapes: "CHEERING (AFFECTIONATE), CHEERING (AROUSED), HOSTILE MOB in an assortment of 22 languages, . . . FIRE-FIGHT (CONVENTIONAL), FIRE-FIGHT (NUCLEAR), FIRE-FIGHT (URBAN), CATHEDRAL ACOUSTICS. . . ." [86]

In insisting that phonograph records have something to do with the Great War, Burgess indicates his sensitivity to the theatrical element that actually was a part of the very machinery that made the war operate. There was a popular record of a successful (that is, a fake) attack, which Alfred M. Hale listened to on a tea-shop gramophone in 1917. This record, he recalls,

> depicted a battle, and troops going into action. We heard the General's preliminary address before the battle, urging the men "to do or to die," and then the prayers of the Chaplain, after which there was a long silence, punctuated with very faint noises—presumably the sounds of the battle—which emerged from out the noise the [gramophone] itself was making, at long intervals. Then at last came the General's address of congratulation to the returning battle-stained army, and that was all. I recollect being much impressed by the truly delightful way in which all the difficulties inseparable from the taking of such a record were thus so easily surmounted. [87]

There is extant a postwar version of such a record, undated but probably issued during the late twenties or early thirties. It is aimed at what today might be called the Sick Nostalgia Market: its audience is supposed to consist of veterans who want to recall the war in dramatic terms, over and over. The actors are designated on the record label only as "Some of the Boys." On one side they enact a dramatic piece called "The Estaminet." A group of British soldiers enters a French bar (slamming door) to be served coffee and "cog-nack" by "Marie." They sing "I Want to Go Home" to piano accompaniment and make jokes over their drinks. Suddenly, a shrill whistle:

OTHER RANK: Blimey, look! Here's the sergeant major!
SERGEANT MAJOR: Fall in!
   (*Sounds of leaving the estaminet; good-byes to Marie; door slams.*)
OFFICER: Battalion, attention!
   Slope arms!
   Forward march!
   (*Fifes and drums, marching feet. Cheerful incoherent grumbling from the troops.*)

But the other side of the record is more interesting. This offers a
three-minute playet called "The Attack." It is played absolutely straight,
with great attention to "realism." The voice of the platoon officer could
be that of Mr. Dollimore himself:

   (*Bird song. Then artillery fire in the distance.*)
OTHER RANK: What a lovely morning!
OTHER RANK 2: . . . Soon it'll be bloody hell! . . .
OTHER RANK: Jerry's got the wind up.
OTHER RANK 2: He knows what's coming, I bet.
   (*The cry "Stretcher Bearers!" is repeated and echoed along the line.*)
OFFICER: (*fruity, upper-class voice*) Wire seems pretty well cut in front,
   Sergeant. . . .
OTHER RANK: Thank God, here's the rum.
OFFICER: Give them a double ration, Sergeant.
   (*Continuous artillery fire, louder and louder.*)
OFFICER: . . . Now remember, boys, you mustn't go beyond [the] sunken
   road in the second line. . . . Now where's the corporal of the bombers?
   Who is he?
CORPORAL: Here, Sir!
OFFICER: . . . Look here, Corporal, don't forget: when you get into the
   Boche trench, you must work your way up to the right and clean out that
   big Boche dugout as you go past. . . . And then remember the important
   thing is the machine-gun nest, which must be put out at all costs.
   (*Airplane motor noise.*)
OFFICER: He's much too high up: you can't touch him.
SERGEANT: How many minutes to go?
OFFICER: Just under a minute. Now get ready, boys. Keep steady. . . . Get
   up here, lad.
SERGEANT: How long now, Sir?
   (*Whining shells and explosions nearby.*)
OFFICER: I'll give you the count: . . . ten, nine, eight, seven, six, five,
   four, three, two, one. (*Blows whistle*).
   (*Confused loud shouting:* Everybody go! Everybody over! *Artillery noises
   grow louder, whistles blow, gradual fade-out.*) [88]

The suspense is well managed, and the effect is surprisingly exciting.
"The Attack" must have constituted a sort of folk war-memoir for many

thousands of households, allowing veterans when the need was on them to re-play vicariously the parts they had once played in actuality. Their instinct, if pitiful, was sound: they sensed that so theatrical a war could well be revisited theatrically.

The more sophisticated could revisit it more obliquely, in literature. By the late thirties the habit of remembering the war in theatrical terms was so widespread among readers that by using the title *Between the Acts* Virginia Woolf could indicate the historical location of her novel between two gigantic pieces of theater, the First and the Second. And in *Jacob's Room* Clara Durrant speaks sympathetically to young Jacob Flanders: " 'Poor Jacob,' said Mrs. Durrant, quietly, as if she had known him all his life. 'They're going to make you act in their play.' " [89] Although she imagines that she is speaking only of the conventional kind of amateur theatricals, Jacob will soon play his part in an unthought-of kind of amateur theater, where he will be destroyed.

VII
Arcadian Recourses

THE BRITISH MODEL WORLD

If the opposite of war is peace, the opposite of experiencing moments of war is proposing moments of pastoral. Since war takes place outdoors and always within nature, its symbolic status is that of the ultimate anti-pastoral. In Northrop Frye's terms, it belongs to the demonic world, and no one engages in it or contemplates it without implicitly or explicitly bringing to bear the contrasting "model world" by which its demonism is measured. When H. M. Tomlinson asks, "What has the rathe primrose to do with old rags and bones on barbed wire?" [1] we must answer, "Everything."

At least we must give that answer so long as we are in the presence of English materials. English writing from the beginning ("Whan that April with his showres soote / The droughte of March hath perced to the roote") has been steeped in both a highly sophisticated literary pastoralism and what we can call a unique actual ruralism. When we go right through the *Oxford Book of English Verse*, we find that half the poems are about flowers and that a third seem somehow to involve roses. The man with the *Oxford Book* in his haversack has with him the main English pastoral poems, from Spenser's *Prothalamion* and *Epithalamion* to Arnold's *Scholar Gypsy*. And if he should also be carrying Bridges's anthology *The Spirit of Man* (bound in olive-green as if designed for trench use) he would have access to a tradition even more resolutely pastoral and floral. There poems like Arnold's *Thyrsis* and Shelley's *Adonais* and Yeats's *The Sad Shepherd* and Milton's *Arcades* are set in a context prolific with roses, sunflowers, lilies, sheep and shepherds, birds, bees, poppies, gardens, "vales," and wood-nymphs. In December, 1917, Wilfred Owen was reading Barbusse's *Under Fire*. But at the same time he was also reading Theocritus, Bion, and Moschus.[2]

In Barbusse he would not have found the model world depicted, as in English writing, in pastoral terms, nor would a later reader find it so depicted in Jünger or Remarque. The British are different. What other nation supports through all the vicissitudes of modern economics and politics a periodical like *Country Life*, devoted very largely to the excitation of rural nostalgia? In France, Germany, Italy, and the United States there's nothing quite like it, and it appears not, say, monthly—that would be conceivable—but weekly. Oliver Lyttelton was reading it in the trenches in September, 1916, and Keith Douglas resorted to it in the desert near El Alamein in 1942.[3] In both times and places it was the same as today, with articles on, for example (I am flipping through the number for September 6, 1973), "The Gardens of Burford House, Shropshire," "Color in the Autumn Border," "Adventures with Bulbs," and "The Surviving Craft of Hedging," together with letters on "A Sansevieria in Flower" and "Killing of Sparrows." In Britain, as Ronald Blythe says, "The townsman envies the villager his certainties and . . . has always regarded urban life as just a temporary necessity. One day he will find a cottage on the green and 'real values.' "[4] Perhaps the special British ruralism is partly the result, as Raymond Williams has suggested, of a tradition of Imperialist exile from home. "From about 1880," he points out,

> there was . . . a marked development of the idea of England as "home," in that special sense in which home is a memory and an ideal. Some of the images of this "home" are of central London. . . . But many are of an ideal of rural England: its green peace contrasted with the tropical or arid places of actual work; its sense of belonging, of community, idealized by contrast with the tensions of colonial rule and the isolated alien settlement. The birds and trees and rivers of England; the natives speaking, more or less, one's own language: these were the terms of many imagined and actual settlements. The country, now, was a place to retire to.[5]

That has been a constant in the modern British sensibility regardless of which exile or what war we're talking about. Briefly on leave in Cairo from the desert warfare of 1942, R. L. Crimp finds that "sweating in the sultry night of this soulless, venal, cynical city, it's refreshing to be with Cobbett's 'Rural Rides' in the changing weathers and seasons of the English countryside."[6]

And perhaps English ruralism is partly the result, as Julian Moynahan suggests, of the early establishment of industrialism in England. The British, he observes, "by and large invented the Industrial Revolution and urbanization as far back as the later eighteenth century. . . ." They have thus needed more than others "what Empson calls 'the permanent tradition of the country,' "

> . . . an idea of themselves as indeed a people with immemorial rural roots and of a spiritual disposition that is at home among tall trees, clear streams, hunting fields and sheepfolds, ricks, roses, and hedgerows. There is a pleasant, superficial side to this continuing illusion of the pastoral in Britain: seen in the persistent English cult of the domestic and the homely; . . . in country sports . . . ; in weekend gardening, which, according to a recent survey, is the only hobby of some 22,000,000 Britons.[7]

Whatever the reason, it is striking that no other modern literature possesses as its most popular bad poem anything like Thomas E. Brown's "The Garden":

> A garden is a lovesome thing, God wot!
> > Rose plot,
> > Fringed pool,
> > Fern'd grot—
> > > The veriest school
> > > Of peace. . . .

What other literature has in it a Herrick, a Cowper, a Clare, a Cobbett, a Wordsworth, or a Blunden? And an important thing about all this Godwottery—as Anthony Burgess calls it—is that all types and classes embrace it. A professional gardener interviewed by Ronald Blythe notes that three-quarters of those registered with the National Gardens Scheme are retired Army and Navy officers,[8] but even Gunner W. R. Price, who can't spell *canvas* or *frantically*, never makes a mistake with the names of flowers like marsh marigolds, cuckoo pint, or golden saxifrage.[9] It somehow seems entirely English that D. H. Lawrence, in the various versions of the Lady Chatterley scene involving the lovers' decorating each others' parts with flowers, should particularize so knowledgeably and precisely: "He came back with a mixed bunch of forget-me-nots, campions, bugle, bryony, primroses, golden-brown oak-sprays."[10]

Once when Stephen Graham's battalion was out of the line, the men set themselves to elaborate, indeed pedantic, competitive gardening, instinct with symbolism:

> Each company . . . marked out the pattern of its formal garden. . . . A
> platoon of A company . . . enclosed a tent in a heart; a border of boxwood
> marks out the pattern of the heart—the plan is that the crimson of many
> blossoms shall blend to give a suggestion of passion and loyalty and suffer-
> ing. . . . Primroses and daffodils and narcissi are soon blossoming in
> plenty. Lilies followed, arums and Solomon's-seal, and then forget-me-nots,
> pansies and violas.

"Gardeners camouflaged as soldiers": that was the RSM's rueful comment.[11] Even in trenches the English idea of domesticity was inseparable
from the image of a well-kept gardening allotment. The only trench that
ever gave Anthony French "any satisfaction," as he says, was one which,
although only forty-seven yards from the enemy, had "walls lined with
basketwork and trellis": "Bottlenecks and junctions had a homely atmosphere with nasturtiums climbing the trellis." [12] One day late in the war
Sassoon, by this time conspicuous as a pacifist, was being harangued by
Winston Churchill, whom he had visited in the office of the Ministry of
Munitions. "War," Churchill told him, "is the normal occupation of
man." Sassoon wonders: "Had he been entirely serious? . . . He had
indeed qualified the statement by adding 'War—and gardening!' " [13]
Those are the poles.

A standing joke for years has been the earnestness of the annual letters
to the *Times* and the *Telegraph* reporting, with due self-praise, the correspondent's sighting of the spring's first swallow and cuckoo. The convention was funny as far back as 1916, when the *Wipers Times* for February
12 printed this parody letter:

> To the Editor.
> Sir,
>   Whilst on my nocturnal rambles along the Menin Road last night, I am
> prepared to swear that I heard the cuckoo. Surely I am the first to hear it
> this season. Can any of your readers claim the same distinction?
>   A LOVER OF NATURE.

That is comic, but it extends as well a more serious invitation to entertain a significant polar memory of "home," where people were calmly
carrying on the old delightful English pastoral usages. Colonel A. C.
Borton, his two sons away at the war, makes these entries in his diary in
1916: "*14 April:* At Maidstone all day as President of a Board of Examination for Volunteer Officers in West Kent. SAW FIRST SWALLOW";
"*18 April:* I to Maidstone Diseases of Animals Committee in morning,
and to Tonbridge afternoon for Medway Conservancy Meeting. HEARD
CUCKOO." [14]

For the English, nature is, as Wordsworth and his Victorian successors instructed them, a "stay" against the chaos of industrial life. It is not
just from his prolonged encounter with Great War writings invoking
pastoral and rural images as a saving contrast that Bernard Bergonzi
derives this very English way of beginning a "Letter from England,"
addressed to American readers in 1973:

> The daffodils are beginning to bloom outside my window and the country
> is on the verge of chaos. . . . Nothing will stop the daffodils from coming
> up in their complacent fashion one March after another; nature is pro-
> foundly reactionary . . . in its blind attachment to the status quo.[15]

Whether those daffodils originate in Wordsworth's "I Wandered Lonely
as a Cloud"—that is, in "pastoral"—or spring from the actual rural earth,
they are unmistakably English, as is the use to which Bergonzi puts
them.

THE USES OF PASTORAL

Recourse to the pastoral is an English mode of both fully gauging the
calamities of the Great War and imaginatively protecting oneself against
them. Pastoral reference, whether to literature or to actual rural localities
and objects, is a way of invoking a code to hint by antithesis at the indescribable; at the same time, it is a comfort in itself, like rum, a deep dugout, or a woolly vest. The Golden Age posited by Classical and Renaissance literary pastoral now finds its counterpart in ideas of "home" and
"the summer of 1914." The language of literary pastoral and that of particular rural data can fuse to assist memory or imagination. Thus the behavior of the exiled pastoralists in Lord Dunsany's sentimental short
story "The Prayer of the Men of Daleswood." Isolated after an attack,
they comfort themselves by recalling the village of Daleswood (Kent?)
"as it used to be." They remember its "deep woods" and "great hills"
and dwell longingly on "the ploughing and the harvest and snaring rabbits in winter, and the sports in the village in summer." Anxious to "fix"
and record their memories of Daleswood, they uncover a large block of
chalk in the trench and consider using their knives to inscribe on it—like
a group of late-eighteenth-century poets writing pastoral "Inscriptions"—a list of bucolic details: the woods; the foxgloves; mowing the
hay with scythes; "the valleys beyond the wood, and the twilight on
them in summer"; and "slopes covered with mint and thyme, all solemn
at evening" (cf. Gray's *Elegy:* "And all the air a solemn stillness holds").
But deciding finally that even details so evocative as these cannot do jus-

tice to their idea of a prelapsarian Daleswood, they carve on the chalk only these words: "Please, God, remember Daleswood just like it used to be." [16]

There the pastoral details are invoked as a comfort. In Wilfred Owen's "Exposure" they are invoked as a standard of measurement. The scene is the very terrible winter of 1917. Owen and his freezing men are huddling in forward holes during a blizzard:

> Pale flakes with fingering stealth come feeling for our faces—
> We cringe in holes, back on forgotten dreams, and stare, snow-dazed,
> Deep into grassier ditches. So we drowse, sun-dozed,
> Littered with blossoms trickling where the blackbird fusses.
>   Is it that we are dying?

Presented with the datum *snow-dazed*, we are then supplied with a measure for it deriving from the pastoral norm—*sun-dozed*.

It was Middleton Murry's complaint that too often Sassoon's poems didn't behave this way. Too many of them, he found, neglected to indicate the pastoral norm by which their horrors were to be gauged. [17] But he would have had no objection to this performance in *Memoirs of an Infantry Officer*, where, after having been subjected to the famous Scottish major's famous lecture on the Spirit of the Bayonet, Sherston says,

> Afterwards I went up the hill to my favorite sanctuary, a wood of hazels and beeches. The evening air smelt of wet mold and wet leaves; the trees were misty green; the church bell was tolling in the town, and smoke rose from the roofs. Peace was there in the twilight. . . . But the lecturer's voice still battered on my brain. "The bullet and the bayonet are brother and sister." "If you don't kill him, he'll kill you." [18]

And sometimes Sassoon's contrasting pastoral details are even more flagrantly literary, as when once at the front he recalls a twilight at Aunt Evelyn's front gate when "John Homeward had just come past with his van, plodding beside his weary horse." [19] It is almost a verbal anagram of Gray's

> The plowman homeward plods his weary way.

Such moments of brief recurrence to the pastoral ideal are like miniatures of those "bucolic interludes" or "pastoral oases" specified by Renato Poggioli as points of illumination or refreshment in late-medieval or Renaissance narrative. [20] The standard Great War memoir generally provides a number of such moments sandwiched between bouts of violence and terror. Sassoon's pattern of going in and out of the line is typical. The same pattern provides the dynamics for Hemingway's *A Farewell to*

Arms (1929) as well as for such memoirs of the Second War as Raleigh Trevelyan's *The Fortress* (1956). There the "pastoral oasis"—in the Hemingway tradition—is provided by the hero's withdrawal not just to Naples but to idyllic Capri to recover from wounds sustained at Anzio. Such a withdrawal makes it all the more awful to return to the line to be wounded again.

Sometimes a pastoral oasis can be found without leaving the line. Speaking of how rarely one caught sight of the enemy, Guy Chapman says,

> Once we heard him. On a morning when light filtered through a bleached world, and a blanket of mist was spread upon the ground so that our eyes could only pierce to the edge of our own wire, the whole company seized the opportunity to repair the damages of shell fire. We worked furiously with muffled mauls and all the coils of wire we could beg. Presently we realized that the enemy was doing as we were; dull thuds and the chink and wrench of wire came from his side. We might perhaps have opened fire, but that suddenly there came out of the blankness the sound of a young voice.

As he goes on, Chapman seems to recall the word *Dorian* in a special way, the way the pastoral context of Arnold's *Thyrsis* gives that word a meaning synonymous with *Doric*—that is, rustic, pastoral—just as he seems to recall the solitariness of Wordsworth's Reaper, who also sings an unidentified folksong so beautifully that the listener is adjured to

> Stop here, or gently pass.

The young voice sounding through the mist, says Chapman,

> was raised in some Dorian-moded folksong. High and high it rose, echoing and filling the mist, pure, too pure for this draggled hill-side. We stopped our work to listen. No one would have dared break the fragile echo. As we listened, the fog shifted a little, swayed and began to melt. We collected our tools and bundled back to our trench. The singing voice drew further off, as if it was only an emanation of the drifting void.

After the oasis, a return to the real world:

> The sun came out and the familiar field of dirty green with its hedges of wire and pickets rose to view, empty of life.[21]

Hedges of wire: there is an image useful to the pastoral ironist. A nation of gardeners was not likely to miss the opportunities for irony in pretending that "thickets" of barbed wire were something like the natural hedges of the English countryside. Reginald Farrer pretends that he is safely botanizing at home as he observes the wire snarled at the bottom of a mine crater near Pozières: it is, he declares, "the characteristic war-flora, . . . represented here as usual by derelict snarls of *Barbedwiria volubilis*. . . ." [22] The *Wipers Times* of December 25, 1917, in its column "In My Garden"—a parody of the weekly column so titled in *Country Life*—amused itself by pretending to confuse the industrial with the natural, the nasty with the organic and domestic:

> Where the soil is damp and heavy, an early planting of gooseberries is attended with some risk. This hardy perennial, being a strong grower, will quickly cover an unsightly patch of waste ground. The best crops of this luscious fruit have been obtained when some support was given by stakes.

Furthermore,

> It must be remembered that the planting of Toffee-apples on the border of your neighbor's allotment will seriously interfere with the ripening of his gooseberries.*

The tradition could be said to go back at least as far as late-Victorian metallic mock-organic Art Nouveau. Thus Richard le Gallienne's reaction to London gas-lighting fixtures in 1892:

> Lamp after lamp against the sky
> Opens a sudden beaming eye,
> Leaping alight on either hand
> The iron lilies of the Strand.

And it persists through the Second War. In "Cairo Jag," after dramatizing the soft dubious pleasures of Cairo, Keith Douglas speaks of the proximity of the desert war:

> But by a day's travelling you reach a new world
> the vegetation is of iron
> dead tanks, gun barrels split like celery
> the metal brambles have no flowers or berries. . . .

That ironic vegetation will make it seem appropriate that the author of the Introduction to Douglas's poems should be his old Merton College tutor in English, the eminent pastoralist and gentle ironist Edmund Blunden.

Useful as pastoral allusion is for purposes of comfort or of implicit description through antithesis, it has a more basic function of assisting ironic perception. But as David Kalstone reminds us, even when it has

---

* Gooseberries: thick balls of barbed wire five or six feet in diameter used to block trenches or fill gaps in wire entanglements. Toffee-apples: globular projectiles about a foot in diameter fired from the British trench mortar.

been regarded ironically, "a clarifying or restorative force . . . has always been associated with pastoral." [23] The ironist thus has it both ways: he marks the distance between the desirable and the actual and at the same time provides an oasis for his brief occupancy, as in Owen's "Anthem for Doomed Youth," where the figure of

> The shrill, demented choirs of wailing shells

both gauges the obscenity of industrialized murder and returns us for a fleeting moment to the pastoral world where the "choirs" consist of benign insects and birds. The last line of *Lycidas*,

> Tomorrow to fresh woods, and pastures new,

was a favorite for ironic operations like Stephen H. Hewett's in a letter to his sister of May 7, 1916. He has been back at an idyllic trench mortar course well behind the line, but he must now leave that oasis: "To-day I must rejoin my Battalion in the firing line. 'Cras ingens interabimus aequor': 'Tomorrow to fresh woods and pastures new'." [24] Gentle wine-bibbing at evening is the setting of Horace's "Tomorrow we'll set out again on the wide sea" (*Odes* I, vii, 32); pastoral twilight is the setting at the end of *Lycidas*. What a pleasure to revisit such scenes sentimentally, if even for a moment, while simultaneously registering a sardonic view of the immediate future. Oliver Lyttelton gestures in the refreshing direction of Isaak Walton and Charles Cotton while recounting the events of an early morning on the Somme: "Then we killed. I have only a blurred image of slaughter. I saw about ten Germans writhing like trout in a creel at the bottom of a shell-hole and our fellows firing at them from the hip. One or two red bayonets." [25] Thank God for that cool creel and those fresh trout! The irony attaching to spring's being the favorite time for launching offensives is still resonating four years after the war in the ironic pastoral of Eliot's

> April is the cruellest month, breeding
> Lilacs out of the dead land. . . .

SHEEP AND BIRDSONG

Strictly speaking, pastoral requires shepherds and their sheep. The war offers both in officers and Other Ranks, especially when the Other Ranks are wearing their issue sheepskin coats with the fur outside. Thus, as Joyce put it in his song "Mr. Dooley," registering his contempt for English cant about "war-aims,"

> "Poor Europe ambles
> Like sheep to shambles. . . ." [26]

But unlike the shepherds in literary pastoral, the shepherds here have ranks; installed in a ghastly defensive position near Ypres, Gladden remembers an odd scene:

> Cooped in our sodden trap it is not easy to exaggerate the weight of depression that fell upon us. After a time the rain did abate, leaving a Scotch mist wrapped like a cloak about the lines.
>   Then among the broken trees not far ahead we saw a tall figure, moving like a shepherd with the aid of a simple staff, a shepherd looking for his lost sheep! We recognized the brigadier-general, alone and quite unconcerned.[27]

Sometimes the shepherd must prod the sheep to get them moving, as R. H. Tawney had to do during the Somme attack when one of his men "buried his head in the ground and didn't move. I think he was crying. I told him I would shoot him, and he came up like a lamb." [28] During the 1917 mutinies in the French army, the *poilus* being marched up to the line frequently made loud *baa*-ing noises "in imitation of lambs led to the slaughter, and their officers were helpless to prevent it." [29]

The properly responsible relation of shepherds to their sheep is much on the mind of C. E. Montague in *Disenchantment*, where he chronicles the troops' transition from initial enthusiasm to final cynicism. "In the first weeks of the war," he says, "most of the flock had too simply taken on trust all that its pastors and masters had said." Later, "they were out to believe little or nothing—except that in the lump pastors and masters were frauds." A happy bucolic image assists his point: "From any English training-camp . . . you almost seemed to see a light steam rising, as it does from a damp horse. This was illusion beginning to evaporate." [30] *Lycidas* seems to be Montague's reference point, as in the title of his chapter dealing with the deficiencies of chaplains and the sad inefficacy of the religious consolations officially provided: "The Sheep That Were Not Fed." When a body of troops in training is marched by its sergeant major directly to a pub with whose landlord he has an understanding, those who decline to enter and who stand around outside all afternoon finally begin to break down into cynicism and to wonder, "Were it not better done as others use?" [31]

There seems to be no action or emotion of the line that cannot be accommodated to some part of the pastoral paradigm. In his poem "Going into the Line" Max Plowman—could there be a more English name?—dominates the shelling by imagining an almost literal pastoral oasis:

> And slowly in his overheated mind
> Peace like a river through the desert flows,
> And sweetness wells and overflows in streams
> That reach the farthest friend in memory.

Now that he has quieted himself he can contemplate his men, and he feels a silent, secret

> dear delight in serving these,
> These poor sheep, driven innocent to death.

In reacting to the death of Sassoon's friend David Thomas, Graves seizes on the happy coincidence between Thomas's Christian name and the name of the Biblical shepherd to produce a bitter poem, "Goliath and David." As the form of the title suggests, Graves reverses the traditional outcome of the story. Here David, the "goodly faced boy so proud of strength," finds his "knotted shepherd staff" no match for Goliath's saber. This is a fight which shepherd boys, no matter how well favored, cannot win. Nor can their silly sheep. In December, 1915, P. H. Pilditch was examining the German line near Auchy through binoculars:

> All that could be seen from our O.P. was a high chalk parapet with the almost green strip of No Man's Land sloping up towards it. When I first looked at this strip . . . , I made out what seemed to be a flock of sheep grazing all over it.

This seemed "very weird." Applying a powerful telescope, Pilditch finally "saw clearly what they were":

> They were hundreds of khaki bodies lying where they had fallen in the September attack on the Hohenzollern [Redoubt] and now beyond the reach of friend and foe alike, and destined to remain there between the trenches till one side or the other advanced, which seemed unlikely for years.[32]

The "almost green strip" between the lines is a telling detail: it is there primarily that the sheep may not safely graze.

In addition to shepherds and sheep, pastoral requires birds and birdsong. One of the remarkable intersections between life and literature during the war occurred when it was found that Flanders and Picardy abounded in the two species long the property of symbolic literary pastoral—larks and nightingales. The one now became associated with stand-to at dawn, the other with stand-to at evening. (Sometimes it is really hard to shake off the conviction that this war has been written by someone.) Such a symmetrical arrangement was already available in the

Oxford Book, where Milton's *L'Allegro* has its lark and *Il Penseroso* its nightingale. Such a symmetry was also a familiar fixture of the aubade, as in the colloquy between Romeo and Juliet (III, v, 1–36). For some the stand-to larks were a cruel reminder of home and safety. Sergeant Major F. H. Keeling writes: "Every morning when I was in the front-line trenches I used to hear the larks singing soon after we stood-to about dawn. But those wretched larks made me more sad than anything else out here. . . . Their songs are so closely associated in my mind with peaceful summer days in gardens or pleasant landscapes in Blighty." [33] But for most, the morning larks were a comfort: they were evidence that ecstasy was still an active motif in the universe. Sergeant Ernest Nottingham admitted to feeling melancholy "at the sound of a train, or a church bell, or the scent and sight of a flower, or the green of a hedge. But not now in the ecstasy of a lark.—That is inseparably connected with 'stand-to' in the trenches.—So often . . . have I heard him singing joyously in the early dawn." [34] Frank Richards sardonically consoles himself by asserting a general theological principle; namely, that "God made both the louse and the lark." [35] A war later, in the Italian campaign of 1944, Alex Bowlby automatically employs lark song as a gauge of absurdity: "In the lulls between explosions I could hear a lark singing. That made the war seem sillier than ever." [36]

What the lark usually betokens is that one has got safely through another night, a night made poignantly ironic by the singing of nightingales. "We heard the guns," Cloete remembers. "There was continuous fire. It was a background to the singing of the nightingales. It has often seemed to me that gunfire makes birds sing, or is it just that the paradox is so great that one never forgets it and always associates the two?" [37] A good question. Sassoon "associates the two" when he remembers a moment in spring, 1918, near the Somme: "Nightingales were singing beautifully. . . . But the sky winked and glowed with swift flashes of the distant bombardments at Amiens and Albert, and there was a faint rumbling, low and menacing. And still the nightingales sang on. O world God made!" [38] The nightingale tradition persists through the Second War just like the lark tradition. Bowlby in Italy is again its beneficiary. Under fire from German 88's, he suddenly hears a bird singing and at first thinks it may be a German signaling by imitating birdcalls. "But as the bird sang on I realized that no human could reproduce such perfection. It was a nightingale. And as if showing us and the Germans that there were better things to do it opened up until the whole valley rang with song. . . . I sensed a tremendous affirmation that 'this would go on.' . . ." [39] It is a point comprised in the final six lines of Eliot's

"Sweeney Among the Nightingales," written two years after the war. The "bloody wood" where his nightingales sing is Nemi's Wood, from folklore, the place where the priest or ruler is slain by his younger successor. But the irony which juxtaposes nightingales and bloody woods would come naturally to one who had passed a night in, say, Mametz Wood.

ROSES AND POPPIES

As Rupert Brooke asserts in "The Soldier," one of England's main attractions is that she provides "flowers to love." As Brooke and his friends knew, a standard way of writing the Georgian poem was to get as many flowers into it as possible. Edward Thomas could make virtually a whole poem by naming flora, as he does in "October." In the same tradition is Brooke's "The Old Vicarage: Grantchester," written in "exile" in Germany before the war. The garden Brooke imagines there is "English" both in the careless variety of its species and in the freedom in which they flourish:

> Just now the lilac is in bloom,
> All before my little room;
> And in my flower-beds, I think,
> Smile the carnation and the pink;
> And down the borders, well I know,
> The poppy and the pansy blow.

The trouble with Germany is that everything is orderly and forced and obedient:

> Here tulips bloom as they are told.

It is different in England, where

> Unkempt about those hedges blows
> An English unofficial rose.

Brooke wrote this poem two years before the war, but already the Georgian focus was appearing to narrow down to the red flowers, especially roses and poppies, whose blood-colors would become an indispensable part of the symbolism of the war. Stephen Graham's gardener-soldiers instinctively grasped the principle when, as we have seen, they massed "the crimson of many blossoms . . . to give a suggestion of passion and loyalty and suffering." The troops who used *rose-colored* as a jocular periphrasis for *bloody* perhaps understood, in their way, the same principle. An ancient tradition associates battle scars with roses, perhaps because a newly healed wound often does look like a red or pink rose. In *The Tribune's Visitation* David Jones's Roman officer exploits this association in his pep talk:

> . . . within this sacred college we can speak *sub rosa* and the rose that seals our confidence is that red scar that shines on the limbs of each of us who have the contact with the fire of Caesar's enemies, and if on some of us that sear burns, then on all, on you tiros no less than on these *veterani*. . . .[40]

The time-honored image wound-rose has found its way into Wallace Stevens's "Esthétique du Mal":

> How red the rose that is the soldier's wound,
> The wounds of many soldiers, the wounds of all
> The soldiers that have fallen, red in blood,
> The soldier of time grown deathless in great size.

Redness: when the news of a soldier's death is brought to his parents' house in Hardy's "Before Marching and After," written in September, 1915,

> . . . the fuchsia-bells, hot in the sun,
> Hung red by the door. . . .

In "Bombardment," Lawrence conceives of a town being bombed as

> . . . a flat red lily with a million petals. . . .

The person who waited outside Buckingham Palace on August 3, 1914, hoping to catch sight of the King, and who testified later to noticing that "the red geraniums . . . looked redder than they had ever looked before," was deepening their red in retrospect.[41]

Roses were indispensable to the work of the imagination during and after the Great War not because Belgium and France were full of them but because English poetry was, and because since the Middle Ages they had connoted "England" and "loyalty" and "sacrifice." As Frye says,

> In the West the rose has a traditional priority among apocalyptic flowers; the use of the rose as a communion symbol in the *Paradiso* comes readily to mind, and in the first book of *The Faerie Queene* the emblem of St. George, a red cross on a white ground, is connected not only with the risen body of Christ and the sacramental symbolism which accompanies it, but with the union of the red and white roses in the Tudor dynasty.[42]

A typical use of the rose during the war was to make it virtually equal to the idea of England. Even though France has her roses, they are somehow not hers but England's. "Life was good out of the line that September," writes Philip Gibbs of the year 1915. "It was good to go into the garden of a French château and pluck a rose and smell its sweetness, and think back to England, where other roses were blooming. England!" [43] For David Jones, the rose means more than England: it means England-in-the-spring. The dog-rose, he says, "moves something in the English-man at a deeper level than the Union Flag" because English poetry, "drawing upon the intrinsic qualities of the familiar and common June rose, has . . . managed to recall and evoke, for the English, a June-England association." [44] For ordinary people roses connoted loyalty to England, service to it, and sacrifice for it. When after Loos Liddell Hart was carried wounded through Charing Cross station, he found roses cast onto his stretcher by the crowd. And "this happened again the following year," he says, "when I was brought back from the battle of the Somme." [45] The "bar" to the Military Cross indicating a second award is not a bar at all: the device added to the ribbon is a little silver rose.

The rose is indispensable to popular sentimental texts of the war, from the song "Roses of Picardy" to

> The roses round the door
> Makes me love mother more,

"the favorite song of the men," as Owen reports from France in January, 1917; "they sing this everlastingly." [46] "Crimson roses" symbolizing sac-rificial love are at the center of one of Stephen Graham's most sentimen-tal anecdotes. A Sergeant Oliver of the Sixth Black Watch was shot, Graham tells us, when he went forward in violation of orders to frater-nize with the enemy on Christmas Day, 1915. (Sentries on both sides had been ordered to kill anyone trying to re-enact the famous Christmas truce of the year before.) This sergeant, Graham found, is buried in "a little old cemetery by the side of the road a mile or so from Laventie." From his grave is growing—this is English literature, we realize—"a tall rose tree with crimson roses blooming even in autumn." It is especially noticeable because "beside him lies one who was both captain and knight, with only a dock rising from his feet." Indeed, "On all graves are weeds except on that of the man who gave his life to shake hands on Christmas Day." [47] And because crimson roses are thus associated with sacrificial love, they have been made a conspicuous feature of the British military cemeteries in France.

In a brilliant essay Erwin Panofsky has discovered that ever since the eighteenth century the English have had an instinct not shared by Conti-nentals for making a special kind of sense out of the classical tag *Et in ar-cadia ego*. Far from taking it as "And I have dwelt in Arcadia too," they take it to mean (correctly) "Even in Arcadia I, Death, hold sway." As Panofsky says, this meaning, "while long forgotten on the Continent, remained familiar" in England, and ultimately "became part of what may be termed a specifically English or 'insular' tradition—a tradition which tended to retain the idea of a *memento mori*." Skulls juxtaposed with roses could be conventionally employed as an emblem of the omnipotence of Death, whose power is not finally to be excluded even from the sequestered, "safe" world of pastoral. "In . . . Waugh's *Brideshead Revisited*," Panofsky points out, "the narrator, while a sophisticated undergraduate at Oxford, adorns his rooms at college with a 'human skull lately purchased from the School of Medicine which, resting on a bowl of roses, formed at the moment the chief decoration of my table. It bore the motto *Et in Arcadia ego* inscribed on its forehead.' " [48] This "specifically English or 'insular' tradition" added a special resonance of melancholy to wartime pastoral in general and to the invocation of roses in particular. "Rose Trench," says Sassoon, "Orchard Alley, Apple Alley, and Willow Avenue, were among the first objectives in our sector [of the Somme attack], and my mind very properly insisted on their gentler associations. Nevertheless, this topographical Arcadia was to be seized, cleared, and occupied. . . . " [49] Speaking of the rude violation by facts of the troops' early illusions, Montague puts it in this very English, indeed Blakean, way: "At the heart of the magical rose was seated an earwig." Those finally saw the earwig who saw "trenches full of gassed men, and the queue of their friends at the brothel-door in Béthune." [50] The English understanding of *Et in Arcadia ego* is enacted when Cloete's Jim Hilton, wandering amidst the rubble at Ypres, comes upon "a little courtyard garden." There, "a rose was in flower. He picked it. . . . There were dead buried in the rubble. You could smell them—sickly, sweet, putrid—an odor that mixed with that of the red rose in his hand." [51] It is like the smell of the aging Billy Pilgrim's breath when late in the nights after drinking he decides to telephone his old Army friends all over the country: on those occasions his breath smells "like mustard gas and roses." [52]

The case of the other red flower inseparable from writings about the Great War, the poppy, is even more complicated and interesting. Properly speaking, the Flanders poppy—it is called that now—is *Papaver rhœas*, to be distinguished from the kind generally known in the United States, the California poppy, or *Platystemon californicus*. The main difference is that the California poppy is orange or yellow, the Flanders poppy bright scarlet. The familiarity of the California poppy to Americans has always created some impediment to their accurate interpretation of the rich and precise literary symbolism of Great War writing. If one imagines poppies as orange or yellow, those that blow between the crosses, row on row, will seem a detail more decorative than symbolic. One will be likely to miss their crucial relation (the coincidence of Mc-Crae's and Milton's rhyme-sound will suggest it) to that "sanguine flower inscrib'd with woe" of *Lycidas* and with the "red or purple flower," as Frye reminds us, "that turns up everywhere in pastoral elegy." [53] Flanders fields are actually as dramatically profuse in bright blue cornflowers as in scarlet poppies. But blue cornflowers have no connection with English pastoral elegiac tradition, and won't do. The same principle determines that of all the birds visible and audible in France, only larks and nightingales shall be selected to be remembered and "used." One notices and remembers what one has been "coded"—usually by literature or its popular equivalent—to notice and remember. It would be a mistake to imagine that the poppies in Great War writings get there just because they are actually there in the French and Belgian fields. In his sentimental, elegiac *The Challenge of the Dead* (1921), Stephen Graham produces a book of 176 pages without once noticing a poppy, although he chooses to notice plenty of other indigenous flowers, including roses and cornflowers. We can guess that he omits poppies because their tradition is not one he wants to evoke in his book, the point of which is that survivors should now imitate the sacrifice of the soldiers, who in turn were imitating the sacrifice of Christ. There is something about poppies that is too pagan, ironic, and hedonistic for his purposes. The same principle of literary selection—as opposed to "documentary" or photography—is visible in a poem of the Second War by Herbert Corby. In "Poem" he is projecting a contrast between pastoral serenity and its sinister opposite, the emblems of which are the flowers near his aerodrome and his own metallic airplane. The flowers he selects for this purpose are not any old flowers; they are specifically the two species already freighted with meaning from the earlier war:

> The pale wild roses star the banks of green
> and poignant poppies startle their fields with red, . . .
> I go to the plane among the peaceful clover,
> but climbing in the Hampden, shut myself in war.

By the time the troops arrived in France and Belgium, poppies had accumulated a ripe traditional symbolism in English writing, where they had been a staple since Chaucer. Their conventional connotation was the blessing of sleep and oblivion (that is, of a mock-death, greatly to be desired), as in Francis Thompson's "The Poppy" or Tennyson's "The

Lotos-Eaters" (both in the *Oxford Book*). But during the last quarter of the nineteenth century the poppy began to take on additional connotations, some of which are discreetly glanced at by W. S. Gilbert in *Patience* (1881). There the aesthete Bunthorne, an exponent of "sentimental passion," is made to sing,

> . . . if you walk down Picadilly with a poppy or a lily in your medieval hand,
> . . . everyone will say,
> As you walk your flowery way,
> . . . what a most particularly pure young man this pure young man must be!

Bunthorne's poppy is *Papaver orientale*, the floppy, scarlet kind. For late Victorians and Edwardians, it was associated specifically with homoerotic passion. In Lord Alfred Douglas's "Two Loves," the allegorical figure who declares that he is "the love that dare not speak its name" is a pale youth whose lips are "red like poppies." For half a century before the fortuitous publicity attained by the poppies of Flanders, this association with homoerotic love had been conventional, in works by Wilde, Douglas, the Victorian painter Simeon Solomon, John Addington Symonds, and countless others. No "poppy" poem or reference emerging from the Great War could wholly shake off that association. When Sassoon notes that "the usual symbolic scarlet poppies lolled over the sides of the communication trench," [54] he is aware, as we must be, that they symbolize something more than shed blood and oblivion.

One of the neatest "turns" in popular symbolism is that by which the paper poppies sold for the benefit of the British Legion on November 11 can be conceived as emblems at once of oblivion and remembrance: a traditional happy oblivion of their agony by the dead, and at the same time an unprecedented mass remembrance of their painful loss by the living. These little paper simulacra come from pastoral elegy (Milton's Arcadian valleys "purple all the ground with vernal flowers"), pass through Victorian male sentimental poetry, flesh themselves out in the actual blossoms of Flanders, and come back to be worn in buttonholes on Remembrance Day. In his "A Short Poem for Armistice Day" Herbert Read is struck by the paper poppy's ironic inability to multiply as well as by its ironic resistance to "fading" and dying. When he sees it as a sad antithesis to something like Milton's unfading amaranthus of *Lycidas*, we are reminded of the war's tradition of ironic wonder that the metal "bramble thicket" is, as R. H. Sauter finds it in his poem "Barbed Wire," "unflowering." Read's poem on the paper poppy enacts the inverse consolation appropriate to an inverse elegy involving inverse flowers.

The most popular poem of the war was John McCrae's "In Flanders

Fields," which appeared anonymously in *Punch* on December 6, 1915. Its poppies are one reason the British Legion chose that symbol of forgetfulness-remembrance, and indeed it could be said that the rigorously regular meter with which the poem introduces the poppies makes them seem already fabricated of wire and paper. It is an interesting poem because it manages to accumulate the maximum number of well-known motifs and images, which it gathers under the aegis of a mellow, if automatic, pastoralism. In its first nine lines it provides such familiar triggers of emotion as these: the red flowers of pastoral elegy; the "crosses" suggestive of calvaries and thus of sacrifice; the sky, especially noticeable from the confines of a trench; the larks bravely singing in apparent critique of man's folly; the binary opposition between the song of the larks and the noise of the guns; the special awareness of dawn and sunset at morning and evening stand-to's; the conception of soldiers as lovers; and the focus on the ironic antithesis between beds and the graves where "now we lie." Not least interesting is the poem's appropriation of the voice-from-the-grave device from such poems of Hardy's as "Channel Firing" and "Ah, Are You Digging on My Grave?" and its transformation of that device from a mechanism of irony to one of sentiment:

> In Flanders fields the poppies blow
> Between the crosses, row on row,
>   That mark our place; and in the sky
>   The larks, still bravely singing, fly
> Scarce heard amid the guns below.
>
> We are the Dead. Short days ago
> We lived, felt dawn, saw sunset glow,
>   Loved and were loved, and now we lie
>   In Flanders fields.

So far, so pretty good. But things fall apart two-thirds of the way through as the vulgarities of "Stand Up! Stand Up and Play the Game!" begin to make inroads into the pastoral, and we suddenly have a recruiting-poster rhetoric apparently applicable to any war:

> Take up our quarrel with the foe:
> To you from failing hands we throw
>   The torch; be yours to hold it high.

(The reader who has responded to the poppies and crosses and larks and stand-to's, knowing that they point toward some trench referents, will wonder what that "torch" is supposed to correspond to in trench life. It suggests only Emma Lazarus.)

> If ye break faith with us who die
> We shall not sleep, though poppies grow
> In Flanders fields.

We finally see—and with a shock—what the last six lines really are: they are a propaganda argument—words like *vicious* and *stupid* would not seem to go too far—against a negotiated peace; and it could be said that for the purpose, the rhetoric of Sir Henry Newbolt or Horatio Bottomley or the Little Mother is, alas, the appropriate one. But it is grievously out of contact with the symbolism of the first part, which the final image of poppies as sleep-inducers fatally recalls.

I have not broken this butterfly upon the wheel for no reason. I have done it to suggest the context and to try to specify the distinction of another "poppy" poem, Isaac Rosenberg's "Break of Day in the Trenches," written seven months after McCrae's poem. I think it superior even to Rosenberg's "Dead Man's Dump," which would make it, in my view, the greatest poem of the war. It is less obvious and literal than "Dead Man's Dump": everything is done through indirection and the quiet, subtle exploitation of conventions of English pastoral poetry, especially pastoral elegy. It is partly a great poem because it is a great traditional poem. But while looking back on literary history in this way, it also acutely looks forward, in its loose but accurate emotional cadences and in the informality and leisurely insouciance of its gently ironic idiom, which is, as Rosenberg indicated to Edward Marsh, "as simple as ordinary talk." [55]

BREAK OF DAY IN THE TRENCHES

The darkness crumbles away—
It is the same old druid Time as ever.
Only a live thing leaps my hand—
A queer sardonic rat—
As I pull the parapet's poppy
To stick behind my ear.
Droll rat, they would shoot you if they knew
Your cosmopolitan sympathies.
Now you have touched this English hand
You will do the same to a German—
Soon, no doubt, if it be your pleasure
To cross the sleeping green between.
It seems you inwardly grin as you pass
Strong eyes, fine limbs, haughty athletes
Less chanced than you for life,
Bonds to the whims of murder,

> Sprawled in the bowels of the earth,
> The torn fields of France.
> What do you see in our eyes
> At the shrieking iron and flame
> Hurled through still heavens?
> What quaver—what heart aghast?
> Poppies whose roots are in man's veins
> Drop, and are ever dropping;
> But mine in my ear is safe,
> Just a little white with the dust.

The filaments reaching back to precedent poetry are numerous and
richly evocative:

> The darkness crumbles away

inverts Nashe's

> Brightness falls from the air,

just as the literal trench *dust* of the last line translates Nashe's

> Dust hath closed Helen's eye

and invokes as well all the familiar *dusts* of Renaissance lyric-elegy which
fall at the ends of lines:

> Leave me, O Love, which reacheth but to dust.
> (Sidney)

> Golden lads and girls all must,
> Like chimney-sweepers, come to dust.
> (Shakespeare)

> Only the actions of the just
> Smell sweet and blossom in their dust.
> (Shirley)

The *only* with which the third line begins recalls the opening dynamics
of another poem which begins by finding a scene looking "normal" only
to discover that there is something disturbing in it that will require some
unusual thought, feeling, or action. I am thinking of "Dover Beach":

> The sea is calm tonight . . .
> Only. . . .

So Rosenberg begins:

> The darkness crumbles away—
> It is the same old druid Time as ever.
> Only a live thing leaps my hand—

"You might object to the second line as vague," he wrote Marsh, "but that was the best way I could express the sense of dawn." [56] Despite the odd way the darkness dissipates in tiny pieces (like the edge of the trench crumbling away), the dawn is the same as in any time or place: gray, silent, mysterious. "Only" on this occasion there's something odd: as the speaker reaches up for the poppy, a rat touches his hand and scutters away. If in Frye's terms the sheep is a symbol belonging to the model— that is, pastoral or apocalyptic—world, the rat is the creature most appropriate to the demonic. But this rat surprises us by being less noisome than charming and well-traveled and sophisticated, perfectly aware of the irony in the transposition of human and animal roles that the trench scene has brought about. Normally men live longer than animals and wonder at their timorousness: why do rabbits tremble? why do mice hide? Here the roles are reversed, with the rat imagined to be wondering at the unnatural terror of men:

> What do you see in our eyes . . .
> What quaver—what heart aghast?

The morning which has begun in something close to the normal pastoral mode is now enclosing images of terror—the opposite of pastoral emotions. It is the job of the end of the poem to get us back into the pastoral world, but with a difference wrought by the understanding that the sympathetic identification with the rat's viewpoint has achieved.

All the speaker's imagining has been proceeding while he has worn— preposterously, ludicrously, with a loving levity and a trace of eroticism—the poppy behind his ear. It is in roughly the place where the bullet would enter if he should stick his head up above the parapet, where the rat has scampered safely. He is aware that the poppies grow because nourished on the blood of the dead: their blood color tells him this. The poppies will finally fall just like the "athletes," whose haughtiness, strength, and fineness are of no avail. But the poppy he wears is safe for the moment—so long as he keeps his head below the parapet, hiding in a hole the way a rat is supposed to. The poppy is

> Just a little white with the dust,

the literal dust of the hot summer of 1916. It is also just a little bit purified and distinguished by having been chosen as the vehicle that has prompted the whole meditative action. But in being chosen it has been "pulled," and its death is already in train. Its apparent "safety" is as delusive as that currently enjoyed by the speaker. (Rosenberg was killed on April 1, 1918.) If it is now just a little bit white, it is already destined to be very white as its blood runs out of it. If it is now lightly whitened by the dust, it is already fated to turn wholly to "dust." The speaker has killed it by pulling it from the parapet. The most ironic word in the poem is the *safe* of the penultimate line.

As I have tried to suggest, the poem resonates as it does because its details point to the traditions of pastoral and of general elegy. As in all elegies written out of sympathy for the deaths of others, the act of speaking makes the speaker highly conscious of his own frail mortality and the brevity of his time. Even if we do not hear as clearly as Jon Silkin the words "Just a little while" behind "Just a little white," we perceive that the whole poem is saying "Just a little while." We will certainly want to agree with Silkin's conclusions about the poem's relation to tradition. The poem pivots on what Silkin calls "the common fantasy" about poppies, that they are red because they are fed by the blood of the soldiers buried beneath them. "It is one thing to invent," says Silkin; it is "quite another to submit one's imagination to another's, or to the collective imagination, and extend it, adding something new and harmonious." [57] "Extend" is just what Keith Douglas does, twenty-seven years later in North Africa. He writes "Desert Flowers," aware that

> . . . the body can fill
> the hungry flowers. . . .

"But that is not new." As he knows,

> Rosenberg I only repeat what you were saying.

Pastoral has always been a favored mode for elegy, whether general or personal, because pastoral contains perennial flowers, and perennials betoken immortality. Especially flowers that are red or purple, the colors of arterial and venous blood. Here is a pastoral elegy by Ivor Gurney associating homoerotic emotion with a literal pastoral setting and bringing into poignant (but still very literary) relation the redness of appropriate memorial flowers and the redness of blood nobly and sacrificially shed. Not the least of its merits is the complex play on the idea of "remembrance" undertaken by the last stanza:

> TO HIS LOVE
> He's gone, and all our plans
> Are useless indeed.
> We'll walk no more on Cotswold
> Where the sheep feed
> Quietly and take no heed.

His body that was so quick
   Is not as you
Knew it, on Severn river
Under the blue
   Driving our small boat through.

You would not know him now . . .
   But still he died
Nobly, so cover him over
   With violets of pride
   Purple from Severn side.

Cover him, cover him soon!
   And with thick-set
Masses of memoried flowers—
   Hide that red wet
   Thing I must somehow forget.

("Cover him over / With violets of pride / Purple from Severn side": one can't help observing that the beauty of those lines might seem to go some small way toward redeeming the ghastliness of the misbegotten events that prompted them. *O felix culpa!*)

Given the tradition, one can write a pastoral elegy in prose, as Cecil Lewis does in *Sagittarius Rising*. For a change he is not flying over the Somme battlefield but walking over it a bit back from the line, observing its desolated, riven trees, its debris and rubble and ravaged cemeteries, its overlapping filthy craters extending for scores of miles. "It was diseased, pocked, rancid, stinking of death in the morning sun":

> Yet (Oh, the catch at the heart!), among the devastated cottages, the tumbled, twisted trees, the desecrated cemeteries, opening, candid, to the blue heaven, the poppies were growing! Clumps of crimson poppies, thrusting out from the lips of craters, straggling in drifts between the hummocks, undaunted by the desolation, heedless of human fury and stupidity, Flanders poppies, basking in the sun.

And suddenly "a lark rose up from among them and mounted, shrilling over the diapason of the guns. . . ." [58]

A HARMLESS YOUNG SHEPHERD: EDMUND BLUNDEN

An extended pastoral elegy in prose is what Blunden's *Undertones of War* (1928) may be called. Whatever it is (G. S. Fraser once called it "the best war *poem*" and printed some of it as free verse in the *London Magazine* [59]), no one disagrees that together with Sassoon's and Graves's "memoirs" it is one of the permanent works engendered by memories of the war. Its distinction derives in large part from the delicacy with which it deploys the properties of traditional English literary pastoral in the service of the gentlest (though not always the gentlest) kind of irony. If Spenser or Milton or Gray or Collins or Clare or the author of *Thyrsis* had fought in the Great War, any one of them could have used Blunden's final image to end a memoir of it. With a due sense of theatrical costume and an awareness of a young subaltern's loving responsibility for the flock under his care, Blunden brings *Undertones of War* to a close by calling himself "a harmless young shepherd in a soldier's coat" (314).[60] That characterization is English to a fault, and beautiful.

This harmless young shepherd was born in the Tottenham Court Road on November 1, 1896, the son of a London schoolmaster. His family soon after moved to Yalding, Kent, where Edmund attended the local Grammar School. From the beginning he was scholarly and sensitive. He went on to a scholarship at Christ's Hospital, and there, conscious of such predecessors as Charles Lamb and Leigh Hunt, plied his Latin and Greek well enough to win a Classics scholarship at Queen's College, Oxford, where he practiced Georgian poetry exactly reflecting his character: shy, modest, accurate, gentle, courteous, and a bit dreamy. Edmund Gosse thought he looked like an intelligent chinchilla, with great sincere poetic eyes.

He was eighteen when the war began. Early in 1915 he joined the Royal Sussex Regiment as one of its youngest, most innocent lieutenants. He was nicknamed "Rabbit" and later, "Bunny." In May, 1916, he was in the trenches. Guy Chapman remembers: "On our third evening in Hedge Street we welcomed a very young, very fair and very shy subaltern from the Royal Sussex, who were to relieve us next day. . . . I showed our incoming tenant from the Sussex over his noxious habitation. As we bade him good-bye, he shyly put a small paper-covered book into my hand. *The Harbingers*, ran the title, 'Poems by E. C. Blunden.' "[61] After two years at the front, during which he was gassed and won the Military Cross (typically, he withholds mention of either event in *Undertones of War*), he was invalided home to a training camp in March, 1918.

He was married in 1918 and finally discharged from the army in 1919. At first he tried to resume his life as an Oxford student, but, wearying of it, in 1920 he joined the *Athenaeum Magazine* as an assistant editor and writer of essays and reviews. The same year he started his own prolific writing career with the publication of his first important book of poems, bucolically titled (and spelled) *The Waggoner*. Before he was done, he published sixteen volumes of poems and four collected volumes; thirty volumes of essays, criticism, and travel pieces; and fifteen edited volumes recommending the work of such fellow spirits as Clare, Christopher Smart, Collins, Lamb, Shelley, and Keats and introducing the war poetry of Wilfred Owen and Ivor Gurney. His second important book of poems, *The Shepherd* (1922), won him the Hawthornden Prize.

He had been fond of distant places ever since a recuperative voyage to South America after the war, and in 1924 he began a three-year stint as Professor of English Literature at the University of Tokyo, an adventure that turned out very differently from Graves's experience at Cairo. In Tokyo, surrounded by unpretentious beauty, gentleness, ceremony, and courtesy, he composed *Undertones of War*. When it was published back in London, the first edition sold out in one day. He continued teaching poetry through the 1930's, but now back at Oxford, as a Fellow of Merton College and Tutor in English. He also wrote copiously for the *Times Literary Supplement* (whose staff he joined in 1943), and after Kipling's death in 1936, he replaced him as Honorary Literary Adviser (that is, rhetorical consultant) to the Imperial War Graves Commission.

During the Second War he taught soldiers at Oxford map-reading—elegantly, it is said—and worked on his biography of Shelley, published in 1946. After the war he returned to Tokyo as Cultural Liaison Officer with the British Mission, lecturing all over Japan on English poetry. During the fifties he taught at the University of Hong Kong, and in 1966 he "defeated" Robert Lowell to win election as Professor of Poetry at Oxford, a position he had to relinquish in about a year because of poor health. He never stopped writing and editing to the very end, which came on January 20, 1974, when he was seventy-seven. He died in the country, at Sudbury, Suffolk. The flowers placed on his coffin were Flanders poppies.

Shortly before his death he wrote, "My experiences in the First World War have haunted me all my life and for many days I have, it seemed, lived in that world rather than this." [62] The prophecy which he had uttered by choosing as an epigraph for *Undertones of War* Bunyan's couplet

> Yea, how they set themselves in battle-array
> I shall remember to my dying day

was abundantly fulfilling itself. Obvious evidence of that is available in much that Blunden wrote, in poems, for example, like "1916 Seen from 1921" or "War Cemetery" (1940). But sometimes the residue of the Great War is more subtly lodged in his work, as in "The Midnight Skaters" of

Edmund Blunden. (Imperial War Museum)

1925, a poem which, for all its exactly observed rural setting, its traditional form and diction, and its precise, even fussy syntax, has at its
center a menace for which the term "Georgian," as commonly understood, would be quite inadequate:

THE MIDNIGHT SKATERS

The hop-poles stand in cones,
    The icy pond lurks under,
The pole-tops steeple to the thrones
    Of stars, sound gulfs of wonder;
But not the tallest there, 'tis said,
Could fathom to this pond's black bed.

Then is not death at watch
    Within those secret waters?
What wants he but to catch
    Earth's heedless son's and daughters?
With but a crystal parapet
Between, he has his engines set.

Then on, blood shouts, on, on,
    Twirl, wheel and whip above him,
Dance on this ball-floor thin and wan,
    Use him as though you love him;
Court him, elude him, reel and pass,
And let him hate you through the glass.

The one word there which would never have entered such a poem before
the war, even if Hardy had written it, is *parapet*. The enemy is "out
there"—or "down there"—always "at watch" like an unsleeping malevo-
lent sentry amply equipped with infernal machines. One is moderately
safe so long as one keeps the parapet between, but the parapet here is not
of earth and sandbags but of *crystal*. Rural life itself is now forever
tainted by awareness of "the enemy"; forever one will be conscious in a
new way of adversary hazards.

   To speak of rural England as "this Arcadia" would be to register considerable pastoral enthusiasm. But to speak of it, as Blunden does quite
solemnly, as "this genuine Arcadia" is to commit oneself to an extraordinary position, even if one is writing an essay titled "The Preservation of
England" and defending the countryside against the commercial de-
spoilers of the 1930's.[63] For to Blunden the countryside is magical. It is
as precious as English literature, with which indeed it is almost identical.
As the author of the critical study *Nature in English Literature* (1929),
Blunden can easily be imagined writing a companion study, *English Liter-*

ature in Nature. Once at Thiepval, he says in *Undertones of War,* "in double gloom the short day decayed"—a perception inverting Tennyson's "the long day wanes" ("Ulysses")—and the interminable night that followed "dragged its muddy length" (168) as if it as well as the author were familiar with the wounded snake of Pope's *Essay on Criticism,* which "drags its slow length along."

To Blunden, both the countryside and English literature are "alive," and both have "feelings." They are equally menaced by the war. And the French countryside is little different. For it to be brutally torn up by shells is a scandal close to murder. The pathos and shock of it are what Blunden returns to repeatedly in *Undertones of War:* "The greensward, suited by nature for the raising of sheep, was all holes, and new ones appeared with great uproar as one passed" (220). As one critic says, "It was the countryman [in Blunden] . . . who saw 'a whole sweet countryside amuck with murder'; for whom the sight of rich and fruitful land, much like his own, laid waste was an additional torment." [64] The naiads and hamadryads that he conceives to be lurking in the as yet unruined countryside west of the Somme early in 1918 are actual to him, more actual, perhaps, than were to Milton the nymphs of *Il Penseroso,* whose woodland haunts are threatened by the "rude axe." The streams and woods of these timorous spirits are in the path of the German attack of March, 1918, and it is their fate, not his own, that prompts Blunden's bitterness: "I might have known the war by this time, but I was still too young to know its depth of ironic cruelty" (314). In Japan later, he found that the Japanese poem he most admired was the one in which the poet, coming to his well to draw water and finding a morning-glory newly twined around the bucket, decides that he must get his water elsewhere. [65]

The literary culture implicit in *Undertones of War* is so ripe, so mellow and mature, that it is a surprise to recall that Blunden was only twenty-eight when he began writing it and only thirty-two when it was published. He is already practiced in the old man's sense of memory as something like a ritual obligation. "I must go over the ground again," he says in his "Preliminary" (xiii): the "ground" is the past imaged as military terrain, spread out for visiting and mapping as his battalion front had been mapped by the younger Blunden only seven years earlier. And the paragraph that follows, in its Baroque mortuary complications, is a startling performance from a young man:

> A voice, perhaps not my own, answers within me. You will be going over the ground again, it says, until that hour when agony's clawed face softens into the smilingness of a young spring day; when you, like Hamlet, your prince of peaceful war-makers, give the ghost a *"Hic et ubique?"* then we'll

change our ground," and not that time in vain; when it shall be the simplest
thing to take in your hands the hands of companions like E. W. T., and
W. J. C., and A. G. V., in whose recaptured gentleness no sign of death's
astonishment or time's separation shall be imaginable (xiii).

Anticipating his own death—how could a gassed asthmatic imagine that
it would be all of a half-century away?—as a form of almost-wished-for
death in battle, he is confident that when he achieves it he will rejoin his
three Christ's Hospital friends who were in his battalion: the young
officers E. W. Tice, W. J. Collyer, and A. G. Vidler, translated into
blessed spirits whose gentleness he will recapture at last. "Going over the
ground again": that is the act of memory conceived as an act of military
reconnaissance, just as later in *Undertones of War* Blunden (the former
"Field Works Officer") will imagine it as resembling the digging of saps
out into No Man's Land. But acts of recall are "saps of retrogression"
(231), diggings not forward but backward, burrowings into a torn landscape both familiar and totally unknown.

Unlike *The Memoirs of George Sherston* and *Good-bye to All That, Undertones of War* is less about its author than about the experience of the 11th
Battalion of the Royal Sussex Regiment, initially "a happy battalion"
(viii), but one whose history could be described as a series of grim disil-
lusions bravely borne. We hear nothing of Blunden's boyhood—partly
because throughout the war he's still a boy. "A schoolboy officer" he
calls himself (114). When he starts his memoir with the words "I was not
anxious to go," he means overseas, and to the front. He's already in the
army, ordered to accompany a draft and to join his battalion near the
line. It is spring of 1916, and the first couple of months, near Festubert,
Cuinchy, and Richebourg, are not so bad. But on June 30 near La Bassée
the battalion undergoes the first of the four terrible encounters which
will finally destroy it and leave it fit only for silent head-shaking and for
training duties. The three others are the attacks at Hamel on September
3, at Thiepval in the Somme during late October and early November,
and finally, in July, 1917, at Ypres, the beginning of the action called
Passchendaele. Blunden experiences a gradual but inexorable loss of innocence not because he's Blunden, or a poet, or shy, or sensitive, but
because he's a member of the battalion.

It all starts off gaily enough, with whimsy reminiscent of Laurence
Sterne but braced by British phlegm: "There was something about
France in those days which looked to me, despite all journalistic en-
chanters, to be dangerous" (1). But at Etaples there are ominous signals:
Blunden's ebony walking-stick, "which . . . was to be my pilgrim's
staff," is stolen; and a rifle-grenade instructor, just after announcing,

"I've been down here since 1914, and never had an accident," blows himself up together with most of his "class" (4–5). Joining C Company of his battalion on the line, Blunden at first can't believe that the pastoral landscape is being violated: "In the afternoon, looking eastwards . . . , I had seen nothing but green fields and plumy grey-green trees and intervening tall roofs; it was as though in this part the line could only be a trifling interruption of a happy landscape" (12). But going up that night, he becomes aware, as his similes suggest, that he is entering a perverse pastoral: "several furious insect-like zips went past my ear" (13) (later the bullets will whizz "like gnats" [28]). And daylight brings him further towards the truth. Although the morning was "high and blue and inspiriting," on the ground there were phenomena not at all like those of the benign bucolic world. Indeed, "at some points in the trench . . . skulls appeared like mushrooms" (15). Very soon he indicates that he has perfected the form of vision with which he will see the war: with ironic mock-wonder ("This is not what we were formerly told"), he will invoke all the beloved details of literary pastoral as a model and measure: "On the blue and lulling mist of evening, proper to the nightingale, the sheep-bell, and falling waters, the strangest phenomena of fire inflicted themselves" (16–17). Some of these juxtapositions are as overt as that, but many are so admirably subtle that we rub our eyes. Like the hated general's criticism of a subaltern's written report on a patrol: "Too flowery for a military report" (23). And it would be a mistake to patronize Blunden by imagining that the tone of these juxtapositions can be predicted. Not far from Festubert Village, he says, was "a red brick wall, to which fruit-trees reached their covert"; but only delusively. Part of the wall, "being but painted wood, presently swung open, and a field battery glaring brutally out would 'poop off' " (25–26). Here an appreciation of theater joins a fondness for pastoral to project the understanding of war as a travesty—comic this time—of nature.

It is while Blunden is luxuriating in the pastoral oasis of a "gas course" behind the line that we are told that before the war his batman was—what else?—a gardener (35). This man has his own form of Arcadian poetic resistance to the war. When threatened with the nastiest circumstances, we learn later, he is fond of singing his theme song, "You Must Sprinkle Me with Kisses If You Want My Love to Grow" (136). With this original Blunden enjoys a few weeks' respite in this "country-rectory quietude and lawny coolness" (37). But soon he must rejoin the battalion in trenches near Cuinchy, and as he goes up he notices the way "evidence of a war began to gnarl the scenery": indeed, the last estaminet has not roses but "shell-holes round the door" (38). It is at Cuinchy that he first speaks of the sun as "gilding" something—here, the sandbags at stand-to (49). Later, the sun will "gild" the barbed wire (130), and in Poperinghe a concert-party (called the "Red Roses") will seem with its strenuous cheerfulness to "gild the clouds of fate" (183). *Gilding* has been an action of tongue-in-cheek pastoral ever since Gay's *Shepherd's Week* (1714), where at dawn

> . . . the sunbeams bright to labor warn,
> And gild the thatch of Goodman Hodge's barn.
>   ("Monday")

Since to gild a thing is merely to surface it with gold—or more often, the gold-like—it is, as it were, to advertise its meretriciousness or (like Goodman Hodge's thatch) its inappropriateness for such treatment. Blunden's acts of gilding, while exhibiting a genuine pastoral motive of ennobling and "raising" low or common objects, at the same time "put them in their place" by illuminating them as unnatural, misplaced, anomalous, cruel, or fatally wrong.

At Cuinchy, Blunden says, "I began to understand the drift of the war" (42). It is a drift—nobody wants it, no one is in charge—toward insensate destruction conducted against a background of the unbearably beautiful:

> Over Coldstream Lane, the chief communication trench, deep red poppies, blue and white cornflowers, and darnel thronged the way to destruction; the yellow cabbage-flowers thickened here and there in sickening brilliance. . . . Then the ground became torn and vile, the poisonous breath of fresh explosions skulked all about, and the mud which choked the narrow passages stank as one pulled through it. . . . Much lime was wanted at Cuinchy (47).

But worse is to come. After the failed attack near Richebourg, Blunden's innocence and the battalion's is at an end: "deep down in the survivors there grew a bitterness of waste" (68).

At this time Blunden is elevated to "Field Works Officer," in which capacity his duty is to perform (largely at night) a parody of rural daily work—puttering, shoring-up, ditching, and draining—"honest labor," he calls it (132). The battalion now moves, in August, 1916, to the Somme. Near Hamel the communications trench sardonically named Jacob's Ladder is so long that it seems to extend all the way from the bucolic past (the rear) to the industrial present (the front):

> Leafy bushes and great green and yellow weeds looked into it as it dipped sharply into the green valley by Hamel, and hereabouts the aspect of peace

and innocence was as yet prevailing. A cow with a crumpled horn, a harvest cart should have been visible here and there.

But as one moves forward, one sees that "the trenches ahead were curious, and not so pastoral":

> Ruined houses with rafters sticking out, with half-sloughed plaster and dangling window-frames, perched on a hill-side, bleak and piteous . . . ; half-filled trenches crept along below them by upheaved gardens, telling the story of wild bombardment (104–105).

The terrible attack on Hamel of September 3 leaves the battalion again "much impaired" (122). Its state is equivalent to that of the town of Englebelmer behind the line, to which the battalion retires to "reorganize." Englebelmer, formerly "a sweet village," is now "entering upon a dark period. Its green turf under trees loaded with apples was daily gouged out by heavy shells; its comfortable houses were struck and shattered, and the paths and entrances gagged with rubble, plaster, and woodwork" (122–23). Another victim is the quiet rural town of Auchonvillers. From a distance its shame is masked by its surrounding hedges and shrubs and orchards, but its center is a cruel mess: "That heart nevertheless bleeds" (125). But even worse lies ahead along the road passing through town—the battlefield of Beaumont Hamel itself. Better to turn back and contemplate even battered Auchonvillers; or as Blunden puts it, "Turn, Amaryllis, turn—this way the tourist's privacy is preserved by ruins and fruitful branches" (126).

After a bad time at Thiepval in October and November, the battalion marches north to Ypres again, but it is "not the same 'we' who in the golden dusty summer tramped down into the verdant valley, even then a haunt of every leafy spirit and the blue-eyed ephydriads, now Nature's slimy wound with spikes of blackened bone" (175–76). But the battalion's recuperative powers are extraordinary, and even though it now knows its destiny, "from company headquarters and cookhouses one heard such cheerful singings and improvisations as seemed to hail the Salient as the garden of Adonis" (189). At Ypres Blunden is transferred to Brigade Headquarters as Intelligence Officer. His duties consist largely of drawing beautiful maps of the countryside, which now comprises millions of adjoining shell holes filled with corpses. But in the spring he returns to the battalion, preparing now to attack Pilckem Ridge.

"The instinct revolted," he says, "against the inevitable punishment to come" (244). The battalion harbors "in readiness among the familiar woods west of Vlamertinghe," but the woods are not the same as when the battalion lived in them during the relatively "Arcadian" days of 1916.

Their wood-spirit has been shelled silly like everyone else, and "the part-
ing genius must have gone on a stretcher" (245). The rural roads are now
made of greasy planks instead of earth, rails criss-cross the Salient, ammunition and stores are piled everywhere, and even the "eels, bream,
and jack" in the streams are dead. They have been shelled and gassed
(248).

The attack of July 31, 1917, goes well at first, and Blunden, passing
the mangled German trenches, regards them with special pity when he
sees that they are "mostly mere hedges of brushwood, hurdles, work for
a sheepfold . . ." (255). But rain soon clogs the attack, and "one more
brilliant hope, expressed a few hours before in shouts of joy, sank into
the mud" (257). Back to Poperinghe, finally, redolent now of gasoline
rather than violets.

He is now obsessed with "the growing intensity and sweep of destructive forces" (270). Taking up the rations at night in a quiet sector used to
be something almost like a pastoral activity, but the scene is now a
demonic anti-pastoral, dominated by such details as "slimy soil," "a
swilling pool of dirty water," "tree-spikes," and the everlasting "plank
road," to leave which is to fall into a swamp up to the armpits, although
to remain on it is "to pass through accurate and ruthless shell-fire" (272).
Even cleverness seems now to belong to earlier years. Along the road
leading to Kemmel "stood the column of tree stumps among which our
sniping authorities had formerly smuggled in one or two steel trees, now
[like the "veteran" in Johnson's *Vanity of Human Wishes*] lagging superflu-
ous on the stage, their green paint and tubular trunks being out of
season" (273). It is too late for wit: the war is now a matter of nothing
but "a dead sea of mud" (285), rain, putrefying flesh, flies, constant bombardment, unremitting fear, and the murder of friends.

Blunden's very worst moment occurs at a position consisting of two
concrete pillboxes south of Ypres. His party of battalion signalers is in
one, the adjutant and the doctor and their men are in the other. The pillboxes come under terrible artillery fire, and immediately the trench joining them is full of bodies. The adjutant delivers a phlegmatic invitation
to dinner at his pillbox, and just after he returns, a shell enters his door,
killing at once adjutant, doctor, wounded, and everyone else. "Don't go
over, sir, it's awful," Blunden's gardener-servant tells him (279). Forty
men were killed and tumbled together. Blunden was haunted by this
moment for the rest of his life, as he indicates in poems like "Third
Ypres" and "The Welcome."

Relieved at last after 280 men have been killed or wounded, the battalion recovers back of the line, and Blunden ironically speculates on which

of these remaining men were ultimately to be killed in March, 1918, on that spot. He does not write, "Some were destined to fight and fall not many months later on that very ground." He writes, "Some were destined to fight and drop . . ." (292). They are not like men anymore. They are like the orchard fruit in *The Merchant of Venice*, where Antonio finds that

> . . . the weakest kind of fruit
> Drops earliest to the ground . . . ,
>   (IV, i, 115–16)

or the fruit in George Meredith's "Dirge in Woods":

> And we go,
> And we drop like the fruits of the tree,
> Even we,
> Even so.

Or they are like the flowers in Fitzgerald's *Rubaiyat*, a poem which—foreshadowing the tradition of the way Great War poppies are nourished—finds that red or purple flowers feed on blood-drops underground:

> I sometimes think that never blows so red
> The rose as where some buried Caesar bled;
>   That every hyacinth the garden wears
> Dropt in her lap from some once lovely head.

Blunden is now used up and inefficient, irritable with his superiors and not at all well: he has inhaled too much gas on top of his asthma. In the spring of 1918 he is returned to England to serve out the rest of the war at a training camp. It is while finally departing from the war and passing through the as yet untouched countryside way behind the line that he proposes his final pastoral contrast, the one that closes his book with complex pathos and irony:

> Could any countryside be more sweetly at rest, more alluring to naiad and hamadryad, more incapable of dreaming a field-gun? Fortunate it was that at the moment I was filled with this simple joy. . . . No conjecture that, in a few weeks, Buire-sur-Ancre would appear much the same as the cataclysmal railway-cutting by Hill 60, came from that innocent greenwood.

Despite the knowledge he has attained, especially at the pillboxes, Blunden is still innocent. Only an innocent man would say something like "that innocent greenwood." But the next sentence, the superb final one, implies what will constitute Blunden's Fall. His Fall will take place out-

Hill 60, near Messines, Ypres Salient, from the air. The large craters were made by two of the nineteen mines set off on June 7, 1917. The pock marks are shell craters. The squiggly lines are German trenches. One follows the rim of the mine crater; those at the lower left exhibit two kinds of traverses, square and zig-zag. Saps extend downward from the zig-zag trench. On the right, between the two mine craters, is Blunden's "cataclysmal railway-cutting," once the route of the railway between Ypres and Comines. The dark patches on the left are woods, now cratered. The area on the right, once equally wooded, is now bare.

The thick light-colored lines marked on the photograph (roughly parallel with the railway-cutting) represent boundary lines between British units which will attack over this ground, advancing from the bottom of the photograph to the top. (Imperial War Museum)

side the confines of the book, and the book will thus maintain its innocence. The Fall is just this: the knowledge that the remaining countryside of Picardy has been ravaged. That will be the fatal, serpent-borne Knowledge of Good and Evil for Blunden. But at the (delusive) present moment, "no destined anguish lifted its snaky head to poison"—not "me," notice, but—"a harmless young shepherd in a soldier's coat" (314). With that objective distancing, that tender withdrawing vision of a terribly vulnerable third-person, the book ends.

Or almost ends. There is an addendum of thirty-one poems (thirty-two in the 1956 edition, which adds "Return of the Native"), titled, in Blunden's old-fashioned, retrogressive way, "A Supplement of Poetical Interpretations and Variations." Almost a fourth of these poems establish pastoral oases as an ironic gauge of what things are like further on up. A literal pastoral oasis is imagined in "The Guard's Mistake." The battalion has left the line and marched back to rest at a miraculously undamaged town, so untouched by the war that even nature has not forgotten that she is supposed to be friendly to man:

> The cherry-clusters beckoned every arm,
> The brook ran wrinkling by with playful foam.
>> And when the guard was at the main gate set,
>> Surrounding pastoral urged them to forget.

So benign and homelike is the setting that one sentry, a former keeper of cows, forgets himself—or remembers himself—and replaces his rifle with a country cudgel:

> . . . out upon the road, gamekeeper-like,
> The cowman now turned warrior measured out
> His up-and-down *sans* fierce "bundook and spike,"
> Under his arm a cudgel brown and stout;
>> With pace of comfort and kind ownership,
>> And philosophic smile upon his lip.

It is too good to last. A minute later

>> . . . a flagged car came ill-omened there;
> The crimson-mottled monarch, shocked and shrill,
> Sent our poor sentry scampering for his gun,
> Made him once more "the terror of the Hun."

A constant element in the poems is the equivalent of the staff officer's reimposition of "the war" on the pastoral sentry: namely, the insistent ravaging of innocent, unpretentious rural houses and trees. "A House at Festubert," apostrophized as "Sad soul," is nothing now but "one wound"; "La Quinque Rue" is lined with "trees bitterly bare or snapped in two." There is a free transfer between feelings evoked by torn-up farmhouses and trees and those evoked by torn-up human bodies. In someone else this might suggest some lack of interest in misery that is specifically human. But not in Blunden: his vision of nature ravaged always in human terms and images indicates what his gentle heart has been attentive to all the while.

When all is said, it is the sheer literary quality of *Undertones of War* that remains with a reader. One does not forget its rhythms ("Musically sounded the summer wind in the trees of Festubert" [27]) nor its shrewd, witty reliance on etymology: "The officer and I, having nothing to do but wait, sat in a trench . . . considering the stars in their courses" (114). Blunden's weakness, on the other hand, is a calculation so precise, even niggling, that on occasion it leads to the archness of the confirmed schoolmaster. Jocularities like "mine host" or mock-pedantic elevated diction like *adumbrated* are Blunden's fatal Cleopatras. His way with language is as archaic as Graves's is "modern": his style is rich with personifications (Fancy, Imagination, War, Detraction, Order, Cheerfulness) which act like "displaced" classical deities and give the prose a mock-epic dimension redolent of the eighteenth century. The constant allusion and quotation reveal a mind playing over a past felt to be not at all military or political but only literary. The style is busy with rhetorical questions and exclamations, with half-serious apostrophes evocative of Sterne. A pleasant room in a farm cottage behind the line is thus addressed: "Peaceful little one, standest thou yet? cool nook, earthly paradisal cupboard with leaf-green light to see poetry by, I fear much that 1918 was the ruin of thee" (82). And the archaism points back even earlier than to Sterne, back to Sir Thomas Browne and Robert Burton and Isaak Walton, who might lament of the peasantry very much the way Blunden ironically does of the infantry: "There is no pleasing your ancient infantryman. Attack him, or cause him to attack, he seems equally disobliging" (307).

But archaism is more than an overeducated tic in Blunden. It is directly implicated in his meaning. With language as with landscape, his attention is constantly on pre-industrial England, the only repository of criteria for measuring fully the otherwise unspeakable grossness of the war. In a world where literary quality of Blunden's sort is conspicuously an antique, every word of *Undertones of War*, every rhythm, allusion, and droll personification, can be recognized as an assault on the war and on the world which chose to conduct and continue it. Blunden's style is his critique. It suggests what the modern world would look like to a sensibility that was genuinely civilized.

Is Blunden "escaping" into the past? If he is, let him. But I don't think he is. He is, rather, engaging the war by selecting from the armory of the past weapons against it which seem to have the greatest chance of withstanding time. In his own shy way, he is hurling himself totally and emotionally into opposition. This is what H. M. Tomlinson seems to recognize when he says of *Undertones of War*, "Something in Blunden's story is more than queer. . . . This poet's eye is not in a fine frenzy rolling. There is a steely glitter in it." [66] We can now begin to understand what Blunden is getting at when in his "Preliminary" he says that he's afraid "no one will read it. . . . No one? Some, I am sure; but not many. *Neither will they understand* . . ." (xii).

Any writer about the war who has recourse to Arcadian contrasts at all runs a terrible risk of fleeing into calendar-art sentiments:

> The roses round the door
> Makes me love mother more.

It is for this reason that Blunden's literary bravery seems so admirable. The distinction of *Undertones of War* emerges most tellingly when we see what it might be but is not. "The Things that Make a Soldier Great," Edgar A. Guest sings in 1918,

> Are lilacs by a little porch, the row of tulips red,
> The peonies and pansies too, the old petunia bed,
> The grass plot where his children play, the roses on the wall:
> 'Tis these that make a soldier great. He's fighting for them all.
>   . . .
> What is it through the battle smoke the valiant soldier sees?
> The little garden far away, the budding apple trees. . . .

To select terms very like these and yet to keep them from saying what these say is what Blunden does. In doing so, he risks everything. I think he wins through in the end.

VIII
Soldier Boys

MARS AND EROS

After considering the matter for centuries, the ancients concluded that one of the lovers of Venus is Mars. And Eros, some held, is their offspring. Since antiquity everyone who has experienced both war and love has known that there is a curious intercourse between them. The language of military attack—*assault, impact, thrust, penetration*—has always overlapped with that of sexual importunity. Seventeenth-century wit, so conscious of its classical inheritance, would be sadly enfeebled if deprived of its staple figure of "dying" on one's "enemy."

War and sexuality are linked in more literal ways as well. That a successful campaign promises rape as well as looting has been understood from the beginning. Prolonged sexual deprivation will necessitate official brothels—in the Great War, "Blue Lights" for officers, "Red Lights" for Other Ranks. The atmosphere of emergency and the proximity of violence will always promote a relaxing of inhibition ending in a special hedonism and lasciviousness. And of course a deeper affection as well. As Auden puts it in *The Age of Anxiety* (1947), "In times of war even the crudest kind of positive affection between persons seems extraordinarily beautiful, a noble symbol of the peace and forgiveness of which the whole world stands so desperately in need." [1] And the gender of the beloved will not matter very much. During the Second War, says Goronwy Rees, "Guy [Burgess] brought home a series of boys, young men, soldiers, sailors, airmen, whom he had picked up among the thousands who thronged the streets of London . . . ; for war, as Proust noticed, provokes an almost tropical flowering of sexual activity behind the lines which is the counterpart to the work of carnage which takes place at the front." [2] On the one hand, sanctioned public mass murder. On the other, unlawful secret individual love. Again, severe dichotomy.

And some relations between warfare and sexuality are more private and secret still. There are numerous testimonies associating masturbation and exhibitionism with the fears and excitements of infantry fighting. Perhaps prolonged threats to the integrity of the body heighten physical self-consciousness and self-love. When Christopher Isherwood began masturbating precociously at the age of six (around 1910), his fantasies took the form of this battlefield vignette: "He imagined himself lying wounded on a battlefield with his clothes partly torn off him, being tended by a woman. . . . The mood of this fantasy was exhibitionistic; Christopher's own nakedness was what excited him." [3] Wilfred Owen seems to hint that there is something ambiguously exhibitionistic about exposing the body to bullets and shellfire when he describes the "sensations of going over the top" to his brother Colin in May, 1917, and uses terms that he refrains from using in describing his feelings to his mother: "There was extraordinary exultation," he says, "in the act of slowly walking forward, showing ourselves openly." [4] What he seems to imply is explored exhaustively and honestly in James Jones's underrated novel about the Guadalcanal campaign, *The Thin Red Line* (1962). After a breathless and risky infantry action against the Japanese, Jones's hero John Bell wonders why he has actually felt pleased to "expose himself" to mortal danger. Immediately he remembers occasions during his adolescence at home when he openly risked discovery while masturbating. One occasion in particular he recalls: he deliberately placed his clothes too far behind him in a woods to recover them safely if he should be spotted; then, standing just behind a screen of leaves, he masturbated only a few feet from a gathering of strangers. Now,

crawling along . . . a ledge on Guadalcanal, . . . John Bell stopped and stared, transfixed by a revelation. And the revelation, brought on by his old memory [of masturbating at the edge of the woods] . . . , was that his volunteering, his climb out into the trough that first time, even his participation in the failed assault, all were—in some way he could not fully understand—sexual, and as sexual, and in much the same way, as his childhood incident. . . .

After this "revelation," he begins wondering: "Could it be that with the others? Could it be that *all* war was basically sexual? . . . A sort of sexual perversion? Or a complex of sexual perversions? That would make a

funny thesis and God help the race." [5] Jones's point and terms are not
essentially different from Graves's, who in "Recalling War" (1938) remembers the exulation he and other young men felt at the beginning of
the Great War, and remembers that exultation in images of youthful
erections and ejaculations:

> we . . . thrust out
> Boastful tongue, clenched fist and valiant yard.
> Natural infirmities were out of mode,
> For Death was young again: patron alone
> Of healthy dying, premature fate-spasm.

Given this association between war and sex, and given the deprivation
and loneliness and alienation characteristic of the soldier's experience—given, that is, his need for affection in a largely womanless
world—we will not be surprised to find both the actuality and the recall
of front-line experience replete with what we can call the homoerotic. I
use that term to imply a sublimated (i.e., "chaste") form of temporary
homosexuality. Of the active, unsublimated kind there was very little at
the front. What we find, rather, especially in the attitude of young
officers to their men, is something more like the "idealistic," passionate
but non-physical "crushes" which most of the officers had experienced at
public school. Sassoon's devotion to Dick Tiltwood we have already
seen, and we have seen how it constitutes a transferral to a new theater
of a passion like Graves's for his "Dick" at Charterhouse. What inspired
such passions was—as always—faunlike good looks, innocence, vulnerability, and "charm." The object was mutual affection, protection, and
admiration. In war as at school, such passions were antidotes against
loneliness and terror.

Do the British have a special talent for such passions? An inquirer
turning over the names of late nineteenth and early twentieth century lit-
erary worthies might be led to think so as he encounters Wilde, Samuel
Butler, Edward Fitzgerald, Housman, Hopkins, Symonds, Strachey,
Edward Marsh, William Johnson Cory (author of the "Eton Boating
Song"), Hugh Walpole, John Maynard Keynes, E. M. Forster, and J. R.
Ackerley. Even such professionally manly figures as Cecil Rhodes and
Sir Richard Burton proved homoerotically excitable, and Strachey says
of General Gordon, the martyr of Khartoum: "He was particularly fond
of boys. Ragged street arabs and rough sailor-lads crowded about him.
They were made free of his house and garden; they visited him in the
evenings for lessons and advice; he helped them, found them employment, corresponded with them when they went out in the world." [6] Of

the sainted Lord Kitchener Queen Victoria once said, "They say he dislikes women, but I can only say he was very nice to me." [7]

J. R. Ackerley, during the war a platoon leader in the 8th East Surreys, is an authority on these matters, and we should listen carefully to his scrupulous testimony in *My Father and Myself*. While in the army he apparently succeeded in sublimating his homosexuality into something purely aesthetic and "ideal." During the war, he says, "I never met a recognizable or self-confessed adult homosexual . . . ; the Army with its male relationships was simply an extension of my public school." [8] Friendships, that is, were chaste. The Other Ranks were equivalent to the younger boys, and as at school, one generally admired them from a distance, as Ackerley did "the younger Thorne," a handsome boy in his platoon.

> The working classes also, of course, now took my eye. Many a handsome farm- or tradesboy was to be found in the ranks of one's command, and to a number of beautiful but untouchable NCO's and privates did I allot an early sentimental or heroic death in my nauseous verse.

Being near male beauty is always important to Ackerley, but it is especially important when one is lonely and in physical danger:

> My personal runners and servants were usually chosen for their looks; indeed this tendency in war to have the prettiest soldiers about one was observable in many other officers; whether they took more advantage than I dared of this close, homogeneous, almost paternal relationship I do not know. [9]

(R. J. E. Tiddy, a lieutenant in the Oxfordshire and Buckinghamshire Light Infantry, confessed to "great self denial" in choosing as his servant "a phenomenally plain" young person. But he could regale himself with the constant sight of the handsome NCO's, whom he found "luckily enchanting." [10]) Ackerley himself was once virtually proposed to by a senior member of the brigade staff, who stopped him just before an attack and offered him a safe job on the staff: "He begged me to accept it. He had always been fond of me, I knew, indeed he had a crush on me, I think, for I was a pretty young man. . . . 'You've done your bit already,' said he gently." [11] Given this atmosphere, it will seem natural that it was on the Western Front early in the war that W. Somerset Maugham encountered the young American ambulance driver Gerald Haxton, who was to become his lifelong companion.

It was largely members of the upper and upper-middle classes who were prepared by public-school training to experience such crushes, who "hailed with relief," as J. B. Priestley remembers, "a wholly masculine

way of life uncomplicated by Woman." [12] Some, as we have seen, generalized the affection and extended their love to whole groups of men.
Montague reports that "some of the higher commanders, as well as the
lower, do really love with a love passing the love of women—'the dear
men' of whom I have heard an officer, tied to the staff and the base by
the results of head wounds, speak with an almost wailing ache of desire,
as horses whinny for a friend. . . ." [13] But the tradition of "buddies"
and "mates" promoted sentimental associations among the Other Ranks
also that in their intensity and chastity were very close to the crushes indulged in by their superiors. Anthony French had such a feeling for his
"Bert," Private Albert William Bradley, "a young fair-haired soldier"
who died with his head cradled in French's arms on September 15, 1916.
When French first met Bert, he says, "I liked his face. It was boyish,
pale and slightly drawn. His forehead was high and broad and his lips
had a whimsical curl that seemed to compel one cheek to dimple." [14] On
Armistice Day, left alone in a deserted camp at Wimbledon, French
decides to write Bert a "letter":

> "This, my dear Bert, is *your* day, and I'm more than ever reminded of
> you. . . ." I wrote on and on telling him all that would interest him. . . .
> As I wrote a deep sadness afflicted me. . . . I drafted some verses of a
> lyric of unadulterated sentimentality that told of a friendship and what it
> had meant to me. It ended with
> > . . . one happy hour to be
> > With you alone, friend of my own,
> > That would be heaven to me. [15]

Although the usual course of protective affection was from superior to
subordinate, sometimes the direction was reversed, with men developing
hero-worshipping crushes on their young officers. Such a situation, reflected in Raleigh's devotion to the older Stanhope in *Journey's End*, constitutes part of a crucial disclosure in Graves's play *But It Still Goes On*
(1931). Curiously, this play is very little known. Graves does not choose
to have its title appear in the list of his works prefixed to his *Collected
Poems* of 1955, although that list includes thirty-five books. Nor is it
mentioned by J. M. Cohen in his critical study of Graves, nor listed in
his bibliography. It is as if Graves wanted us not to know about it at all.
His desire to forget it is perhaps not unconnected with his deletion from
Good-bye to All That of the sentence indicating that he did not "recover"
from his school homosexuality until he was twenty-one.

But It Still Goes On is a Shavian comedy about two nice young people,
David and Charlotte, who have been affecting love for each other in the

standard postwar house-party way but who finally confess their disingenuousness: each is actually homosexual and secretly in love elsewhere. When David reveals that he was a little minx at school and a
practiced homosexual at Cambridge, Charlotte, in some disbelief, says,

> Well, then the War. Surely in the War—?
>
> DAVID: That's another part of life that isn't generally known. Do you know
> how a platoon of men will absolutely worship a good-looking gallant
> young officer? If he's a bit shy of them and decent to them they get a
> crush on him. He's a being apart: an officer's uniform is most attractive
> compared with the rough shapeless private's uniform. He becomes a sort
> of military queen-bee.
>
> CHARLOTTE: And his drilling them encourages the feeling?
>
> DAVID: (*nods*) Of course, they don't realize exactly what's happening, neither
> does he; but it's a very strong romantic link.[16]

It's worth noticing that the boy with whom David finds himself currently in love is named Dick. As the name especially of a beloved boy
who is dead (cf. Sassoon's "The Effect": "When Dick was killed last
week he looked like that, / Flapping along the fire-step like a fish"),
"Dick" may derive in part from Housman's "The Night is Freezing
Fast":

> . . . winterfalls of old
> Are with me from the past;
> And chiefly I remember
> How Dick would hate the cold.

We will notice too how often the "Dick" figure is blond, even when
his name is "Bert," or when, sufficiently distanced, he appears as David
Jones's "flaxen-haired" Lieutenant Jenkins. To be fair-haired or (better)
golden-haired is, in Victorian iconography, to be especially beautiful,
brave, pure, and vulnerable. For Victorian painters of Arthurian themes,
the convention was that Galahad, Tennyson's "bright boy-knight," was
golden-haired. "I glanced at Dick," says Sassoon, "and thought what a
Galahad he looked." [17] Victorian pederastic poetry swarms with adored
lads like the one in Wilde's "Wasted Days":

> A fair slim boy not made for this world's pain,
> With hair of gold thick clustering round his ears. . . .

In this tradition, beloved acolytes and boy-saints are most often blond so
that the poet can treat their hair as a golden halo. Or use it to evoke
images of exotic, libertine southern places. Thus Wilfred Owen celebrates the good looks of a "navy boy" met in a train compartment:

> His head was golden like the oranges
> That catch their brightness from Las Palmas sun.[18]

During the war those who found beauty even in German corpses tended to find it in blond ones. One of C. E. Carrington's most vivid memories of the whole war is that of "a tall, pale, plump, blond, young Teuton of my own age or a little more":

> He lay as he had fallen, in an attitude of running, struck by three shrapnel bullets in the back—not running away but carrying a message, for in his hand was a dirty scrap of paper on which three words were scribbled: *'die Engländer kommen.'* His grey eyes were open, and his mouth showed strong white teeth. I looked on him and loved him. . . .[19]

A dead blond equally fascinates Sassoon. Coming upon some German bodies, he experiences "an impulse" to lift up a blond one and prop it against a bank. "He didn't look to be more than eighteen. . . . I thought what a gentle face he had. . . . Perhaps I had some dim sense of the futility which had put an end to this good-looking youth." [20]

The equation of blondness with special beauty and value helps explain the frantic popularity of Rupert Brooke, whose flagrant good looks seemed an inseparable element of his poetic achievement. His features were available to everyone in the famous bare-shouldered photograph by S. Schell which served as a frontispiece to his volume of poems. Apparently no one was immune to his golden beauty. D. H. Lawrence speaks of "his general sunniness," [21] A. C. Benson found that his complexion had "a tinge of sun-ripened fruit," [22] and Frances Cornford, while ridiculing his romanticism, immortalized his person as that of "A young Apollo, golden-haired." We can agree with Christopher Hassall that the famous photograph of Brooke provided "a visual image that met the needs of a nation at a time of crisis," [23] although we would want to suggest that the "needs" were as deeply homoerotic as they were patriotic.

To the degree that front-line homoeroticism was sentimental it can be seen to constitute another element of pastoral. To the degree that it seems conscious of the pathetic fate of the pretty, it seems identifiably British: the British understanding of *Et in Arcadia ego* is never far away. It seems to surface in Robert Nichols's memory of his medical examination at enlistment:

> I remember very well the face of a kind, keen major . . . amid a crowded room. . . . He smiled at each of us in turn, but his eyes were sad. "How old are you?" he asked an applicant who was blowing out a bare chest to fill the tape around it. "Nineteen, sir." "You seem in a great hurry to be killed,

my boy." The applicant, disconcerted, stammered, "I only want to do my bit, sir." "Very well, so you shall, so you shall; and good luck to you." But . . . as the major laid his head to my bare chest—I was next—I experienced a curious sensation: his eye-lashes were wet.[24]

The characteristic "pastoral" homoerotic tenderness of Great War British male love is specifically contrasted by Pynchon with something superficially resembling it in the Second War and after. *Gravity's Rainbow*—and this is one of the most notable things about it—depicts all modern sex as aggression, hatred, selfishness, and cynicism. After exhibiting the preposterous, mincing Sir Marcus Scammony's bitchery in a colorful scene, Pynchon lays aside the jokes, steps to the front of his stage, and speaks in his own serious, measured voice:

> It wasn't always so. In the trenches of the First World War, English men came to love one another decently, without shame or make-believe, under the easy likelihoods of their sudden deaths, and to find in the faces of other young men evidence of otherworldly visits, some poor hope that may have helped redeem even mud, shit, the decaying pieces of human meat. . . . While Europe died meanly in its own wastes, men loved.

But that was only for a moment. What succeeded has been far different: "The life-cry of that love has long since hissed away into no more than this idle and bitchy faggotry." [25]

On the last day he ever spent in England, August 31, 1918, Wilfred Owen went swimming at Folkestone. He had hoped to read Shelley while drying on the beach, "But I was too happy," he wrote Sassoon the next day, "or the Sun was too supreme." Besides, he says,

> there issued from the sea distraction, in the shape, Shape I say, but lay no stress on that, of a Harrow boy, of superb intellect and refinement: intellect because he hates war more than Germans; refinement because of the way he spoke of my Going, and of the Sun, and of the Sea there; and the way he spoke of Everything. In fact, the way he spoke—.

Owen has fallen in love, and it is in this state that he has rejoined his inarticulate needy men back in France: "And now I am among the herds again, a Herdsman; and a Shepherd of sheep that do not know my voice." [26] The beach represents the scene of pastoral romance as delineated by Frye, "a world of magic or desirable law, [tending] to center on a youthful hero, still overshadowed by parents, surrounded by youthful companions." In that pastoral world, "the archetype of erotic innocence is less commonly marriage than the kind of 'chaste' love that precedes marriage; the love of brother for sister, or of two boys for each other." [27]

Frye points to the sixth book of *The Faerie Queene* as a repository for the imagery of such a world. But a more stimulating and less innocent example was, for Owen and his contemporaries, closer to hand: namely, Virgil's *Second Eclogue*, in which Corydon, who has fallen in love with the boy Alexis, complains of his boyish disdain of love. Every young officer from a public school had read this poem: it recommended itself especially by virtue of usually being skipped by teachers making assignments in Virgil. Acquaintance with it was a great help in promoting sentimental friendships between older and younger boys at school and in training up what Graves calls "pseudo-homosexuals," [28] that is, innocent ones. It must have been a complex and poignant experience to remember this poem in the trenches, for among the flowers with which Corydon hopes to bribe Alexis are poppies, and it is sundown (cf. evening stand-to) that temporarily lays to rest Corydon's "raging heats" (as Dryden translates). But the really British thing about the *Second Eclogue* is that Corydon does not attain his desire. Alexis remains at a "Platonic" distance, an image of *formose puer* forever unenjoyed.

One of the pre-existing motifs available for "translation" and "purification" by Great War writers was provided by the tradition in Victorian homosexuality and homoeroticism that soldiers are especially attractive. What makes them so is their youth, their athleticism, their relative cleanliness, their uniforms, and their heroic readiness, like Adonis or St. Sebastian, for "sacrifice." Many of Whitman's poems of "adhesiveness"—that is, sublimated male love—in *Drum Taps* run parallel with this tradition. For example, "Vigil Strange I Kept on the Field One Night," a vigil not just for any soldier but a

> Vigil for boy of responding kisses. . . .

The poems of Hopkins likewise take place in this atmosphere. In "The Soldier" he asks,

> . . . Why do we all, seeing of a soldier, bless him? bless
> Our redcoats, our tars?

And in "The Bugler's First Communion" Hopkins dwells with extraordinary affection on the innocent but manly soldierliness of the "bugler boy from barrack" who "knelt . . . in regimental red" to receive his first communion, among whose blessings, Hopkins prays, shall be "bloom of a chastity in mansex fine." (And speaking of the infusion of the homoerotic by the pastoral, we remember that it was of "Harry Ploughman" that Hopkins wrote defensively to Bridges: "When you read it let me know if there is anything like it in Walt Whitman; as perhaps there may be, and I should be sorry for that." [29])

When we turn from these gentle literary fantasists to active prewar pederasts, we find soldiers specifically the focus of desire. The Other Ranks of H.M. Brigade of Guards had of course been notoriously employable as sexual objects since early in the nineteenth century, and the first German edition of Symonds's *Sexual Inversion* (Leipzig, 1896) contained an appendix, "*Soldatenliebe und Verwandtes*" (The Love of Soldiers and Related Matters), examining the pursuit of soldiers as a well-known special taste. We are told that it was with a soldier that Symonds shared his "first collaborative orgasm," [30] and in his arguments for pederasty he habitually turns to the love of soldiers for his readiest examples, as in *A Problem in Modern Ethics* (1891), where he says,

> It would not be easy to maintain that a curate begetting his fourteenth baby on the body of a worn-out wife is a more elevating object of mental contemplation than Harmodius in the embrace of Aristogeiton, or that a young man sleeping with a prostitute picked up in the Haymarket is cleaner than his brother sleeping with a soldier picked up in the Park.[31]

In Section D ("Pederasty") of the "Terminal Essay" in his 1885 translation of *The Arabian Nights*, Sir Richard Burton resorts to Plato and advances Socrates's argument in the *Symposium* that "a most valiant army might be composed of boys and their lovers; for . . . of all men they would be most ashamed to desert one another." [32] Achilles and Patroclus come to mind as well as the famous Theban band, described by Graves as "a regiment of mature soldiers, each paired with a younger homosexual companion, who fought to the death in defense of their country and in honor of their love bond." [33]

When Aubrey Beardsley contemplates a certain painting of St. Sebastian, he can't help noticing that "there is a charming soldier in the background picking up the arrows that have missed the saint." [34] For the Victorian pederast, uniform was so attractive in itself that even quasi-military uniform would do, like the uniforms of elevator operators and post office telegraph boys. As John Gambril Nicholson puts it in his book *A Garland of Ladslove* (1911):

> Smart-looking lads are in my line;
> The lad that gives my boots a shine,
> The lad that works the lift below,
> The lad that's lettered G. P. O.[35]

THE BRITISH HOMOEROTIC TRADITION

No one turning from the poetry of the Second War back to that of the First can fail to notice there the unique physical tenderness, the readiness to admire openly the bodily beauty of young men, the unapologetic recognition that men may be in love with each other. "War poetry," observes Richard Fein, "has the subversive tendency to be our age's love poetry." [36] That seems strikingly true about poetry of the Great War. From the Second War we find nothing like, say, Gurney's "To His Love" or Graves's sensuous little ode, "Not Dead," to the memory of David Thomas:

> Walking through trees to cool my heat and pain,
> I know that David's with me here again.
> All that is simple, happy, strong, he is.
> Caressingly I stroke
> Rough bark of a friendly oak.
> A brook goes bubbling by: the voice is his.
> Turf burns with pleasant smoke;
> I laugh at chaffinch and at primroses.
> All that is simple, happy, strong, he is.
> Over the whole wood in a little while
> Breaks his slow smile.

From the Second War there is nothing like Robert Nichols's "Plaint of Friendship by Death Broken," with its loving list of male physical features, or Herbert Read's "My Company," which at one point seems to recover the world of Whitman's *Calamus* and *Drum Taps:*

> A man of mine
>   lies on the wire;
> And he will rot
> and first his lips
> the worms will eat.
> It is not thus I would have him kissed,
> but with the warm passionate lips
> of his comrade here.

No one's warm passionate lips come anywhere near those of Jarrell's Ball Turret Gunner. The very point of Ewart's "When a Beau Goes In" is that "nobody says 'Poor lad.' " Were writers of the Second War sexually and socially more self-conscious than those of the First? Were they more sensitive to the risks of shame and ridicule? Had the presumed findings of Freud and Adler and Krafft-Ebing and Stekel so diffused themselves down into popular culture that in the atmosphere of strenuous "democratic" uniformity dominating the Second War, one was careful now not to appear "abnormal"?

It would take another book to consider such questions. Less problematic are some of the reasons for the homoerotic motif in Great War writing. Chief among them is the war's almost immediate historical proximity to such phenomena as the Aesthetic Movement, one of whose most powerful impulses was the rediscovery of the erotic attractiveness of young men. Aestheticism was an offshoot of the kind of warm late-Romanticism that makes it seem appropriate that Tennyson should be fond of Arthur Hallam, Whitman of Peter Doyle, and Housman of Moses Jackson.

What can be called the main prewar tradition of homoerotic poetry runs from Whitman to Hopkins to Housman. Whitman's homoeroticism was publicized in England by Edward Carpenter, who visited Whitman in 1877 and from 1883 to 1905 issued the four parts of a long prose-poem, *Towards Democracy*, a celebration of "comradeship" among pastoral youths uncorrupted by industrialism and undeformed by "education." The ideal type, Carpenter finds, is to be discovered among what Robert Nichols, writing about his soldiers thirty-five years later, will call "men rough." One of Carpenter's examples is "the thick-thighed hot coarse-fleshed young bricklayer with the strap round his waist." [37] There in embryo is Hopkins's Harry Ploughman, and set forth in what is very much like Sprung Rhythm.

The homoerotic impulses of Hopkins were largely subdued by his conversion in 1866, but images of attractive young men continue to surface in his poems throughout his career. His early infatuation with Digby Mackworth Dolben is thought to be the implicit subject of his 1865 sonnet "Where Art Thou, Friend," and the three sonnets of the same year titled "The Beginning of the End" record the termination of a crush with such warmth that Bridges noted of two of them, "these two sonnets must never be printed." [38] "The Loss of the *Eurydice*," written late in the seventies, provides Hopkins with an occasion for dwelling on the physical merits of the "lads" and "boldboys" drowned when this vessel foundered. The "Mortal Beauty" celebrated in "To What Serves Mortal Beauty?" (1885) is that of "lovely lads," and admiration of it is found to be permitted by God since it is "heaven's sweet gift." Blonds are preferred throughout, as in the fragment "The furl of fresh-leaved dogrose down," a description of a flower-decked rural youth:

> His locks like all a ravel-rope's-end,
>   With hempen strands in spray—
> Fallow, foam-fallow, hanks. . . .

And Harry Ploughman himself is notably blond. His wind-laced lily-locks are whitish gold, and on his sinewy forearms grows the memorable "broth of goldish flue. . . ."

But most of Hopkins was not published until 1918. The book that was "in every pocket" just before the war, says Robert Nichols, was *A Shropshire Lad* (1896).[39] It did not sell very well at first. As Roy Fuller says, "Housman himself believed that the start of [its popularity] was the outbreak of the First World War, and doubtless he was right, for the book was on the spot (as was Rupert Brooke) to take advantage of the increase of interest in poetry brought about by 1914."[40] But it might seem that the "increase of interest" was less in poetry than in the theme of beautiful suffering lads, for which the war sanctioned an expression more overt than ever before. Homoeroticism was now, as it were, licensed. *A Shropshire Lad*, Brian Reade observes, "is like a beautiful ruin built over an invisible framework, and Housman obscured the framework so well that until recently not many readers of the poems seemed to guess that it was *l'amour de l'impossible* which haunted many of them. . . ."[41] Whether or not readers were really that naive—I think they were not—it is remarkable the way *A Shropshire Lad*, like Hardy's *Satires of Circumstance*, anticipates, and in my view even helps determine, the imaginative means by which the war was conceived. "One feels Housman foresaw the Somme," says Robert Lowell.[42] The headstone inscription there, "If love could have saved him, he would not have died," is virtually a précis of the first stanza of one of Housman's most popular poems:

> If truth in hearts that perish
>   Could move the powers on high,
> I think the love I bear you
>   Should make you not to die.

(Interestingly, we find that the allusion of the Somme headstone inscription is erotic, like the other favorite inscription, "Until the day break, and the shadows flee away": this is spoken by the love-smitten girl in the *Song of Solomon* (2:17) and constitutes part of her invitation to her lover to browse on her body all night long.)

Perhaps Housman's greatest contribution to the war was the word *lad*, to which his poems had given the meaning "a beautiful brave doomed boy." In Great War diction there are three degrees of erotic heat attaching to three words: *men* is largely neutral; *boys* is a little warmer; *lads* is very warm. In the autumn of 1914 Lytton Strachey is to be found knitting mufflers for, as he writes Clive Bell, "our soldier and sailor lads," just as R. J. E. Tiddy, writing a friend about what trench life has done to his company, says, "We have lost some very adorable lads. . . ."[43] Blond hair seems to have been characteristic of "lads." Alfred M. Hale defines the term: "Lt. Ossington . . . was one of those fair-

haired young men whom one speaks of as a 'Lad.' . . . [He was] young
and lusty and full-blooded. . . ." [44] As *men* grow more attractive, they
are seen as *boys*, until finally, when conceived as potential lovers, they
turn into *lads*. Thus Philip Gibbs on the men coming back from the
Somme front for a few hours of leave in Amiens: "To these men—boys,
mostly—who had been living in lousy ditches under hell fire, Amiens
was Paradise. . . ." There were shop-girls in Amiens, but the boys'
leaves were so brief that "there could be no romantic episode, save of a
transient kind, between them and these good-looking lads in whose eyes
there were desire and hunger. . . ." [45] The lads who populate the
poems and memoirs of the Great War have about them both the doom of
Housman's lads and the pederastic allure of John Gambril Nicholson's.
All these associations reverberate in lines like Owens's

> . . . blunt bullet-leads
> Which long to nuzzle in the hearts of lads,

or Wilfrid Gibson's

> Lads who so long have stared death in the face,

or the last stanza of Sassoon's Housman-like "Suicide in Trenches":

> You smug-faced crowds with kindling eye
> Who cheer when soldier lads march by,
> Sneak home and pray you'll never know
> The hell where youth and laughter go.

A less respectable but no less influential prewar tradition of homoeroticism was that of the so-called Uranians, a body of enthusiastic pedo-
phils who since the late eighties had sent forth from Oxford and London
a stream of pamphlets, poems, drawings, paintings, and photographic
"art studies" arguing the attractions—and usually the impeccable mora-
lity—of boy-love. Among the Uranians can be numbered such writers,
schoolmasters, scholars, divines, and Roman Catholic converts as William Johnson Cory, Symonds, Carpenter, Frederick Rolfe ("Baron
Corvo"), Wilde, Lord Alfred Douglas, Aleister Crowley, Montague
Summers, Leonard Green, Beverley Nichols, Sholto Douglas, and
Gerard Hamilton (Isherwood's "Mr. Norris"). Their ambition and
achievement have been exhaustively studied by Timothy d'Arch Smith,
to whose elegant *Love in Earnest* (1970) I am deeply indebted. The periodical outlets of the Uranians were mainly two: *The Artist and Journal of
Home Culture*, edited by Charles Kains-Jackson, and *The Quorum: A Mag-*

azine of Friendship. Their official organization was the British Society for the Study of Sex Psychology, organized in London in July, 1914. Its most active members included John Gambril Nicholson, Edward Carpenter, Montague Summers, and A. E. Housman's brother Laurence—later, it should be noted, the devoted editor of *War Letters of Fallen Englishmen* (1930). At its most pure, the program of the Uranians favored an ideal of "Greek love" like that promulgated in Walter Pater's essay on Winckelmann (1867), stressing the worship of young male beauty without sex. But very frequently such highmindedness was impossible to sustain, and earnest ideal pedophilia found itself descending to ordinary pederastic sodomy.

It is impossible to say how widely known the work of the Uranians was. Although some of it circulated "privately," most was quite public. If it could not be said that their performances created the atmosphere favorable to the wider homoerotic conception of soldiering, they are at least an indication of what was in the air. The erotic and military career of one of them, Ralph Nicholas Chubb (1892–1960) is instructive here. He was an eccentric, and ultimately prophetic and lunatic, prose-poet who published a series of quasi-Blakean lithographed and hand-lettered works recommending a sincere and strenuous pederasty. What led him to this position, it is said, were five unforgettable experiences in his early life. Timothy d'Arch Smith lists them in significantly parallel and climactic form: "The sight . . . at an early age of some village boys bathing naked; the discovery of masturbation; a passionate attachment to an unknown choir-boy; a love-affair when he was eighteen with a boy three years his junior; and his witnessing the death of a young soldier in battle." Chubb enlisted in 1914, became a captain, fought at Loos, and was discharged as neurasthenic in September, 1915. In the Loos fighting, says Smith, Chubb

> watched the slaughter of a boy, a creature such as those he had always mentally, and once physically, loved. He was the curly-haired, seventeen-year-old son of a blacksmith. . . . His death symbolized for Chubb all the horrors and taboos of society. The boy, a beloved object, was not only forbidden by law to be loved by an adult male but was legally sacrificed by the same laws in the service of his country. In Chubb's neurasthenic condition, the images of boyhood impinged on his mind at that moment of bloodshed: the sight and touch of beautiful lads ending with their frightful death in a wanton slaughter.[46]

His response to this irony Chubb set down in *The Sacrifice of Youth: A Poem* (1924).

But long before the war the Uranians were producing poems that were at first glance indistinguishable from poems of the Great War. This they did by placing before themselves and their readers the most moving image they could think of—the sudden death of boys. E. M. Forster was also excited by this "period" motif: witness the number of boys who meet inexplicable sudden deaths in the novels Forster wrote from 1905 to 1910. In their search for subjects the Uranians were stimulated by the proposals of *The Artist*, which in July, 1889, recommended a number of good ideas for homoerotic paintings. One was for a picture of Hyacinthus and Apollo, with Apollo holding "the dearest of mortal heads on his knee, and gazing for the last time on the loving eyes now closing." Another affecting picture could show the whippings of boys which took place at the annual festival of Diana Orthia, goddess of the Spartans. A challenge to any artist would be "the expression of steady and determined endurance under pain in a young face," as well as the depiction of "the complex emotions of pride, compassion and love in that of the father, elder brother or lover who would be the lad's natural aid or attendant at the sacrifice." [47]

The theme of the sacrificial martyrdom of lads (largely naked) prompted hundreds of verses on St. Sebastian, like Frederick Rolfe's, published in *The Artist* in June, 1891:

> A Roman soldier-boy, bound to a tree,
>   His strong arms lifted up for sacrifice,
>   His gracious form all stripped of earthly guise,
> Naked, but brave as a young lion can be,
> Transfixed by arrows he gains the victory. . . .

These lines by another author,

> Fair as the Boy that Mary loved was he
>   . . . nor has the beauty fled
> From his still form with blood-stained limbs outspread,

are not the result of Wilfred Owen's contemplating a corpse on the Somme. They are from "St. William of Norwich," a homoerotic-pious exercise (1892) by Rolfe and Nicholson.[48] In the same way, the lines

> Thy lips are still and pale,
> Pale from Death's icy kiss

are not from a Great War elegy on a dead soldier but from a homoerotic elegy on a dead acolyte by Montague Summers published in *Antinous and Other Poems* (1907). With the poems of Owen, Ivor Gurney, and Robert Nichols we can compare "In Memoriam, E. B. F.," by G. G. Gillett. It first appeared in March, 1893, in *The Spirit Lamp*, an Oxford magazine

edited by Lord Alfred Douglas. It is typical of hundreds of Uranian
elegies on dead boys that were stylish for twenty years before 1914:

> Brave boy with the bright blue eyes,
>   Faithful and fair and strong!
> Dead now—when the short day dies
>   Like a broken song,
> And the night comes dark and long.
>
> Friend and more than a friend,
>   Brother and comrade true,
> We are come to the dim sad end
>   Of the way we knew:
> I bleed in the dark for you.[49]

Some of these are so apposite to the circumstances of 1914–18 that a very
little alteration would turn them into Great War poems. Like Rennell
Rodd's "Requiescat," written in 1881:

> He had the poet's eyes,
>   —Sing to him sleeping,—
> Sweet grace of low replies,
>   —Why are we weeping?
>
> He had the gentle ways,
>   —Fair dreams befall him!—
> Beauty through all his days,
>   —Then why recall him?—
>
> That which in him was fair
>   Still shall be ours:
> Yet, yet my heart lies there
>   Under the flowers.[50]

That last stanza, with *here* substituted for *there* in the third line, would
do admirably as a headstone inscription at Ypres or the Somme.

THE HOMOEROTIC SENSUOUSNESS OF WILFRED OWEN

It is most conspicuously in the poetry of Wilfred Owen that these impulses of Victorian and early-twentieth-century homoeroticism con-
verge, and it is there that they are transfigured and sublimated with little
diminution of their emotional warmth. If the boys of Owen's early imag-

ination begin as interesting "lads," ripe for kissing, they end as his "men" in France, types not just of St. Sebastian but of the perpetually sacrificed Christ. The route Owen negotiates leads him from a world something like Baron Corvo's to one resembling that of Hopkins. The tradition of Victorian homoeroticism teaches him how to notice boys; the war, his talent, and his instinct of honor teach him what to make of them.

Owen was shy, sensitive, and intense. He was born March 18, 1893, at Oswestry, Shropshire—a genuine Shropshire lad, as John H. Johnston reminds us.[51] His father was a minor railway official whose salary sufficed to maintain the family in genteel poverty. His mother, to whom Wilfred cleaved with what will strike post-Freudians as an abnormal devotion, was pious, puritanical, and strong-willed, entirely dedicated to her four children, but especially to Wilfred, the eldest. She execrated smoking, drinking, and the theater, but approved of "poetry," watercolors, and clergymen, especially those of the evangelical wing of the Established Church. After a brief residence in Shrewsbury, the family moved to Birkenhead, where Wilfred studied at the Birkenhead Institute for seven years. Already he was a dignified, serious boy with a marked scholarly bent who read constantly. The family returned to Shrewsbury in 1907 and Wilfred was sent to Shrewsbury Technical School.

During his adolescence his mother encouraged him to find a vocation in the church, but he was more interested in poetry, which he read and wrote apparently day and night. As he grew closer to his mother he became distant from his father, who never tired of exhorting him to find a trade since, as everyone knows, poetry doesn't pay. Wilfred's early model for his life as well as for his writing was Keats, and long before the war he pursued his literary studies with an urgent intensity as if mindful of the shortness of his time. If he was to attend a university he would have to win a scholarship. But when he took the matriculation examination at London University he failed to win the required First-Class Honors. As a way of continuing to read poetry and of considering at the same time his fitness for the church, he became a private student and lay-assistant to the Reverend Mr. Herbert Wigan, vicar of Dunsden, near Reading. For two years he read theology and assisted Wigan in services and parish work.

Wilfred's pre-war psychological state can be estimated from his poem "Maundy Thursday," which registers the tension he is beginning to feel between Establishment theology and homoerotic humanism. The kissing which is the leitmotif of the poem became one of Owen's favorite images:

Between the brown hands of a server-lad
The silver cross was offered to be kissed.
The men came up, lugubrious, but not sad,
And knelt reluctantly, half-prejudiced.
(And kissing, kissed the emblem of a creed.)
The mourning women knelt; meek mouths they had,
(And kissed the Body of the Christ indeed.)
Young children came, with eager lips and glad.
(These kissed a silver doll, immensely bright.)
Then I, too, knelt before that acolyte.
Above the crucifix I bent my head:
The Christ was thin, and cold, and very dead:
And yet I bowed, yea, kissed—my lips did cling
(I kissed the warm live hand that held the thing.)

It would be hard indeed to ignore the ways in which that poem allies itself with the tradition of Symonds, Wilde, Rolfe, Charles Edward Sayle,
John Francis Bloxam, and other writers of warm religio-erotic celebra-
tions of boy-saints, choirboys, acolytes, and "server-lads." Hopkins's
"The Handsome Heart: At a Gracious Answer" is in the tradition. As
early as January, 1916, it was available in Bridges's anthology *The Spirit
of Man*. That Owen knew some of Hopkins seems likely from certain
moments in his diction and meter. For example,

    But all limped on, blood-shod. All went lame; all blind
                 (*"Dulce et Decorum Est"*)

and, from "Spells and Incantation,"

                          . . . wrathful rubies
        You rolled. I watched their hot hearts fling
        Flames. . . .

Hopkins's affectionate physical notice of the boy who has helped him in
the sacristy is like that at which Owen will become adept:

                       . . . more than handsome face—
        Beauty's bearing or muse of mounting vein,
        All, in this case, bathed in high hallowing grace. . . .

But Owen did not have to wait until 1916. Before he had seen any of the
war he had already made his own the sentimental homoerotic theme
which his greatest poems of the war would proceed to glorify.

Sensing (as we may gather) his unsuitability for a religious vocation,
despairing of entering a university, and warned by some apparent lung
trouble that a warmer climate might do him good, he moved to Bordeaux

in 1913 and went to work as a teacher of English in the Berlitz School there. About a year later he quit to become a private tutor with a French family. He was in France when the war broke out, and for over a year he was tempted to ignore it, to consider it an irrelevancy and a distraction from his work of writing poems. But in September, 1915, he returned to England and a month later joined the Artists' Rifles—he was surprised to find no artists in the unit—as an officer cadet. In London he trained in happy proximity to Harold Monro's Poetry Bookshop, where he could be seen browsing for long hours after drill. On June 4, 1916, he was commissioned as a second lieutenant in the Manchester Regiment and almost immediately became an excellent officer—conscientious, efficient, and sympathetic.

Up to now, for all his disappointment about missing the university and his frustrations over money, he had been a strikingly optimistic, cheerful young man, skilled in looking on the bright side and clever at rationalizing minor setbacks. But with his first experience of the trenches in the middle of January, 1917, everything changed. What he encountered at the front was worse than even a poet's imagination could have conceived. From then on, in the less than two years left to him, the emotions that dominated were horror, outrage, and pity: horror at what he saw at the front; outrage at the inability of the civilian world—especially the church—to understand what was going on; pity for the poor, dumb, helpless, good-looking boys victimized by it all. He was in and out of the line half a dozen times during the first four months of 1917, but what finally broke him was an action in late April, when he had to remain in a badly shelled forward position for days looking at the scattered pieces of a fellow officer's body. No one knows exactly how he reacted or what he did, but he was evacuated and his condition was diagnosed as neurasthenia. He wrote his sister from the Casualty Clearing Station: "I certainly was shaky when I first arrived. . . . You know it was not the Bosche that worked me up, nor the explosives, but it was living so long by poor old Cock Robin (as we used to call 2/Lt. Gaukroger), who lay not only near by, but in various places around and about, if you understand. I hope you don't!" [52] Soon he was being treated at Craiglockhart, where he read Barbusse's *Le Feu* with enthusiasm and developed a poetic crush on the patient who lent it to him, Siegfried Sassoon. It was here, while visiting Sassoon, that Graves met Owen and perceived, or thought he perceived, that he was "an idealistic homosexual with a religious background." Graves goes on to say: "It preyed on his mind that he had been unjustly accused of cowardice by his commanding officer." [53] If this is true, Owen nowhere mentions it.

Wilfred Owen with seven-year-old Arthur Newboult, son of friends in Edinburgh whom Owen visited while at Craiglockhart. (Imperial War Museum)

Released from the hospital in the autumn of 1917, he was assigned to light duty in Scarborough, where he managed the Officers' Mess of the 5th Manchesters. He longed to return to the front although he knew he was going to be killed. Having seen the suffering of the men, he had to be near them. As the voice of inarticulate boys, he had to testify on their behalf. After a year of home duty he was pronounced fit to return to France, and in September, 1918, he was back in the line. In an attack during the first days of October he won the Military Cross. In another attack across the Sambre Canal near Ors on November 4, he was machine-gunned to death. At noon on November 11 the Armistice bells had been pealing for an hour in Shrewsbury when the telegram arrived at his parents' house. When he was killed he was twenty-five years old.

Over the years his mother kept going by presiding over his memory. Her way of doing this was to treasure up and arrange the letters he had sent her since he was five—she had kept every one—and to oversee publication of his poems, only four of which had appeared while he was living. She was as possessive of Wilfred dead as of Wilfred alive. She permitted Sassoon to bring out a selection of twenty-three poems in 1920. An edition by Blunden in 1931 contained fifty-nine. Susan Owen died in 1942, and in 1964 C. Day Lewis published a new edition and included for the first time a number of interesting juvenilia, including "Maundy Thursday." His edition contains seventy-nine poems. Twenty-five of these—almost a third—refer to either *boys* or *lads.*

Owen's poetic response to the war is unique. Unlike Sassoon, Blunden, and Graves, with their "university" bent toward structured general ideas, Owen, as Bergonzi has noted, "rarely attempts a contrast, nostalgic or ironic, between the trenches and remembered English scenes." [54] Rather, he harnesses his innate fondness for dwelling on the visible sensuous particulars of boys in order to promote an intimate identification with them. His method is largely a de-sexed extension of his kissing "the warm live hand" of the "server-lad." With a most tender intimacy he contemplates—"adores" would perhaps not be too strong a word—physical details like eyes, hair, hands, limbs, sides, brows, faces, teeth, heads, smiles, breasts, fingers, backs, tongues. Loving these things, he arrives by disciplined sublimation at a state of profound pity for those who for such a brief moment possess them. He seems skilled in the deployment of the sympathetic imagination as defined by Keats. He is practiced in the art of throwing himself into another thing. The physical attributes of others, if sufficiently handsome, become like his own.

Whether he acts to identify himself with a homogeneous group of men ("Insensibility," *"Apologia Pro Poemate Meo,"* "Greater Love," "Anthem for Doomed Youth," "The Send Off," "The Last Laugh," "Mental Cases") or more warmly with one single unfortunate young male ("Arms and the Boy," *"Dulce et Decorum Est,"* "Asleep," "Futility," "The Sentry," "Conscious," *"A Terre,"* "Disabled," "The Dead-Beat," "S.I.W.," "Inspection"), it is the features of the palpable body that set him off. This is noticeable even in so "ideal" a poem as "Anthem for Doomed Youth," where the generalized accessories of funeral services (anthems, prayers, bells, "choirs," candles, and palls) are made to seem supremely irrelevant next to "the hands of boys" and—most tellingly—"their eyes."

Owen's favorite sensuous device is the formula "his ———," with the blank usually filled with a part of the body. Thus "Disabled," with its numerous echoes of "To an Athlete Dying Young," a poem in which Housman also has made certain that we see and admire the boy's eyes, ears, foot, head, and curls. Owen's former athlete, both legs and one arm gone, sits in his wheelchair in a hospital convalescent park listening to the shouts of "boys" playing at sunset. He can't help recalling the excitements of former early evenings in town before the war, back then "before he threw away his knees." His attributes, now all changed, move into focus:

> There was an artist silly for his face,
> For it was younger than his youth, last year.
> Now, he is old; his back will never brace;
> He's lost his color very far from here,
> Poured it down shell-holes till the veins ran dry,
> And half his lifetime lapsed in the hot race,
> And leap of purple spurted from his thigh.

(In one draft these lines were preceded by the following, which may remind us of Hopkins's Harry Ploughman. Their autoerotic tendency would seem to assure us that the artist who was silly for this boy's face is not being imagined as female:

> Ah! he was handsome when he used to stand
> Each evening on the curb or by the quays.
> His old soft cap slung half-way down his ear;
> Proud of his neck, scarfed with a sunburn band,
> And of his curl, and all his reckless gear,
> Down to the gloves of sun-brown on his hand.)

As the final draft resumes, it resumes Housman:

> One time he liked a blood-smear down his leg,
> After the matches, carried shoulder-high.

Why, in the midst of such triumphs, did he enlist? Physical vanity:

> Someone had said he'd look a god in kilts,
> That's why; and may be, too, to please his Meg;
> Aye, that was it, to please the giddy jilts
> He asked to join. He didn't have to beg;
> Smiling they wrote his lie; aged nineteen years.

But after his disaster, his return home was not as triumphal as the traditional athlete's. When Housman's athlete returned to his town,

> Man and boy stood cheering by,

but when Owen's came back,

> Some cheered him home, but not as crowds cheer Goal.

Indeed his main welcomer was one curious not about his face, his youth, his back, or his color, but about a sexless attribute not at all visible:

> Only a solemn man who brought him fruits
> *Thanked* him; and then inquired about his soul.

After quoting C. Day Lewis as saying, "Owen had no pity to spare for the suffering of bereaved women," Bergonzi quite correctly notes, "Male fellowship and self-sacrifice is an absolute value, and Owen celebrates it in a manner that fuses the paternal with the erotic . . . : Owen's attitude to the 'boys' or 'lads' destined for sacrifice has some affinities with Housman's." [55] And as the verbal echoes of Housman in "Disabled" suggest, he derives from Housman more than an attitude. In fact, Owen's early poem on the noble sacrifice of soldiers, "Ballad of Purchase Moneys," suggests that he learned to write by imitating Housman as well as Keats. Here is the last of the ballad's three stanzas:

> Fair days are yet left for the old,
>   And children's cheeks are ruddy,
> Because the good lads' limbs lie cold
>   And their brave cheeks are bloody.

If the eighth line of "Anthem for Doomed Youth" were a foot shorter (by, say, the omission of *for them*), it could be Housman's:

> And bugles calling for them from sad shires;

and the *sad shires* might seem a not entirely unconscious inversion of Housman's joyous *colored counties* in "Bredon Hill." Whatever his verbal borrowings, Owen could have derived from *A Shropshire Lad* a whole set of tender emotions about young soldiers, their braveries, their deaths, their agonies, and even their self-inflicted wounds. Owen was not likely to forget the lovely lads residing in a Housman stanza such as this:

> East and west on fields forgotten
>   Bleach the bones of comrades slain,
> Lovely lads are dead and rotten;
>   None that go return again.

Complicating, dramatizing, and almost literally fleshing out this sort of thing became Owen's wartime business.

Owen's other poems about individual victims ground themselves likewise in physical attributes. In "Arms and the Boy" the final stanza, which explains the outrageous instructions of the first two, does so by making us intimates of *his teeth*, *his fingers*, *his heels*, and *his curls*. In *"Dulce et Decorum Est"* it is *his face*. Indeed, until a late stage of revision, the lines

> If you could hear, at every jolt, the blood
> Come gargling from the froth-corrupted lungs,
> Obscene as cancer, bitter as the cud
> Of vile, incurable sores on innocent tongues,

presented a considerably more attractive picture:

> If you could hear, at every jolt, the blood
> Come gargling from the froth-corrupted lungs,
> And think how, once, his head was like a bud,
> Fresh as a country rose, and keen, and young.

Again, as in "Disabled," the effect of the revision is to efface indications of the poem's original Uranian leanings, to replace the pretty of 1913 with the nasty of 1917. In "Asleep," we are offered his *brow*, *chest*, *arms*, *head*, and *his hair*. In "Futility," his *limbs* and *sides*. In "Conscious," *his fingers*, *his eyes*, and *his head*. In "A Terre," *hands*, *fingers*, *eyes*, *back*, *legs*, and *arms' tan*. In "The Dead-Beat," *his feet* and *hand*. The final two lines of "S.I.W." tell us how the victim of the self-inflicted wound was buried and how his mother was informed, but not without a lingering on attributes and on kissing:

> With him they buried the muzzle his teeth had kissed,
> And truthfully wrote the Mother, "Tim died smiling."

Another young male smile at the moment of death—this time more painful-ecstatic and hence Swinburnean—is that which concludes "Has Your Soul Sipped?" Eight stanzas develop a sequence of lush comparatives of "sweetness" (what could be imagined sweeter than the nightingale's song? the smell of leaves? the martyr's smile?) in order to arrive at an answer. Sweeter than them all, says Owen,

> was that smile,
> Faint as a wan, worn myth,
> Faint and exceeding small
> On a boy's murdered mouth.

> Though from his throat
> The life-tide leaps
> There was no threat
> On his lips.
>
> But with the bitter blood
> And the death-smell
> All his life's sweetness bled
> Into a smile.

And in "Inspection" the "dirt" on a man's tunic which the officer-speaker unimaginatively reprehends proves to be *blood, his own*. The *his* naturally becomes *their* in the poems depicting groups of the pitiable. Thus the opening of "Insensibility":

> Happy are men who yet before they are killed
> Can let their veins run cold.
> Whom no compassion fleers
> Or makes their feet
> Sore on the alleys cobbled with their brothers.

(Is *their brothers* an intimate attribute? Owen would insist that it be seen as such—that is his whole point.)

Owen's extraordinary talent for imagined sensuous immediacy can be measured by contrasting the abstractions offered by Herbert Read when, in "Ode Written During the Battle of Dunkirk, May, 1940," he recalls the rhetorical structure ("Happy . . . happy . . . but") of Owen's "Insensibility" and ventures to bring it up to date:

> Happy are those who can relieve
> suffering with prayer
> Happy those who can rely on God
> to see them through.
>
> They can wait patiently for the end.
>
> But we who have put our faith
> in the goodness of man
> and now see man's image debas'd
> lower than the wolf or the hog—
>
> Where can we turn for consolation?

Owen's mind goes in the opposite direction, feeling always towards male particulars. To speak of "sufferings" is not enough; one must see and feel the bloody head cradled dead on one's own shoulder. In early October, 1918, he writes his mother to explain why he has had to come to France again: "I came out in order to help these boys—directly by leading them as well as an officer can; indirectly, by watching their sufferings that I may speak of them as well as a pleader can." And then, sensing that "their sufferings" is too abstract to do the job, he indicates what he's really talking about: "Of whose blood lies yet crimson on my shoulder where his head was—and where so lately yours was—I must not now write." But he does write about it to a less shockable audience, Sassoon: "The boy by my side, shot through the head, lay on top of me, soaking my shoulder, for half an hour." [56] *His head. My shoulder.* An improvement over *watching their sufferings.*

If contrast with Herbert Read can suggest some of the power of Owen's sensuous immediacy, contrast with Robert Nichols can indicate some of the artistic sophistication with which Owen complicated and finally transcended his leanings towards mere homoeroticism. Born in 1893, Nichols was educated at Winchester and at Trinity College, Oxford, which he left in 1914 to become a 2nd lieutenant in the Royal Field Artillery. After three weeks in the line, he was sent back as a neurasthenic, and after five months in hospital, he was released from the army. Edward Marsh encouraged his poetic ambitions, and his book of war poems, *Ardours and Endurances*, was one of the hits of 1917. He toured the United States after his discharge giving histrionic lectures and poetry readings. From 1921 to 1924 he was Professor of English Literature at the University of Tokyo. He married in 1922, enjoyed a literary and journalistic career in London, and died in Cambridge in 1944. Reactions to him by his more talented contemporaries are what we would expect. Owen, who by 1917 had mastered the art of intimate identification with the objects of his sympathy, found Nichols deplorably "self-concerned and *vaniteux* in his verse. . . ." [57] Graves thought him very close to a phony. [58]

It was the homoerotic warmth of some of the poems in *Ardours and Endurances* that earned the volume a place among the 454 titles in Francis Edwin Murray's *Catalogue of Selected Books from the Private Library of a Student of Boyhood, Youth and Comradeship* (1924), a sixteen-page list issued by this well-known publisher and bookseller specializing in Uranian materials. [59] In Murray's checklist Nichols's poems joined such pederastic curiosa as John Leslie Barford's *Ladslove Lyrics*, Edwin Emmanuel Bradford's *Passing the Love of Women and Other Poems*, "Edmund Edwinsons' " *Men and Boys: An Anthology*, Nicholson's *A Garland of Ladslove*, and Arnold W. Smith's *A Boy's Absence. By a Schoolmaster*. Nichols, whose early poetic career would tempt most readers to associate him with the Uranians, was one of the reviewers recommending Owen's *Poems* in 1920. Another was C. K. Scott-Moncrieff, later to distinguish himself as the translator of Proust but when a young man known in some circles as the author of *Evensong and Morwesong*, a bravely obscene story of adolescent fellatio which caused the suppression of the Winchester student magazine where it first appeared in 1908. It was later issued as a pamphlet by Murray.[60] Scott-Moncrieff, an invalided captain with a Military Cross, thought so well of Owen that during Owen's convalescence he tried to get him a safe military job in England to prevent his having to return to France. It is like the brigade staff-officer's plea to Ackerley.

Despite their gross incompetence Nichols's poems are worth noticing because they were so popular. It was clearly an audience much larger and more innocent than Murray's furtive pederastic book buyers that agreed to find Nichols's achievement significant. This larger audience— and Owen was not unconscious of it—took pleasure in images of the war that featured fated, beautiful soldier boys mourned sentimentally and "romantically" by their intimate male friends. The separate poems in *Ardours and Endurances* are arranged to imply a narrative. One who reads between the lines can piece together this story: before the war, the poet has been implicated in both love and "passion," but has wearied of them. "Passion" is equated with lust and is found to be unclean, a cause of shame. The "summons" to war he regards, like Brooke, as an almost evangelical invitation to a new clean life. During the war he contracts a deep friendship with two soldiers in turn: first with Richard Pinsent, an Oxford friend killed at Loos in 1915, and then with Harold Stuart Gough, killed at Ypres in 1916. The deaths of first one and then the other leave him devastated, and an irredeemable loneliness dominates the rest of the volume.

What the audience of *Ardours and Endurances* apparently found gratifying were effects like the reiterated sound *boy* in Nichols's elegy for Gough ("A Boy I friended with a care like love") titled "The Burial in Flanders (H. S. G., Ypres, 1916)":

> . . .
>
> Shoulder-high, khaki shoulder by shoulder,
> They bear my Boy upon his last journey.
>
> . . .
>
> Even as Vikings of old their slaughtered leader
> Upon their shoulders, so now bear they on
> All that remains of Boy. . . .
>
> . . .

Sudden great guns startle, echoing on the silence.
Thunder. Thunder.
HE HAS FALLEN IN BATTLE.
(O Boy! Boy!)

. . .

"Dick" Pinsent is mourned in terms more fleshly, more suggestive of
Owen's way with physical attributes. Thus Nichols writes in "Plaint of
Friendship by Death Broken (R. P., Loos, 1915)":

His eyes were dark and sad, yet never sad;
In them moved sombre figures sable-clad;
They were the deepest eyes man ever had,
    They were my solemn joy—*now my despair.*

. . .

His face was straight, his mouth was wide yet trim;
His hair was tangled black, and through its dim
Softness his perplexed hand would writhe and swim—
    Hands that were small on arms strong-knit yet spare.

. . .

His voice was low and clear. . . .

. . .

The only relief for the speaker's despair, the last stanza indicates, would
be for Christ to arrive as the ultimate dark-eyed Ideal Friend:

God, if Thou livest, and indeed didst send
Thine only Son to be to all a Friend,
Bid His dark, pitying eyes upon me bend,
    And His hand heal, or *I must needs despair.*

But it would be cruel to quote anymore. This is enough to indicate
that as long as "passion" was visible in war elegies, and as long as handsome eyes, smiles, faces, and arms were adverted to, an inept poet was
largely immune to derision. To appreciate Owen's tact—which is not, to
be sure, faultless—we need only place his work near Nichols's. It is then
that we will not just excuse but may actively relish performances like
"To My Friend (With an Identity Disc)," which absorbs homoerotic passion into technique and tempers intimate excitement until it becomes an
ally of English literary tradition. Keats is occupying Owen's consciousness as he writes this sonnet, and so is Poets' Corner, and so is his dubious prospect of long life in March, 1917. All are passed through his opu-
lent homoerotic imagination:

If ever I had dreamed of my dead name
High in the heart of London, unsurpassed

By Time forever, and the Fugitive, Fame,
    There seeking a long sanctuary at last,—
Or if I onetime hoped to hide its shame,
    —Shame of success, and sorrow of defeats,—
Under those holy cypresses, the same
    That shade always the quiet place of Keats,
Now rather thank I God there is no risk
    Of gravers scoring it with florid screed.

And now he turns—like Gray in the *Elegy*—to the opposite, the rough, irregular lettering of his name (all capitals, and crude ones at that) on the coarse "composition" disc hanging around his neck on the usual dirty string:

But let my death be memoried on this disc.
    Wear it, sweet friend. Inscribe no date nor deed.

And his final couplet—how typical it is!—focuses once more on a physical feature, and one capable of "kissing":

But may thy heart-beat kiss it night and day
Until the name grow blurred, and wear away.[61]

SOLDIERS BATHING

Watching men (usually "one's own" men) bathing naked becomes a set-piece scene in almost every memory of the war. And this conventional vignette of soldiers bathing under the affectionate eye of their young officer recurs not because soldiers bathe but because there's hardly a better way of projecting poignantly the awful vulnerability of mere naked flesh. The quasi-erotic and the pathetic conjoin in these scenes to emphasize the stark contrast between beautiful frail flesh and the alien metal that waits to violate it.

For Nichols, the soldiers-bathing scene becomes a type of the pastoral oasis or a rare "idyllic moment":

Such moments could at best be but brief and perhaps for the officer, subject to the pressure of necessity arising from the responsibility of command, they were even briefer or at any rate less complete than they were for his men. Lying on the grass, watching his men bathing, he could not but wonder how many would be alive in two months' time and whether he would still be alive to watch such as were. So it was that these idyllic moments, sweet though they might be, resembled nothing as much as . . . interims in nightmare when we feel and hope that we are about to awake.[62]

In Sassoon's version the emphasis is less voyeuristic. His soldiers-bathing scene occurs at sunset, his traditional moment, as we have seen, for momentous events. Recalling the warm days just before the Somme attack, he invokes an image of the men of his battalion swimming in a tributary of the Ancre at Bussy. "I thought how young and light-hearted they looked," he writes, "splashing one another and shouting as they rocked a crazy boat under some lofty poplars that shivered in a sunset breeze." Polar images are quick to offer themselves: "How different to the trudging figures in full marching order; and how difficult to embody them in the crouching imprisonment of trench warfare!" [63]

Blunden's version, located at Givenchy, constitutes one of the memories that help establish the "Arcadian quality" of that place. In a canal near the support line, one might see "among the weeds the most self-satisfied pike in numbers, who had almost forgotten the fact of anglers."

> Moreover, a heavy concrete lock, barring the canal . . . , afforded protection . . . to bathing parties of our men, who were marched down in the afternoon, and chaffed and splashed and plunged, with the Germans probably aware but unobjecting a few hundred yards along.

But as always the pastoral is threatened by the industrial, and soft bare flesh feels its appalling vulnerability to damage by "iron":

> Outside the stopped electrical machinery in this place was an old notice, "Danger de Mort"—exactly. The usual nuisance was the wires which generations of field telephonists had run through the bathing pool. . . . on my last occasion there, sudden shelling on the high south bank scattered unwelcome jags of iron in the still lapping water.[64]

In the hands of Brigadier F. P. Crozier, the familiar genre-piece turns sensuous and meditative at once. One of Crozier's companies near the Ancre is commanded by "a warrior of twenty-two," assisted by "three subalterns all under nineteen":

> One day when at Martinsart, where the nightingales keep me awake at night, within a few yards of a heavy gun, I ride up to have tea with these four boys. They had nothing much to do. The bathing pool is good. As I arrive they are all standing stark naked on the improvised spring board, ready to jump in for a race. How wonderful they look, hard, muscular, fit, strong and supple, yet devoid of all coarseness. They ask me to start them and I comply with their request. . . . I think as I watch them ducking each other in the water . . . , "What a pity they are not married in order that they might plant their seed."

But it can never happen: "Mankind has ordained that they shall shortly die." [65] And when the conventional group of bathing soldiers dwindles to just one, the effect can be even more ironic. Walking over the deserted battlefields of the Somme, Reginald Farrer comes upon the ruins of Thiepval—not much left but a scattering of brick powder in the mud. There he contemplates the remains of a burnt-out tank. And then he sees something which stabs him:

> A little way off a naked Tommy was standing under a spout of water. . . . And the beauty of that tiny frail fair thing, vividly white in the sunshine upon that enormous background of emptiness and dun-colored monotony of moorland was something so enormous in itself, that it went straight through me like a violent lance of pain. So minute a little naked frog, hairless and helpless, to have made the earth such a place of horror, and itself, incidentally, in the making, a thing so infinitely great. . . . All those complicated bedevilments of iron and dynamite, got together at so vast an expense of thought and money and labor, to destroy just—*that*. [66]

One of the period's most vigorous exponents of naked bathing was Rupert Brooke, who professed himself to be, as Virginia Woolf reported, "very keen on living 'the free life.' " She remembered his saying to her one day, "Let's go swimming, quite naked." Asked whether she did, she answered, "Of course." [67] Brooke was by no means immune to homoerotic feelings, as he seems to recognize in writing to Lady Eileen Wellesley from his Royal Naval Brigade camp in Kent in October, 1914. The spectacle of soldiers bathing he associates with Aestheticism and wryly indicates that the delights of the pastoral-erotic must be set aside now that he has assumed a disciplined new life:

> My skin is brown and hard. I think of nothing at all, hour after hour. Occasionally I'm faintly shaken by a suspicion that I might find incredible beauty in the washing place, with rows of naked, superb men bathing in a September sun . . .—if only I were sensitive. But I'm not. I'm a warrior. So I think of nothing. . . . [68]

In his famous sonnet "Peace," one of the five of the group "1914," it is the image of "swimmers into cleanness leaping" that betokens the delights of the enthusiastic enlistment that liberates one from "a world grown old and cold and weary" as well as from types (like Lytton Strachey?), "half-men, and their dirty songs and dreary. . . ."

If the war had actually been written by Hardy it could scarcely offer a bolder irony than that by which Brooke's "swimmers" of 1914 metamorphose into the mud-flounderers of the Somme and Passchendaele sinking beneath the surface. Perhaps both have some relation to Phlebas the Phoenician in *The Waste Land*, who finds death by water and who "was once handsome and tall as you." (It has even been suggested that Phlebas may memorialize an actual soldier beloved by Eliot and drowned at

Gallipoli in 1915, the young Parisian medical student and poet Jean Verdenal.[69]) The boyish swimmers of Brooke's "Peace" were a valuable gift to imaginations which survived to experience what the war turned into. Tinkering with a line about attacking troops in his poem "Spring Offensive," Owen arrives finally at the wording "Leapt to swift unseen bullets," but not before he has imagined the same men plunging into waves on a beach: "Breasted the surf of bullets. . . ."[70] His mind is with swimming again when he depicts death by gas as death by drowning in *"Dulce et Decorum Est."* And Sassoon—who ends "Counter-Attack" with an image of a soldier drowning in his own blood—recalls also that one shell-burst resembled nothing so much as a swimmer turned blind, gross, clumsy, and "dingy":

> Against the clear morning sky a cloud of dark smoke expands and drifts away. Slowly its dingy wrestling vapors take the form of a hooded giant with clumsy expostulating arms. Then, with a gradual gesture of acquiescence, it lolls sideways, falling over into the attitude of a swimmer on his side. And so it dissolves into nothingness.[71]

Such imagery seems to fit a landscape which after a rain resembled, as one man recalls, "just one big sea for miles and miles"; indeed, "when the wind blew it made waves just like the real sea."[72] A few of the "swimmers" of 1914 who were lucky revisited the battlefields after the war, as Stephen Graham did in 1920. He found that "in a grey haze of autumn sunshine the battlefields stretch like a sea; green waves [i.e., grass-grown shell craters] to the limit of eyes' view. . . . You see gleaming above the green main-flood . . . a white Ionic cross shining afar and make it your landmark. You reach it as a swimmer coming from some ship to a white buoy on the sea."[73] A later visitor, Henry Williamson, noticed that the craters of the mines at Messines, now filled with water, were used as swimming holes by the local youths.[74]

Naked bodies need not be imagined swimming to dramatize the pathetic vulnerability of flesh. The homely ceremony of louse-hunting provides an occasion for reminding a viewer that the hunters bear an ironic, "displaced" relation to those nude soldiers surprised by attack so popular among Florentine painters and engravers near the end of the fifteenth century.[75] Herbert Read recognized that the title of his 1919 volume of war poems, *Naked Warriors,* pointed in the direction of Renaissance Florence even though at that time, he says, he had not yet seen such a work as Antonio Pollaiuolo's famous engraving of naked soldiers.[76] Frederic Manning's poem "A Shell" is classical in the way it depicts the direct menace of shellfire to naked flesh:

> Here we are all, naked as Greeks,
> Killing the lice in our shirts:
> Suddenly the air is torn asunder,
> Ripped as coarse silk . . . ,

and the shell arrives. Rosenberg's "Louse Hunting" is a variant.

But the favorite theater for exhibiting soldiers' nudity was the sea or the bathing-pool, and it is doubtful that so many scenes of soldiers bathing would have been noticed and fondly recalled as significant if a half-century earlier the Uranians had not established the Boys Bathing poem as a standard type. During the nineties especially the boys-bathing theme was popular in "respectable" painting as well, as it was, of course, in less respectable photography. Henry Scott Tuke, R.A., was one of the best-known serious painters of boys bathing. Others were Frederick Walker and William Scott of Oldham. Tuke's *August Blue* (now in the Tate) is typical of the many popular pictures he sent up to London from Falmouth, where he found his swimming and sun-bathing young models. This painting, dating from 1894, is of four boys in a rowboat out in the bay: one lounges nude in the stern after a swim, one stands nude at the bow with a towel, one is in the water holding onto the gunwale, and one—clothed—is at the oars. Alan Stanley's Uranian poem "August Blue" addresses, one gathers, the boy standing in the bow:

> Stripped for the sea your tender form
> Seems all of ivory white,
> Through which the blue veins wander warm
> O'er throat and bosom slight,
> And as you stand, so slim, upright,
> The glad waves grow and yearn
> To clasp you circling in their might,
> To kiss with lips that burn.[77]

Another Uranian celebration of a painting by Tuke might be seen as a virtual model for many Great War poems. This is Charles Kains-Jackson's "Sonnet on a Picture by H. S. Tuke," which appeared in *The Artist* on May 1, 1889:

> Within this little space of canvas shut
>   Are summer sunshine, and the exuberant glee
>   Of living light that laughs along the sea,
> And freshness of kind winds; yet these are but
> As the rich gem whereon the cameo's cut;
>   The cameo's self, the boyish faces free
>   From care, the beauty and the delicacy

Of young slim frames not yet to labor put.
The kisses that make red each honest face
    Are of the breeze and salt and tingling spray.
So, may these boys know never of a place
    Wherein, to desk or factory a prey,
That color blanches slowly, nature's grace
    Made pale with life's incipient decay.[78]

Here *desk* and *factory* establish a peacetime paradigm for

The hell where youth and laughter go

of Sassoon's "Suicide in Trenches," just as they might be said to
foreshadow the *bayonet-blade*, *steel*, and *bullet-leads* of Owen's "Arms and
the Boy."

Owen was among those who regarded the image of boys bathing as an
attractive constituent of poems. "From My Diary, July 1914" (written
in 1917) lists among numerous other pleasant memories of the French
rural scene

Boys
    Bursting the surface of the ebony pond.
Flashes
    Of swimmers carving thro' the sparking cold.
Fleshes
    Gleaming with wetness to the morning gold.

And in some drafts towards a sardonic-tender poem on the new meaning
the war has given the term *beauty* (a Blighty wound is now "a beauty"),
he sees to it that the shrapnel which wounds a soldier shall arrive while
he is bathing:

A shrapnel ball
Just where the wet skin glistened when he swam
Like a full-opened sea-anemone.
We both said "What a beauty! What a beauty, lad!" . . .[79]

The Uranian and Great War boys-bathing tradition can be seen to
stem in large part from the passage (section 11) in Whitman's *Song of
Myself* (1855) beginning

Twenty-eight young men bathe by the shore,

where the bathers are joined, in fantasy, by the "twenty-ninth bather,"
the constrained woman who owns "the fine house" overlooking the beach
and who hides

> . . . handsome and richly drest aft the blinds of the window.

Whitman's image of young men bathing, so rich with homoerotic potential, would seem to be one of the stimuli for Hopkins's "Epithalamion," written in 1888, which describes

>   . . . boys from the town
> Bathing . . . ,

and bathing so attractively that an older observer plunges in too, but at some distance away. Like Whitman's twenty-eight bathers, Hopkins's boys are not aware of the one who has "joined" them. Under Hopkins's loving solicitude the poem grows so warm that he is impelled to end it "allegorically": the bathing-pool, he concludes, is really a "symbol"—of wedlock; and the water is an image of "spousal love."

During the nineties, as Brian Reade observes, pederastic bathing verses became wildly popular.[80] That they could still seem innocent, romantic, and sentimental—homoerotic rather than riskily homosexual— is suggested by Frederick Rolfe's signing his name as "The Rev. F. W. Rolfe" at the end of his "Ballade of Boys Bathing," published in *The Art Review* for April, 1890. S. S. Saale's "Sonnet," in *The Artist* for September 1, 1890, draws the contrast between present industrial "griminess" and past cleanliness and freedom that would become familiar in the scenes of soldiers bathing during the war:

> Upon the wall, of idling boys a row,
>   The grimy barges not more dull than they
>   When sudden in the midst of all their play
> They strip and plunge into the stream below.

Their transformation is as miraculous as that of Brooke's "swimmers into cleanness leaping":

> Changed by a miracle, they rise as though
>   The youth of Greece burst on this later day,
>   And on their lithe young bodies many a ray
> Of sunlight dallies with its blushing glow.[81]

The contrast between the constraints of uniform and the freedom of nakedness was established as a staple theme of homoerotic poetry well before 1914, as these lines by the Uranian Horatio Forbes Brown, written in 1900, indicate:

>   . . . with comely, capless head,
> With a light, elastic tread

> Came a trooper of some summers twenty-three;
>   With his jacket all unlaced,
>   And his belt about his waist,
> And a ruddy golden color from his bathing in the sea.[82]

Three years before the war a reader of Nicholson's *A Garland of Ladslove* would have encountered this jingling little celebration of the idyllic:

> There is a Pond of pure delight
>   The paidophil adores,
> Where boys undress in open sight
>   And bathers banish drawers.
>
> There youth may flaunt its naked pride
>   Unscathed by withering Powers,—
> Convention's narrow laws divide
>   That swimming-bath from ours! [83]

Youth unscathed: that is the concept bequeathed by the Uranian boys-bathing poets to the Great War writers of soldiers-bathing scenes. The concept was also conveyed by such a novel as Forster's *A Room with a View* (1908). As Samuel Hynes has noticed, in this novel "no physical scene between the lovers is treated as vividly as the all-male bathing scene, so reminiscent of the pederastic bathing of Victorian homosexual writing and photography. In it, male nakedness liberates George [Emerson] and Mr. Beebe from their conventionality, and the women, when they appear, are a confining and depressing end to the affair." [84] One legatee of these scenes is John Dos Passos. In the middle part of *Three Soldiers* (1921), titled "Machines," Privates Chrisfield and Andrews undergo a rigid inspection in the heat during which Chrisfield resents the tightness of his tunic and feels the desire to "strip himself naked." The inspection over, the two wander down the road to a pool. But their swim is spoiled by a stuffy YMCA man—the equivalent of Forster's "women"—who suggests that since they are in view of two French girls they should cover themselves. Andrews says:

> "And why should they not look at us? Maybe there won't be many people who get a chance."
>   "What do you mean?"
>   "Have you ever seen what a little splinter of a shell does to a feller's body?" asked Andrews savagely.[85]

The efficacy of such scenes in dramatizing the irreducible dichotomies was not lost even on writers of the Second War, who can be said to maintain the essential tradition while cautiously trimming away the homoerotic element. Keith Douglas's poem "Mersa" implies the standard contrast between flesh and the iron threats to it, but here, as the speaker and the other "cherry-skinned soldiers" swim on the North African shore, it is the mortality of flesh alone, rather than its sensual attractiveness, that constitutes the focus:

> I see my feet like stones
> underwater. The logical little fish
> converge and nip the flesh
> imagining I am one of the dead.

And F. T. Prince, in "Soldiers Bathing," employs recollections of Michelangelo's and Pollaiuolo's naked soldiers to arrive not at a celebration of youthful beauty and potency but at a renewed appreciation of the Crucifixion, with Christ "murdered, stripped, upon the Cross. . . ." When Lincoln Kirstein revised and reissued his *Rhymes of a PFC* (1964) as *Rhymes and More Rhymes of a PFC* (1966), he added—as if he had at first forgotten one of the obligatory "kinds"—a soldiers-bathing poem. "Bath" enforces a contrast; not between living flesh and sinister metal, but rather between clean flesh and the "crutty layers of uniform wardrobe" which the soldiers resume after their inadequate showers, donning again "stinky shorts, dead shoes, sodden shirt." If the sexuality is gone, the sympathy remains.

As it does in the unforgettable soldiers-bathing scene in Heller's *Catch-22*. We have already seen Heller recovering from Sherriff's *Journey's End* a crucial ironic Great War action. Here he recovers another and adapts it to the more advanced technological actualities of the Second War and to their even more distressing implications for frail flesh and blood. His famous scene of soldiers bathing has its ending not in the appearance of women or a YMCA man or even in the arrival of a few monitory shell fragments, but in a spectacular horror. His scene typifies Second War vision both because the homoerotic element has been purged away and because understatement and subtlety and shrewd allusion to a living literary and artistic past have been displaced by a frantic, guilt-ridden hyperbole of image which will seem close to madness, for all the clearsightedness which the implicit cynical humor would seem to imply.

Heller prepares for this scene very early in the novel while making the point that "McWatt was crazy":

He was a pilot and flew his plane as low as he dared over Yossarian's tent as often as he could, just to see how much he could frighten him, and loved to go buzzing with a wild, close roar over the wooden raft floating on empty oil

drums out past the sand bar at the immaculate white beach where the men
went swimming naked.[86]

Twenty-eight chapters and over 300 pages later, Heller is ready to unite
and animate these data. One day the men of the squadron are swimming
naked around the raft, watched idly from the beach by a lazy crowd of
sun-bathers and card-players, including Yossarian, his sex-object Nurse
Duckett, and the disapproving Nurse Cramer. One of the young soldiers, blond Kid Sampson, is standing on the raft when McWatt's plane

> came blasting suddenly into sight out of the distant stillness and hurtled
> mercilessly along the shore line with a great growling, clattering roar over
> the bobbing raft on which blond, pale Kid Sampson, his naked sides
> scrawny even from so far away, leaped clownishly up to touch it at the exact
> moment some arbitrary gust of wind or minor miscalculation of McWatt's
> senses dropped the speeding plane down just low enough for a propeller to
> slice him half away.
>
> Even people who were not there remembered vividly exactly what hap-
> pened next. There was the briefest, softest *tsst!* filtering audibly through the
> shattering, overwhelming howl of the plane's engines, and then there were
> just Kid Sampson's two pale, skinny legs, still joined by strings somehow at
> the bloody truncated hips, standing stock-still on the raft for what seemed a
> full minute or two before they toppled over backward into the water finally
> with a faint, echoing splash and turned completely upside down so that only
> the grotesque toes and the plaster-white soles of Kid Sampson's feet re-
> mained in view.

An image of the vulnerability of flesh to metal could hardly go further.
The frailty of flesh in its encounter with machine is emphasized by Kid
Sampson's attributes of paleness, scrawniness, and whiteness, all observed by Yossarian, whose contrasting "wide, long, sinewy back with
its bronzed, unblemished skin" has been called to our attention just
before. After the little *tsst!* sound,

> all hell broke loose. . . . Everyone at the beach was screaming and running,
> and the men sounded like women. They scampered for their things in panic,
> stopping hurriedly and looking askance at each gentle, knee-high wave bub-
> bling in as though some ugly, red, grisly organ like a liver or a lung might
> come washing right up against them. Those in the water were struggling to
> get out, forgetting in their haste to swim, wailing, walking, held back in
> their flight by the viscous, clinging sea as though by a biting wind. Kid
> Sampson had rained all over. Those who spied drops of him on their limbs
> or torsoes drew back with terror and revulsion, as though trying to shrink
> away from their own odious skins.[87]

If there is horror on the beach, there is deeper horror in McWatt's cockpit. When he sees what he has done, he flies his plane into a mountainside.

Blond but scrawny Kid Sampson is the latest of the naked swimming soldiers contrived by the post–Great War imagination to register the supreme pathos of flesh menaced by hurtling iron. These latter-day soldier boys of Douglas and Prince and Kirstein and Heller are objects of concern not because they are beautiful and sexually attractive—it is too late for that, their delineators seem to be saying—but because they are surely doomed. Brigadier Crozier's bathing soldiers look "wonderful"— "hard, muscular, fit, strong and supple." Heller's Kid Sampson is pale, skinny, and white, "even from so far away." Heller seems to be remembering something from a long time ago.

IX
Persistence and Memory

THE LITERARY STATUS OF GREAT WAR MEMOIRS

As we have seen, the memoir is a kind of fiction, differing from the "first novel" (conventionally an account of crucial youthful experience told in the first person) only by continuous implicit attestations of veracity or appeals to documented historical fact. It is with a work like Frank Conroy's *Stop-Time* (1967) that we perceive the impossibility of ever satisfactorily distinguishing a memoir from a first-person novel. The elements comprising *Stop-Time* appeared originally as "short stories"; when collected, they presented themselves as a memoir. It is finally up to the bookseller alone to determine whether he will position such a book on the table marked Fiction or on that marked Autobiography. Lady Gregory was being not at all thick when, enthusiastic over Joyce's *Portrait of the Artist as a Young Man*, which most readers would approach as a fiction and specifically as a "first novel," she said that it was "a model autobiography." [1] In considering what these Great War memoirs really are, it will help to remember that Sassoon's, Graves's, and Blunden's books were the first works in prose that these authors published.

The further personal written materials move from the form of the daily diary, the closer they approach to the figurative and the fictional. The significances belonging to fiction are attainable only as "diary" or annals move toward the mode of memoir, for it is only the ex post facto view of an action that generates coherence or makes irony possible. Lillian Hellman testifies that during 1944 she kept copious and detailed diaries, hoping to register important experience. But reading them later, she found that they did not include "what had been most important to me, or what the passing years have made important." [2] Robert Kee, an RAF flyer in the Second War, found the same of the diaries in which he hoped he was recording his wartime experience:

> From all the quite detailed evidence of these diary entries I can't add up a very coherent picture of how it really was to be on a bomber squadron in those days. There's nothing you could really get hold of if you were trying to write a proper historical account of it all. No wonder the stuff slips away mercury-wise from proper historians. No wonder they have to erect rather artificial structures of one sort or another in its place. No wonder it is those artists who re-create life rather than try to recapture it who, in one way, prove the good historians in the end. [3]

"Rather artificial structures": that points to Sassoon's exaggerated antitheses, Graves's farcical dramaturgy, Blunden's unremitting literary pastoralism. The fictions of these three are subtle enough to be credible. Not so some others in which every rift is loaded with portentousness. In *The Old Front Line* John Masefield solemnly asks us to believe that when he saw a pocket Bible lying open next to the body of a British soldier after the Somme attack, "it was open at the eighty-ninth Psalm, and the only legible words were, 'Thou hast broken down all his hedges; thou has brought his strong holds to ruin.' " [4] A structure rather too artificial, but an example of the necessity of fiction in any memorable testimony about fact.

"Fictions," says Frye, "may be classified . . . by the hero's power of action, which may be greater than ours, less, or roughly the same." As he continues to set forth his "Theory of Modes," Frye indicates that the modes in which the hero's power of action is greater than ours are myth, romance, and the "high mimetic" of epic and tragedy; the mode in which the hero's power of action is like ours is the "low mimetic," say, of the eighteenth- and nineteenth-century novel; and the mode in which the hero's power of action is less than ours is the "ironic," where "we have the sense of looking down on a scene of bondage, frustration, or absurdity. . . ." In literature a complete historical "cycle," Frye implies, comprises something like this sequence of prevailing modes (a sequence betokening, in part, a progressive secularization): myth and romance in the early stages; high and low mimetic in the middle stages; ironic in the last stage. Thus the course of ancient literature from Hebrew scriptures to Roman comedy, or of modern European literature from medieval through Renaissance (myth, romance, and high mimetic) to bourgeois (low-mimetic) to modern (ironic). And the interesting thing is the way the ironic mode (in the modern cycle, Joyce, Kafka, and Beckett are exemplars) again "moves steadily towards myth, and the dim outlines of sacrificial rituals and dying gods begin to reappear in it." Our five modes, Frye concludes, "evidently go round in a circle." [5]

Throughout this book we have been in the presence of literary characters (including narrators) who once were like us in power of action but who now have less power of action than we do, who occupy exactly Frye's "scene of bondage, frustration, or absurdity." Conscription is bondage ("It was a 'life-sentence,' " says Hale's Private Porter); and trench life consists of little but frustration (Sassoon, Blunden) and absurdity (Graves). The passage of these literary characters from prewar freedom to wartime bondage, frustration, and absurdity signals just as surely as does the experience of Joyce's Bloom, Hemingway's Frederick Henry, and Kafka's Joseph K. the passage of modern writing from one mode to another, from the low mimetic of the plausible and the social to the ironic of the outrageous, the ridiculous, and the murderous. It is their residence on the knife-edge between these two modes that gives the memoirs of the Great War their special quality, a quality often overlooked because too few readers have attended to their fictional character, preferring to confound them with "documentary" or "history." These memoirs are especially worthy of the closest examination because, for all the blunt violence they depict, they seem so delicately transitional, pointing at once in two opposite directions—back to the low mimetic, forward to the ironic and—most interestingly—to that richest kind of irony proposing, or at least recognizing, a renewed body of rituals and myths.

"In the late phase in which it returns to myth," Frye notes, the ironic mode seizes upon "demonic" imagery regardless of the "world" it observes: the divine world, the human, the animal, the vegetable, or the mineral. Frye's description of the domain of demonic imagery, "the world that desire totally rejects," is intended to imply a (mere) literary classification. Yet it comes close to delineating the literal Western Front:

> Opposed to apocalyptic symbolism is the presentation of the world that desire totally rejects: the world of the nightmare and the scapegoat, of bondage and pain and confusion; . . . the world . . . of perverted or wasted work, ruins and catacombs, instruments of torture and monuments of folly. . . . just as apocalyptic imagery in poetry is closely associated with a religious heaven, so its dialectic opposite is closely linked with an existential hell, like . . . the hell that man creates on earth. . . .

Demonic aspects of the "divine world" include "the inaccessible sky," with its suggestions of "inscrutable fate or external necessity." In the human world the demonic aspects seem very like those apparent in a unit of "the army": "The demonic human world," Frye says, "is a society held together by a kind of molecular tension of egos, a loyalty to the group or the leader which diminishes the individual, or, at best, contrasts his pleasure with his duty or honor." Seen in its demonic aspect, the animal world swarms with the opposite of doves and sheep, the creatures of apocalypse and pastoral. We find noisome beasts of prey, like lice and rats, and wild dogs. Moving down to the demonic vegetable world, we notice how closely it resembles what was to be seen from the trenches and what memory has decided to preserve as significant from that perspective. It features the sinister forest (we will think of Trones or Mametz Wood), wilderness, or waste land. The tree of death is there, and the Cross (we will think of the roadside calvary). And in the mineral world the demonic aspect shows us the whole apparatus of "perverted work" characteristic of a tour of duty in the trenches: "engines of torture, weapons of war, armor, and images of a dead mechanism . . . which is unnatural as well as inhuman." That would include everything from shovels, twisted iron wire-pickets, and rolls of barbed wire, to the bizarre cylinders of the "accessory" and the perverse toffee-apple. Another characteristic image of the demonic mineral world is "the labyrinth or maze, the image of lost direction," like the trench system. Add to this universal demonic image-system the inevitable burning cities (Sodom, Bapaume) and the "water of death, often identified with spilled blood" (fetid shell craters will come to mind), and we have a virtually complete canvass of the features of the Western Front. And yet Frye is not talking about actuality at all, only about conventional and as it were "necessary" literary imagery.[6] He is not even accepting memoirs of historical events as literature. Yet everything he specifies as belonging to the universal literary and mythic demonic world can be found in memories of the Great War.

In the ironic phase of the literary cycle, a standard character is the man whom things are done to. He is Prufrock, Jake Barnes, Malone, Charlie Chaplin. Or he is George Sherston, "Robert Graves," or "Edmund Blunden." It is perhaps the stylistic traditionalism of most Great War writing that has prevented our perceiving such similarities. The reader in search of innovation will find it in few places other than Owen's near-rhymes, Read's *vers libre*, and Jones's Eliotic and Poundian juxtapositions. The roster of major innovative talents who were not involved with the war is long and impressive. It includes Yeats, Woolf, Pound, Eliot, Lawrence, and Joyce—that is, the masters of the modern movement. It was left to lesser talents—always more traditional and technically prudent—to recall in literary form a war they had actually experienced. Sassoon, Graves, and Blunden are clearly writers of the second rank. But their compulsion to render the unprecedented actualities they had experienced brought them fully to grips with the modern theme which we now recognize as the essence of Frye's ironic mode.

A corollary to the technical traditionalism of these memoirists is the kind of backward-looking typical of war itself—it is a profoundly conservative activity, after all—and of any lifelong imaginative obsession with it. Every war is alike in the way its early stages replay elements of the preceding war. Everyone fighting a modern war tends to think of it in terms of the last one he knows anything about. The tendency is ratified by the similarity of uniform and equipment to that used before, which by now has become the substance of myth. To illustrate from the American Second War: first-aid packets, gas-masks, web equipment, the rifles used for training, the heavy machine gun, the screw-pickets for barbed wire—all were either identical with or very like those of the Great War. The act of fighting a war becomes something like an unwitting act of conservative memory, and even of elegy. The soldier dwells not just on the preceding war but on the now idyllic period before the present war as well. For him, the present is too boring or exhausting to think of, and the future too awful. He stays in the past. Thus the personal style found appropriate for the earlier war will usually be the one adopted for the new one, at least during its early stages. Lincoln Kirstein seems to notice this in his poem "Tudoresque," where he speaks of his pre-invasion billet near Manchester and of his landlord's snapshots of his army son:

Since his boy's been absent from home five years, we drink deep to his luck—
      An *officer*, no less—
A tall fair youth with starry eyes, like Raleigh or Rupert Brooke
      (His snapshots show me this).
The *Illustrated London News* taught me that specific look
      From World War No. 1. . . .

And in an attempt to make some sense out of the wars that have succeeded the Second, the contemporary imagination likewise turns back about a quarter of a century. I am thinking about a photograph which appeared in the *Listener* not long ago. It showed Indian infantry attacking in the Rajasthan Desert in 1971 with flat helmets and fixed bayonets. It is captioned "Shades of Tobruk." [7] We have seen Blunden recovering the apostrophic rhetoric of Sterne, an act that propels him backwards 168 years. Similar retrogressive and traditional imaginative tendencies had for the commanders and staffs large tactical consequences, as Tomlinson points out: "The battles they were prepared to direct were of the classic order, centers to be pierced, flanks turned, communications cut. They were still thinking of victory in the field, complete and sublime. . . . they thought it was around 1870." [8]

BUT IT STILL GOES ON

The whole texture of British daily life could be said to commemorate the war still. It is remembered in the odd pub-closing hours, one of the fruits of the Defense of the Realm Act; the afternoon closing was originally designed, it was said, to discourage the munitions workers of 1915 from idling away their afternoons over beer. The Great War persists in many of the laws controlling aliens and repressing sedition and espionage. "D"-notices to newspapers, warning them off "national-security matters," are another legacy. So is Summer Time. So are such apparent universals as cigarette-smoking, the use of wristwatches (originally a trench fad), the cultivation of garden "allotments" ("Food Will Win the War"). So is the use of paper banknotes, entirely replacing gold coins. The playing of "God Save the King" in theaters began in 1914 and persisted until the 1970's, whose flagrant cynicisms finally brought an end to the custom.

Every day still the *Times* and the *Telegraph* print the little "In Memoriam" notices—"Sadly missed," "Always in our thoughts," "Never forgotten," "We do miss you so, Bunny"—the military ones dignified by separation from the civilian. There are more on July 1 than on other days, and on that date there is always a traditional one:

> 9th AND 10th BNS., K.O.Y.L.I.—To the undying memory of the Officers and Men of the above Battalions who fell in the attack on Fricourt (Somme) on July 1, 1916.
> "Gentlemen, when the barrage lifts."

B. H. Liddell Hart, who was in the 9th Battalion of the King's Own Yorkshire Light Infantry, explains. Just before the Somme attack,

> the officers assembled in the headquarters mess, in a typical Picardy farm-house. Recent strain between the commanding officer and some of the others led to an embarrassing pause when the senior company commander was called on to propose a toast to the C.O. On a sudden inspiration, he

raised his glass and gave the toast with the words: "Gentlemen, when the barrage lifts." [9]

The battalion attacked with some 800 men. Twenty-four hours later its strength was 80 men and four officers. And because of the annihilation of the Ulster Division on the same date, "1 July," as Brian Gardner notes, "is still a day of deep mourning in Ulster." [10] In remote, self-contained towns like "Akenfield" it is Armistice Day that draws everyone to church. By contrast, says Ronald Blythe, "Good Friday is barely observed at all, everybody playing football then." [11]

Even cuisine commemorates the war. Eggs and chips became popular during the war because both bacon and steak were scarce and costly. It became the favorite soldiers' dish off duty, and to this day remains a staple of public menus not just in England but in France and Belgium as well.[12] Stephen Spender's thrifty grandmother was not the only user "right up to the Second World War" of the "little squares of sweetened paper" popularized during the Great War as a sugar substitute.[13] After the war women dramatically outnumbered men, and a common sight in the thirties—to be seen, for some reason, especially on railway trains— was the standard middle-aged Lesbian couple in tweeds, who had come together as girls after each had lost a fiancé, lover, or husband. A sign of the unique persistence of the war in England is literally a sign, above a large section of shelves in Hatchard's Bookshop, Piccadilly. I have seen nothing like it in any other country. It reads: "Biography and War Memoirs," in recognition of a distinct and very commonly requested English genre.

A lifelong suspicion of the press was one lasting result of the ordinary man's experience of the war. It might even be said that the current devaluation of letterpress and even of language itself dates from the Great War. Speaking of the Somme, Montague observed in 1922:

> The most bloody defeat in the history of Britain . . . might occur . . . on July 1, 1916, and our Press come out bland and copious and graphic, with nothing to show that we had not had quite a good day—a victory really. Men who had lived through the massacre read the stuff open-mouthed. . . . So it comes that each of several million ex-soldiers now reads . . . with that maxim on guard in his mind—"You can't believe a word you read." [14]

No one can calculate the number of Jews who died in the Second War because of the ridicule during the twenties and thirties of Allied propaganda about Belgian nuns violated and children sadistically used. In a climate of widespread skepticism about any further atrocity stories, most people refused fully to credit reports of the concentration camps until ocular evidence compelled belief and it was too late.

The current economic bankruptcy of Britain is another way it remembers. From 1914 to 1918 its gold reserve diminished dramatically. The beneficiary was the United States, which emerged an undisputed Great Power by virtue of manufacturing and shipping matériel. Indeed, the United States, as Marc Ferro observes, "could rightly be considered the only victor of the war, since their territory was intact, and they became creditors of all the other belligerents." [15] Writing his father in May, 1917, Aldous Huxley foresaw something of the future: "France is already a first class power no longer. We shall merrily go on [rejecting a compromise peace] till on a level with Haiti and Liberia." [16] And the economic ruin uncompleted by the Great War was finished by the Second, which necessitated a replay, but much magnified, of immense indebtedness to the United States. The Americanization of Europe from 1945 to the late sixties was the result.

The effect of the war is visible likewise in subsequent military policy. The conduct of the Second War on "the Western Front" was influenced everywhere by memories of the deadly frontal assaulting of the Great War, and it could be said that the cautious use of infantry only after elaborate air and artillery bombardments and the reliance wherever possible on elaborate technology prolonged the war seven or eight months and, ironically, permitted behind the German border the murder of about a million more Jews. On the other hand, Hitler's suspicion of maneuver and distrust of tactical innovations have been traced to the "trench perspective" he picked up a quarter-century earlier, the view that where you found yourself emplanted, there you had to stay.[17] When Hitler did determine on bold maneuver, as in the Ardennes counterattack of December, 1944, he liked to imitate successful German tactics of the Great War, in this case the attack of March, 1918, an offensive aimed at Amiens and at the junction between the British and the French; the Ardennes offensive was aimed at Antwerp and the junction between British and Americans. As Jacques Nobécourt has said, "Even the Gotha raids on Paris and the firing of 'Big Bertha' were reproduced in the form of the V_1 and V_2 attacks on Antwerp, London, and Brussels." One of Hitler's generals confirms Nobécourt's point. "When the documents of this period are carefully studied," says Blumentritt, "it seems likely that Hitler will be seen to have been thinking in terms of the great March offensive of 1918 in the First World War." [18]

The way the data and usages of the Second War behave as if "thinking in terms of" the First is enough almost to make one believe in a single continuing Great War running through the whole middle of the twentieth century. Churchill and the Nazi Alfred Rosenberg had their differences, but both found it easy to conceive of the events running from

1914 to 1945 as another Thirty Years' War and the two world wars as virtually a single historical episode.[19] In *Men at Arms* (1952) Waugh implies as much when he equips Guy Crouchback with the awareness that his predicament in the Second War is like a protracted reprise of his older brother Gervase's experience in the First, an awareness strengthened by Guy's wearing Gervase's Catholic medal around his neck. In his own day Gervase "went straight from Downside into the Irish Guards and was picked off by a sniper his first day in France, instantly, clean and fresh and unwearied, as he followed the duckboard across the mud, carrying his blackthorn stick, on his way to report to company headquarters." [20] It is notably the irony of this earlier experience that establishes irony as the prevailing mode of the superannuated Crouchback's experiences in the Second War, and Waugh never permits us to forget the domination of the Second War by the First. Billeted at Kut-al-Imara House, normally a second-rate preparatory school, Crouchback finds the boys' rooms named after Great War "battles": Loos, "Wipers," "Paschendael." Arriving in these billets, "he . . . dragged his bag . . . to Paschendael." Appropriately, "his spirit sank":

> The occupation of this husk of a house . . . was a microcosm of that new world he had enlisted to defeat. Something quite worthless, a poor parody of civilization, had been driven out; he and his fellows had moved in, bringing the new world with them; the world that was taking firm shape everywhere all about them, bounded by barbed wire and smelling of carbolic.
>
> His knee hurt more tonight than ever before. He stumped woefully to Paschendael. . . .[21]

He might be stumping woefully toward the following classic dénouement:

> Right on time the barrage bursts, the whole line leaps into life. . . . It's a record effort: 800 guns of all calibers along the whole Front, one gun every dozen yards, each gun with hundreds of shells. It's certainly a rousing display. The guns nearby crash incessantly, one against another, searing the darkness with gashes of flame, and those farther up and down the line rumble wrathfully and convulse the northern and southern horizons with ceaseless flashing and flickering. Groups of . . . infantry . . . with bayonets fixed begin filtering forward through the gap. Poor devils—I don't envy them their night's work.

It's all there, including the familiar geographical orientation of the attack eastward. But it was written not of the Western Front in 1915 but of the North African desert war in 1942.[22] It helps illustrate what Keith Douglas says of the all-but-literary "conventionality" of the Second War:

"The behavior of the living and the appearance of the dead were so accurately described by the poets of the Great War that every day on the battlefields of the western desert . . . their poems are illustrated." [23] This is one reason Vernon Scannell, in his poem "The Great War," can write,

> Whenever war is spoken of
> I find
> The war that was called Great invades the mind. . . .

Although wounded in the Second War, what he "remembers," Scannell says, is

> Not the war I fought in
> But the one called Great
> Which ended in a sepia November
> Four years before my birth.

In the same way, when London was being bombed in 1940, the young Colin Perry tried to make sense of what he saw and felt by reading Vera Brittain's *Honourable Estate* (1936), a novel about the earlier war. "I am beginning to look upon this war," he says, "with the interest of a reader of a novel. . . . Sometimes I feel as though I joined up with Dad . . . and fought with him in 1914–1918. The feeling is real and like the novel." Under intense bombing one night, Perry keeps crawling up to the top of the shelter steps to see what's going on. He finds "a ruddy glowing sky. . . , the blinding flash of guns, and the firework effect of the bursting shells." His instinct is like Scannell's, only, at this moment, livelier: "I found myself saying 'B Company over' and we rushed over the barbed wire and attacked the German lines. . . ." [24]

At about the same moment another urgent imagination steeped in the images and dynamics of Great War infantry fighting was that of young Norman Mailer at Harvard, aged sixteen. " 'War,' " says Richard Poirier, "was the determining form of his imagination long before he had the direct experiences of war that went into his first big novel." [25] During his first few weeks at Harvard in 1939 he wrote an extremely short "short story." Like Perry, the war he conceived was the First, even though the Second had just begun, and for him as well as for Perry the action most economically encapsulating "war" was attacking through barbed wire in the old 1914–1918 way. The story, titled "It," is complete as follows:

> We were going through the barbed-wire when a machine-gun started. I
> kept walking until I saw my head lying on the ground.

> "My God, I'm dead," my head said.
> And my body fell over.[26]

Quoting this, Poirier says, "What we discover looking back is that war has always been his subject." It is "the prior condition of his experience," and "it determines the aspects of experience that are to be recorded, and therefore the form of [Mailer's] books and of his career." [27] Conceiving human consciousness as a form of lifelong warfare, he insists, like David Jones, that man acting as an artist must choose the part of an infantryman rather than a staff wallah; careless of bruises and menaces, singlemindedly he must surge forward to do what must be done, even if he destroy himself in the process. "Man's nature, man's dignity," says Mailer, "is that he acts, lives, loves, and finally destroys himself seeking to penetrate the mystery of existence, and unless we partake in some way, as some part of this human exploration (and war), then we are no more than the pimps of society and the betrayers of our Self." [28] Such penetrations across No Man's Land into adversary territory are the business especially of novelists, "those infantrymen of the arts." [29] But if war provides Mailer with an image of a world where dignity may be earned, it also offers him less consoling conclusions. As early as *The Naked and the Dead* he was perceiving that "modern society" is largely a continuation of the army by other means. As General Cummings tells Lt. Hearn, "You can consider the Army, Robert, as a preview of the future." [30] Mailer has asked himself the question, What is the purpose of technological society?, and the only answer modern history has offered him is the one Alfred Kazin proposes: "War may be the ultimate purpose of technological society." [31] It is the business of Pynchon's *Gravity's Rainbow* to enact that conclusion.

"The parapet, the wire, and the mud," Tomlinson posited in 1935, are now "permanent features of human existence." [32] Which is to say that anxiety without end, without purpose, without reward, and without meaning is woven into the fabric of contemporary life. Where we find a "parapet" we find an occasion for anxiety, self-testing, doubts about one's identity and value, and a fascinated love-loathing of the threatening, alien terrain on the other side. In *City of Words*, his study of modern fiction, Tony Tanner begins his analysis of Mailer's achievement by juxtaposing two quotations as epigraphs. The first is from *The Naked and the Dead*: "Every hundred yards Cummings steps up on the parapet, and peers cautiously into the gloom of No Man's Land." No one can miss the anachronisms here: there are no *parapets* in the Second War, and there is, properly speaking, no *No Man's Land*. Such anachronisms anchor Cummings's action within the mythical trench landscape of the Great War rather than his own and imply a powerful imaginative continuity between that war and the one he is fighting. Tanner's other quotation is from *An American Dream:* ". . . and I was up, up on that parapet one foot wide, and almost broke in both directions, for a desire to dive right on over swayed me out over the drop, and I nearly fell back to the terrace from the panic of that." [33] This is a foot-wide parapet running around a balcony at the Waldorf Towers apartment of Barney Kelly, thirty floors above Park Avenue. Like the parapets of the Great War, it marks a total line of division between protection ("home") and death. And Stephen Rojack has elected to teeter along it, plucked at by gusts of wind, both to test his manhood and to expiate his crime. That is, to make himself worthy of going on living again. One could trace a continuous line from the encounter with wire and machine-gun fire in Mailer's "It" all the way to this terribly self-conscious encounter with the destructive principle twenty-five years later. And one could derive both images finally from the adversary atmosphere of the Great War, just as one could trace their tone to the terror attending that atmosphere for those who read about it young. Constantly in Mailer

> The war that was called Great invades the mind . . . ,

and that war detaches itself from its normal location in chronology and its accepted set of causes and effects to become Great in another sense— all-encompassing, all-pervading, both internal and external at once, the essential condition of consciousness in the twentieth century. Types more different than Mailer and Blunden could hardly be imagined: the one American, brassy, self-advertising, manic, brilliant, urban; the other English, shy, gentle, uncertain, retiring, rural. But they have in common a total grasp of what the word *parapet* must now imply. We recall "The Midnight Skaters":

> With but a crystal parapet
> Between, [Death] has his engines set.

Indeed, a striking phenomenon of the last twenty-five years is this obsession with the images and myths of the Great War among novelists and poets too young to have experienced it directly. They have worked it up by purely literary means, means which necessarily transform the war into a "subject" and simplify its motifs into myths and figures expressive of the modern existential predicament. These writers provide for the "post-modern" sensibility a telling example of the way the present influences the past. In eschewing the Second War as a source of myth and in-

stead jumping back to its predecessor, these writers have derived their
myth the way Frye notes most critics derive their principles, not from
their predecessors but from their predecessors' predecessors. "There may
be noticed," he says, "a general tendency to react most strongly against
the mode immediately preceding, and . . . to return to some of the standards of the modal grandfather." [34] An example in criticism is the neglect of the immediately preceding New Criticism by post-modern
critics, who have jumped back to Wilde, the New Criticism's "grandfather," to re-erect their theories of non-representative art.

Some British novelists who have gone to the Great War as the "modal
grandfather" of modern myth are John Harris, author of *Covenant with
Death* (1961), William Leonard Marshall, author of *The Age of Death*
(1970), Derek Robinson, author of *Goshawk Squadron* (1971), and Jennifer
Johnston, who in *How Many Miles to Babylon?* (1974) imagines the trench
scene so authentically that she gets even the homoeroticism right. These
authors' command of the imagery of the Great War is as striking as
Burgess's, and indeed their details are so convincing that it is sometimes
hard for readers to realize that they are being offered "fiction" rather
than documentary or factual testimony. Thus an elderly veteran of the
Royal Flying Corps writes to the *Listener* objecting to the drunken brutal-
ity of Major Stanley Woolley, the squadron commander in *Goshawk
Squadron:*

> I was a pilot in the RFC and this book made me angry. . . . No command-
> ing officer could have behaved like Woolley—and kept his command. His
> brutality, verbal and otherwise, would have wrecked squadron morale. . . .
> Nor did we treat mechanics the way they did in this book. We respected
> them, since our lives and our machines depended on their expertise. . . .[35]

But Robinson must make Major Woolley brutal, for, like Harris and
Marshall and Burgess, he is at pains to locate fictionally in the Great War
the paradigm of that contempt for life, individuality, and privacy, and
that facile recourse to violence that have characterized experience in the
twentieth century.

The younger poets since the Second War have been even more assidu-
ous in seeking out the modal grandfather. We recall "MCMXIV," by
Larkin, born in 1922. Another poet born in 1922, Vernon Scannell,
recalls the "almost obsessive interest in World War One" that marked his
boyhood.[36] Growing up, he fought his own war in North Africa and
Normandy, where he was wounded near Caen and spent the rest of the
war in hospitals. His first poems in the late forties were the usual mildly
ironic domestic notations mixed with an occasional allusion to his own

war. But in the fifties and sixties, in poems like "Remembrance Day,"
"The Great War," and "Uncle Edward's Affliction," he turned back to
"remember" the Great War, the origin of it all. And in the series
"Epithets of War" he conflates moments and images of the two wars,
suggesting, like Churchill and Blumentritt, that they are really one.

A counterpart transatlantic searcher for the modal grandfather is Louis
Simpson, born in 1923. Like Scannell, he was acquainted early with
twopenny weeklies and school annuals, which fed his Jamaican boyhood
with fantasies of flying the dawn patrol and jumping the bags on the
Western Front.[37] In due course he was caught in the Second War, and in
due course remembered it in poems like the haunting ballad "Carentan O
Carentan." But in the late fifties he shifted focus, remembering now a
war that seemed somehow to have more "authenticity"; and in "I
Dreamed that in a City Dark as Paris" projected a dream of joining history in the person of a representative *poilu* of the Great War, whom
Simpson addresses:

> My confrere
> In whose thick boots I stood, were you amazed
> To wander through my brain four decades later
> As I have wandered in a dream through yours?

What has made the mutual wandering possible is the identity of their
victimhood. What has promoted that identity is the violence formerly
characteristic only of warfare but now welling thickly from every opening fissure of modern life:

> The violence of waking life disrupts
> The order of our death. Strange dreams occur,
> For dreams are licensed as they never were.

Unlike Scannell and Simpson, Ted Hughes was too young for the
Second War, but in the early fifties he performed his National Service as
a ground wireless operator in the RAF. By means of early poems like
"Bayonet Charge" and "Out" he explored the topographical and psychological landscape not of the Second War but of the First. So firm was his
imaginative grasp of the Great War material that he instinctively perceived the principle of "threes" and invited it to dominate poems like
"Six Young Men" (of whom the fate of only three is known definitely)
and "Griefs for Dead Soldiers" (which conceives the kinds of grief as
three). Some years later, as if peeling back the layers of Great War
imagery—trenches, wire, mass graves—to see what they really represent, he arrived at the violent simplicities of the poems constituting the

volume *Crow* (1971), poems whose sentimental primitivism and mechanized emotional exhaustion speak precisely to the late-twentieth-century condition. It was his early instructive detour through the landscapes of the Great War that brought Hughes finally to this ultimate waste land.

Another to whom the Great War establishes an archetype for subsequent violence—as well as a criticism of it—is Michael Longley. He too derives his images not from experience but from mythic narrative. The occasion of his poem "Wounds" is the burial in Northern Ireland in 1972 of three teenage British soldiers and a bus conductor, meaninglessly done to death. But only the last half of the poem deals with these absurd murders. The first half prepares a measure for them and a comment on them by recalling two scenes from Longley's father's anecdotes of his moments in the Great War:

> First, the Ulster Division at the Somme
> Going over the top with "Fuck the Pope!"
> "No surrender!"; a boy about to die,
> Screaming "Give 'em one for the Shankill!"
> "Wilder than Gurkhas" were my father's words
> Of admiration and bewilderment.
> Next comes the London Scottish padre
> Resettling kilts with his swagger-stick,
> With a stylish back-hand and a prayer.
> Over a landscape of dead buttocks
> My father followed him for fifty years.
> At last, a belated casualty,
> He said—lead traces flaring till they hurt—
> "I am dying for King and Country, slowly."
> I touched his hand, his thin head I touched.

Only after this gauge has been established does Longley proceed with the matter of the four more recent burials of the "fallen" in another kind of war:

> Now . . . I bury beside him
> Three teenage soldiers, bellies full of
> Bullets and Irish beer, their flies undone. . . .
> Also a bus-conductor's uniform—
> He collapsed beside his carpet-slippers
> Without a murmur, shot through the head
> By a shivering boy who wandered in
> Before they could turn the television down
> Or tidy away the supper dishes.
> To the children, to a bewildered wife
> I think "Sorry Missus" was what he said.[38]

As if there weren't enough irony there, the irony always associated with the Somme attack remains to shade that conclusion. But at least the Somme attack had some swank and style: one could almost admire, if afterward one had to deplore.

Once the war was over, many assumed they could simply leave it behind. As Stanley Casson says of himself in 1920, "I was deep in my [archeological] work again, and had, as I thought, put the war into the category of forgotten things." He soon found how wrong he was: "The war's baneful influence controlled still all our thoughts and acts, directly or indirectly." [39] Chronologically close to the war as he was, we are not surprised to see Mann bring *The Magic Mountain* (1924) to a close with a "memory" of something very much like the shocking mines at Messines. All the rarefied theoretical dialectics of the sanitorium end in "the historic thunder-peal [which] made the foundations of the earth to shake." This thunder-peal is what "fired the mine beneath the magic mountain, and set our sleeper [Hans Castorp] ungently outside the gates." [40] Castorp goes on to be obliterated on the Western Front, and *The Magic Mountain* ends in a pitying if ironic elegy on him, who like Woolf's Jacob Flanders represents a well-meaning, trusting generation. Nor are we particularly surprised to see Eliot as late as 1940 recalling the terror associated with the term *raid* and exploiting that association in *East Coker* to underline the supreme risks of poetic making:

> And so each venture
> Is a new beginning, a raid on the inarticulate
> With shabby equipment always deteriorating
> In the general mess of imprecision of feeling,
> Undisciplined squads of emotion. . . .
>   . . .
> There is only the fight to recover what has been lost.

But we may be surprised to find in what untraditional places, unlike those frequented by Mann and Eliot, the Great War can be perceived still to go on. In the world of Ginsberg's *Howl*, for example, where we encounter this nightmare of the beautiful innocent young men destroyed on the battlefields of Making It:

> . . . burned alive in their innocent flannel suits on Madison Avenue amid blasts of leaden verse & the tanked-up clatter of the iron regiments of fashion & the nitroglycerine shrieks of the fairies of advertising & the mustard gas of sinister intelligent editors. . . .

We may conclude that, as Francis Hope has said, "In a not altogether rhetorical sense, all poetry written since 1918 is war poetry." [41] Seen in its immediate postwar context, a work like *The Waste Land* appears much more profoundly a "memory of the war" than one had thought. Consider its archduke, its rats and canals and dead men, its focus on fear, its dusty trees, its conversation about demobilization, its spritiualist practitioners reminding us of those who preyed on relatives anxious to contact their dead boys, and not least its settings of blasted landscape and ruins, suggestive of what Guy Chapman recalls as "the confluent acne of the waste land under the walls of Ypres." [42] It was common to identify "the waste land" that modern life seemed to resemble with the battlefields of the war rather than with the landscape of Eliot's poem. Philip O'Connor does this in speaking of the stupidities of "school" in the 1930's: "No intelligent, brave and honest social prospect," he says, "lay ahead for the teachers to inculcate. After school the wasteland prognosticated by this euphemistic education was not surprising; the depth of its craters, however, was unexpected and terrifying." [43]

"I had not thought death had undone so many," says one of the voices of *The Waste Land* as the crowds cross London Bridge on a winter dawn. They are City workers, commuters, but they are like the masses of reinforcements going from England daily to the line. Bertrand Russell recalls the early months of the war:

> After seeing troop trains departing from Waterloo, I used to have strange visions of London as a place of unreality. I used in imagination to see the bridges collapse and sink, and the whole great city vanish like a morning mist. Its inhabitants began to seem like hallucinations. . . .

And he adds: "I spoke of this to T. S. Eliot, who put it into *The Waste Land*." [44] Four years after *The Waste Land* Pound published *A Draft of XVI Cantos for the Beginning of a Poem of Some Length*. Hugh Kenner is right to point to the numerous evocations of war in these Cantos: "Wars, ruins, destructions—a crumbling wall in Mantua, smoke over Troy— these are never far out of mind. . . . This is postwar poetry, as much so as *The Waste Land*." [45]

THE RITUAL OF MILITARY MEMORY

Everyone who remembers a war first-hand knows that its images remain in the memory with special vividness. The very enormity of the proceedings, their absurd remove from the usages of the normal world, will guarantee that a structure of irony sufficient for ready narrative recall will attach to them. And the irony need not be Gravesian and extravagant: sometimes a very gentle irony emerging from anomalous contrasts will cause, as Stephen Hewett finds, "certain impressions [to] remain with one—a sunrise when the Huns are quiet, a sunset when they are raising a storm, a night made hideous by some distant cannonades, the nightingales in the warm darkness by a stagnant weedy river, and always the march back from the trenches to reserve-billets in some pretty village full of shady trees." [46] One remembers with special vividness too because military training is very largely training in alertness and a special kind of noticing. And one remembers because at the front the well-known mechanisms of the psychology of crisis work to assign major portent to normally trivial things like single poppies or the scars on a rifle-stock or "the smell of rum and blood." When a man imagines that every moment is his next to last, he observes and treasures up sensory details purely for their own sake. "I had a fierce desire to rivet impressions," says Max Plowman, "even of commonplace things like the curve of a roof, the turn of a road, or a mere milestone. What a strange emotion all objects stir when we look upon them wondering whether we do so for the last time in this life." [47] Fear itself works powerfully as an agent of sharp perception and vivid recall. Oliver Lyttelton understands this process in highly mechanical terms:

> Fear and its milder brothers, dread and anticipation, first soften the tablets of memory, so that the impressions which they bring are clearly and deeply cut, and when time cools them off the impressions are fixed like the grooves of a gramophone record, and remain with you as long as your faculties. I have been surprised how accurate my memory has proved about times and places where I was frightened. . . .

By contrast, "How faded are the memories of gaiety and pleasure." [48] Subsequent guilt over acts of cowardice or cruelty is another agent of vivid memory: in recalling scenes and moments marking one's own fancied disgrace, one sets the scene with lucid clarity to give it a verisimilitude sufficient for an efficacious self-torment.

Revisiting moments made vivid for these various reasons becomes a moral obligation. Owen registers it in an extreme form, but everyone who has shared his circumstances shares his obsession to some degree. He writes his mother in February, 1918: "I confess I *bring on* what few war dreams I now have, entirely by *willingly* considering war of an evening. I do so because I have my duty to perform towards War." [49] Revisiting the battlefields in memory becomes as powerful a ritual obligation as visiting the cemeteries. Of the silent battlefields of Vimy and Souchez, Reginald Farrer says, "They draw and hold me like magnets: I have never had enough." [50] "I still loaf into the past," says Tomlinson, "to the Old Front Line, where now there is only silence and thistles. I like it; it is a phase of my lunacy." [51]

The quality and dimensions of this lunacy of voluntary torment have never been more acutely explored and dramatized than by Thomas Pynchon in *Gravity's Rainbow*. Here for almost the first time the ritual of military memory is freed from all puritan lexical constraint and allowed to take place with a full appropriate obscenity. Memory haunts Pynchon's novel. Its shape is determined by a "memory" of the Second War, specifically the end of it and its immediate aftermath, when it is beginning to modulate into the Third. Just as Harris, Marshall, Robinson, and Burgess have worked up the Great War entirely from documents and written fictions, so Pynchon, who was only eight years old in 1945, has recovered the Second War not from his own memory but from films and from letterpress—especially, one suspects, the memoirs and official histories recalling that interesting British institution of the Second War, the Special Operations Executive (SOE). Like its American counterpart, the Office of Strategic Services, the SOE performed two functions, espionage and sabotage. The unmilitary informality of the SOE's personnel has become proverbial: like an eccentric club it enrolled dons, bankers, lawyers, cinema people, artists, journalists, and pedants. Of its executive personnel the *TLS* observed, "A few could only charitably be described as nutcases." [52] Its research departments especially enjoyed a wide reputation for sophomoric bright ideas and general eccentricity. From the laboratories of its "inventors" issued a stream of cufflink compasses, counterfeit identity documents, plastic explosives, and exploding pencils and thermos bottles and loaves of bread and nuts and bolts and cigarettes. Among booby traps its masterpieces were the exploding firewood logs and coal-lumps designed to be introduced into German headquarters fireplaces, as well as the explosive animal droppings (horse, mule, camel, and elephant) for placement on roads traveled by German military traffic. One department did nothing but contrive "sibs"—bizarre and hair-raising rumors to be spread over the Continent. It was said that even more outré departments staffed by necromancers, astrologers, and ESP enthusiasts worked at casting spells on the German civil and military hierarchy. Euphemized as "The Firm" or "The Racket" by those who worked for it, the SOE had its main offices at 62–64 Baker Street, and established its departments and training centers in numerous country houses all over Britain.

It is an organization something like the SOE, wildly and comically refracted, to which the American Lieutenant Tyrone Slothrop is assigned in *Gravity's Rainbow*. His initial duties, in September, 1944, are to interpret and if possible to predict the dispersal pattern of the V-2 missiles falling on London, a task for which, it is thought, he has a curious

physiological talent. This work brings him into contact with numerous
colleagues pursuing lunatic researches in aid of Victory: behaviorist psy-
chologists experimenting on (abducted) dogs; parapsychologists persuaded that the application of Schrödinger's *psi* function will win the
war; spirit mediums; statistical analysts; clairvoyants; and other modern
experts on quantification, technology, and prediction. Their place of
work is The White Visitation, an ancient country house and former
mental hospital on the southeast coast. The commander of the motley
unit at The White Visitation is Brigadier Ernest Pudding, a likable but
senile veteran of the Great War called back to preside over this fancy enterprise in the Second War. (In choosing his name Pynchon may be
echoing the name of the onetime Director of Operations of the actual
SOE, Brigadier Colin Gubbins [born 1896], a badly wounded winner of
the Military Cross in the Great War.)

The presence of Brigadier Pudding in the novel proposes the Great
War as the ultimate origin of the insane contemporary scene. It is where
the irony and the absurdity began. Pudding's "greatest triumph on the
battlefield," we are told, "came in 1917, in the gassy, Armageddonite
filth of the Ypres salient, where he conquered a bight of no man's land
some 40 yards at its deepest, with a wastage of only 70% of his unit."
After this satire of circumstance he remained in the army until he retired
in the thirties to the Devon countryside, where "it occurred to him to
focus his hobby on the European balance of power, because of whose
long pathology he had once labored, deeply, all hope of waking lost, in
the nightmare of Flanders." In his retirement he pursued the hopeless
hobby of writing a massive "book" titled *Things That Can Happen in European Politics*. But events defeated him and his resolution dissolved:
" 'Never make it,' he found himself muttering at the beginning of each
day's work—'it's changing out from under me. Oh, dodgy—very
dodgy.' " At the beginning of the Second War he volunteered and was
further disappointed:

> Had he known at the time it would mean "The White Visitation" . . . not
> that he'd expected a combat assignment you know, but wasn't there some-
> thing mentioned about intelligence work? Instead he found a disused hospi-
> tal for the mad, a few token lunatics, an enormous pack of stolen dogs,
> cliques of spiritualists, vaudeville entertainers, wireless technicians,
> Couéists, Ouspenskians, Skinnerites, lobotomy enthusiasts, Dale Carnegie
> zealots . . . (77).[53]

His present situation is especially disappointing because the quality of
the personnel seems to have fallen off between wars. The nutty dog ex-

perimenter at The White Visitation, Ned Pointsman, happens to be the
son of an able medical officer Pudding knew at Ypres. Ned is "not as tall
as his father, certainly not as wholesome looking. Father was M.O. in
Thunder Prodd's regiment, caught a bit of shrapnel in the thigh at Polygon Wood, lay silent for seven hours before they"—but the scene clouds
over in Pudding's memory and begins to fade: "without a word before,
in that mud, that terrible smell, in, yes Polygon Wood . . . or was
that—who *was* the ginger-haired chap who slept with his hat on? ahhh,
come back. Now Polygon Wood . . . but it's fluttering away." No use:
Pudding is losing the image he has been soliciting, and only detached,
incoherent details remain: "Fallen trees, dead, smooth gray, swirling-
grainoftreelikefrozensmoke . . . ginger . . . thunder . . . no use, no
bleeding use, it's gone, another gone, another, oh dear . . ." (76).

Sometimes Pudding's memories surface publicly during his prolonged
and rambling weekly briefings of the staff, occasions designated by Pynchon as "a bit of ritual with the doddering Brigadier." Mingled with "a
most amazing volley of senile observations, office paranoia, gossip about
the war," as well as with such detritus as "recipes for preparing beets a
hundred tasty ways," are detached, incoherent "reminiscences of Flanders," little sensual spots of time and images engraved on the failing
memory:

> . . . the coal boxes in the sky coming straight down on you with a roar . . .
> the drumfire so milky and luminous on his birthday night . . . the wet sur-
> faces in the shell craters for miles giving back one bleak autumn sky . . .
> what Haig, in the richness of his wit, once said at mess about Lieutenant
> Sassoon's refusal to fight . . . roadsides of poor rotting horses just before
> apricot sunrise . . . the twelve spokes of a standard artillery piece—a mud
> clock, a mud zodiac, clogged and crusted as it stood in the sun in its many
> shades of brown.

Brown triggers in Pudding a little lyric on human excrement as the dominant material of Passchendaele: "The mud of Flanders gathered into the
curd-crumped, mildly jellied textures of human shit, piled, duck-
boarded, trenches and shell-pocked leagues of shit in all directions, not
even the poor blackened stump of a tree . . ." (79–80).

Such discrete images take on coherence and something like narrative
relationships only during Pudding's secret fortnightly "rituals," his sur-
real nocturnal visits to the "Mistress of the Night," played by Katje
Borgesius, a Dutch girl who has somehow become attached to this mock-
SOE. The passage in which Pynchon presents one of these ritual visits is
one of the most shocking in the novel: it assumes the style of classic English pornographic fiction of the grossly masochistic type, the only style,

Passschendaele. (Imperial War Museum)

Pynchon implies, adequate to memories of the Great War, with its "filth" and "terrible smell." The language and the details may turn our stomachs, but, Pynchon suggests, they are only the remotest correlatives of the actuality. Compared with the actual sights and smells of the front, the word *shit* is practically genteel.

It is not easy to specify exactly the "allegorical" meaning of this "Mistress of the Night" to whom the aging brigadier repairs so regularly and punctually: she is at once Death, Fear, Ruined Youth, and the memory of all these. She is what one must revisit ritually for the sake of the perverse pleasant pain she administers. She is like the Muse of Pudding's war-memory. She waits to receive him in a setting combining bawdy-house with theater, and indeed the scene of Pudding's visit constitutes the climax of the tradition that the Great War makes a sort of sense if seen as a mode of theater.

It is a cold night. Pudding slips out of his quarters "by a route only he knows," quietly singing, to keep his courage up, the soldiers' song he recalls from the Great War:

> Wash me in the water
> That you wash your dirty daughter,
> And I shall be whiter than the whitewash on the wall.

He tiptoes along the sleeping hallways and through a half-dozen rooms, the passage of which, like the traditional approach to the Grail Chapel, constitutes a "ritual": each contains "a test he must pass." In the third room, for example, "a file drawer is left ajar, a stack of case histories partly visible, and an open copy of Krafft-Ebing." The fourth room contains a skull. The fifth, a Malacca cane, which apparently reminds him of the one he carried in 1917 and sets him thinking: "I've been in more wars for England than I can remember . . . haven't I paid enough? Risked it all for them, time after time . . . Why must they torment an old man?" In the sixth and last room the memory of the war becomes more vivid:

> In the sixth chamber, hanging from the overhead, is a tattered tommy up on White Sheet Ridge, field uniform burned in Maxim holes black-rimmed. . . , his own left eye shot away, the corpse beginning to stink . . . no . . . no! an overcoat, someone's old coat that's all, left on a hook in the wall . . . but couldn't he *smell* it? Now mustard gas comes washing in, into his brain with a fatal buzz as dreams will when we don't want them, or when we are suffocating. A machine-gun on the German side sings *dum diddy da da*, an English weapon answers *dum dum*, and the night tightens coiling around his body, just before H-hour. . . .

It is H-hour now, for he has arrived at the seventh room, where the Mistress of the Night waits for him. He knocks, the door unlocks "electrically," and he enters a room lighted only by "a scented candle." Katje Borgesius sits in a throne-like chair, wearing nothing but the black cape and tall black boots of the traditional female "disciplinarian" in British pornography. She is made up to resemble "photographs of the reigning beauties of thirty and forty years ago," and the only jewelry she wears is "a single ring with an artificial ruby not cut to facets" but resembling a convex bloody wound. She extends it, and Pudding kneels to kiss it before undressing.

> She watches him undress, medals faintly jingling, starch shirt rattling. She wants a cigarette desperately, but her instructions are not to smoke. She tries to keep her hands still. "What are you thinking, Pudding?"
> "Of the night we first met." The mud stank. The Archies were chugging in the darkness. His men, his poor sheep, had taken gas that morning. He was alone. Through the periscope, underneath a star shell that hung in the sky, he saw her . . . and though he was hidden, she saw Pudding. Her face

was pale, she was dressed all in black, she stood in No-man's Land, the machine guns raked their patterns all around her, but she needed no protection. "They knew you, Mistress. They were your own."

"And so were you."

"You called to me, you said, 'I shall never leave you. You belong to me. We shall be together, again and again, though it may be years between. And you will always be at my service.' "

Pudding now undergoes his ritual of humiliation. He creeps forward to kiss her boots. He excites her by reminding her of an incident in the Spanish Civil War when a unit of Franco's soldiers was slaughtered. She is pleased to remember. "I took their brown Spanish bodies to mine," she says. "They were the color of the dust, and the twilight, and of meats roasted to a perfect texture . . . most of them were so very young. A summer day, a day of love: one of the most poignant I ever knew." "Thank you," she says to Pudding. "You shall have your pain tonight." And she gives him a dozen blows with the cane, six across the buttocks, six across the nipples. "His need for pain" is gratified; he feels momentarily rescued from the phony paper war he's now engaged in, reinstalled in his familiar original world of "vertigo, nausea, and pain." The whipping over, she obliges him to drink her urine. And then the climax of the ritual: he eats her excrement. It is an act reminiscent of their first encounter at Passchendaele: "The stink of shit floods his nose, gathering him, surrounding him. It is the smell of Passchendaele, of the Salient. Mixed with the mud, and the putrefaction of corpses, it was the sovereign smell of their first meeting, and her emblem." As he eats, "spasms in his throat continue. The pain is terrible." But he enjoys it. Finally she commands him to masturbate before her. He does so and departs regretfully, realizing anew, when he reaches his own room again, that "his real home is with the Mistress of the Night . . ." (231–36).

It is a fantastic scene, disgusting, ennobling, and touching, all at once. And amazingly rich in the way it manages to fuse literal with figurative. The woman is both Katje and the Mistress of the Night, credible for the moment in either identity; she is both a literal filthy slut in 1945 and the incarnation of the spirit of military memory in all times and places. As allegory the action succeeds brilliantly, while its literal level is consistent and, within the conventions, credible: Pudding, we are told later, died in June, 1945, "of a massive *E. coli* infection" (533) brought about by these ritual coprophagic sessions. And yet what he was "tasting" and "devouring" all the time was his memories of the Great War. Pynchon is not the only contemporary novelist to conceive a close relation between perverse sexual desire, memories of war, and human excrement. In *An American*

Dream Mailer offers a scene analogous to that of Pudding and Katje, only here the memories are of the Second War. It is after recalling the sadistic, bloody details of his shooting of four Germans in Italy that Stephen Rojack proceeds to sodomize the "Kraut" Ruta. It is this scene that Kazin has in mind when he observes, "As always in reading Mailer's descriptions of intercourse, one is impressed by how much of a war novelist he has remained." [54] As we perceive in the work of Mailer and Pynchon and James Jones, it is the virtual disappearance during the sixties and seventies of the concept of prohibitive obscenity, a concept which has acted as a censor on earlier memories of "war," that has given the ritual of military memory a new dimension. And that new dimension is capable of revealing for the first time the full obscenity of the Great War. The greatest irony is that it is only now, when those who remember the events are almost all dead, that the literary means for adequate remembering and interpreting are finally publicly accessible.

We began with Hardy's poems and their relation to a war still hidden below the horizon. Let us end by returning to Hardy's poems and glancing at a relation Vernon Scannell once hinted between history and one's personal memories of war. In June, 1970, two anniversaries coincided almost exactly: the date of Hardy's birthday 130 years earlier and the date of the invasion of Normandy 26 years earlier. As an ex-soldier Scannell was implicated in that invasion; as a poet, he is involved in Hardy's vision and language. On the evening of June 2, he must journey to Dorchester to place a wreath on Hardy's monument and then address the Hardy Society on the subject of his poetry. He doesn't much want to do this because he knows that the members of the Hardy Society aren't very literary and don't really care about any poetry, Hardy's included. "Still," he says, "I am quite glad to be asked to give the talk . . . because, in the preparation of it, I have reread a lot of Hardy's work. . . ." He has found it "quite magnificent." Did he reread *Satires of Circumstance?* Did he reread "After a Journey"? "Your Last Drive"? He doesn't say. But perhaps he reread "Channel Firing," for the next thing he is talking about is the thunderous invasion across that Channel over a quarter of a century ago. And now he thinks not about Hardy but about the way his memories of his own war have reduced to a few repeated incoherent images. "I realize that I do not remember it so clearly after all. History remembers it, and I remember it as history. . . ." Perhaps his own psyche has providentially erased a host of memories that might do him harm: "It is almost as if the inner recording eye had fixed itself on phenomena that the mind would be able to live with in the future without too much distress." [55] But "history remembers it." Ex post facto, literary narrative has supplied it with coherence and irony, educing the pattern: innocence savaged and destroyed. Or if not destroyed, transformed utterly into what Frye calls "the total cultural form of our present life." "The culture of the past," he says, "is not only the memory of mankind, but our own buried life." And "study of it leads to a recognition scene, a discovery in which we see, not our past lives, but the total cultural form of our present life." [56]

In this study of a small bit of that culture of the past, I have tried to present just a few such recognition scenes. My belief is that what we recognize in them is a part, and perhaps not the least compelling part, of our own buried lives.

Notes

Place of publication is London unless otherwise indicated. The abbreviation IWM designates manuscript or typescript material in the archives of the Imperial War Museum.

I. *A Satire of Circumstance*

1. Lytton Strachey, *Literary Essays* (New York, 1949), pp. 220–25.
2. Philip Longworth, *The Unending Vigil* (1967), p. 34.
3. *Undertones of War* (1928; 1956), pp. 58–59.
4. *Ibid.*, p. 147.
5. *Siegfried's Journey* (1945), pp. 13, 29.
6. *Ibid.*, p. 19.
7. *Memoirs of an Infantry Officer* (1930; New York, 1937), p. 64.
8. *The Letters of Henry James*, ed. Percy Lubbock (2 vols., New York, 1920), II, 384.
9. *Now It Can Be Told* (New York, 1920), p. 131.
10. "What Really Happened," in *Promise of Greatness*, ed. George A. Panichas (1968), p. 386.
11. William Moore, *The Thin Yellow Line* (1973), pp. 63–64.
12. *War Letters of Fallen Englishmen*, ed. Laurence Housman (1930), p. 221.
13. *History of the First World War* (1934; 1970), p. 196.
14. *Good-bye to All That* (1929; New York, 1957), pp. 145 ff.
15. Quoted by Liddell Hart, *History*, p. 267.
16. Pages i–iv.
17. *The Mind's Eye* (1934), p. 38.
18. A. J. P. Taylor, *The First World War* (1963; Harmondsworth, Eng., 1966), p. 158.
19. *Blasting and Bombardiering* (1937; Berkeley, Calif., 1967), p. 151.

20. Quoted in *My Warrior Sons: The Borton Family Diary, 1914–1918*, ed. Guy Slater (1973), p. 164.

21. John Terraine, *Impacts of War, 1914 and 1918* (1970), p. 174.

22. *A Fatalist at War*, trans. Ian F. D. Morrow (New York, 1929), pp. 209–10.

23. *The First World War*, p. 22.

24. New York, 1929, p. 191.

25. Christopher Isherwood, *Kathleen and Frank* (New York, 1971), pp. 136, 144.

26. "Bourbon County," in *Prose* VI (Spring, 1973), p. 155.

27. *Kathleen and Frank*, pp. 389, 400.

28. *The Early Years of Alec Waugh* (1962), p. 55.

29. *Impacts of War*, p. 40.

30. *Great Morning!* (Boston, 1947), p. 199.

31. *Letters of Rupert Brooke*, ed. Geoffrey Keynes (New York, 1968), p. 625.

32. Quoted by Patrick Howarth, *Play Up and Play the Game* (1973), p. 7.

33. New York, 1917, p. 86.

34. *Ibid.*, p. 94.

35. *S.O.S. Stand To!* (New York, 1918), p. 32.

36. Slater, ed., *My Warrior Sons*, p. 137.

37. *My Father and Myself* (New York, 1969), p. 57.

38. Private L. S. Price, quoted by Martin Middlebrook, *The First Day on the Somme* (1971), p. 124.

39. *Disenchantment* (1922; 1968), pp. 13–14.

40. Duff Cooper, *Haig* (2 vols., 1935), I, 327.

41. Quoted by Middlebrook, *First Day on the Somme*, pp. 80–81.

42. *Memoirs of an Infantry Officer*, p. 71.

43. *The Wet Flanders Plain* (1929), pp. 15–16.

44. IWM.

45. IWM.

46. Letter from G. Bricknall, Sept. 18, 1971.

47. Panichas, ed., *Promise of Greatness*, pp. 182–83.

48. "A Doctor's War," *ibid.*, pp. 193–94.

49. Page 315.

50. *Passchendaele and the Somme* (1928), p. 73.

51. Pages 73–74.

52. New York, 1928, pp. 130–31, 141–42.

53. New York, 1948, p. 705.

54. 1929, pp. 81–82.

55. *Bright Book of Life* (Boston, 1973), p. 84.

56. (New York, 1961; 1966), pp. 426–30.

II. *The Troglodyte World*

1. H. H. Cooper, IWM.

2. *Under Fire*, trans. W. Fitzwater Wray (1926), p. 25.

3. *Steady Drummer* (1935), pp. 49–50.

4. *Gallipoli to the Somme* (1963), p. 97.

5. *Boy in the Blitz* (1972), p. 210.
6. Quoted by A. H. Farrar-Hockley, *The Somme* (1964), p. 188.
7. George Coppard, *With a Machine Gun to Cambrai* (1969), p. 80.
8. John Masefield, *The Old Front Line* (1917), p. 15.
9. *Memoirs of an Infantry Officer*, p. 228.
10. *Old Soldiers Never Die* (1933), pp. 44–45.
11. *Memoirs of an Infantry Officer*, p. 151.
12. *Letters to His Wife* (1917), p. 112.
13. *Wilfred Owen: Collected Letters*, ed. Harold Owen and John Bell (1967), p. 429.
14. *The First Hundred Thousand* (New York, 1916), p. 97.
15. *1914*, pp. 88, 91.
16. *With a Machine Gun to Cambrai*, p. 87.
17. *Ibid.*
18. *Copse 125*, trans. Basil Creighton (1930), pp. 18–19.
19. C. E. Carrington, *The Life of Rudyard Kipling* (New York, 1955), p. 338.
20. *Air with Armed Men* (1972), p. 114.
21. H. H. Cooper, IWM.
22. Quoted by Moore, *The Thin Yellow Line*, p. 203.
23. Owen and Bell, eds., *Collected Letters*, p. 426.
24. R. W. Mitchell, IWM.
25. P. H. Pilditch, IWM.
26. *Behind the Scenes at the Front* (New York, 1915), pp. 92–93.
27. *Old Soldiers Never Die*, pp. 31, 42–43.
28. *Memoirs of a Fox-Hunting Man* (1928; New York, 1937), p. 331.
29. *The Contrary Experience* (New York, 1963), p. 95.
30. IWM.
31. *Hindoo Holiday* (1952), p. 16.
32. Isherwood, *Kathleen and Frank*, p. 426.
33. E. Norman Gladden, *Ypres, 1917: A Personal Account* (1967), p. 65.
34. *In Parenthesis* (New York, 1963), p. 202.
35. *Passchendaele and the Somme*, pp. 87–88.
36. *Memoirs of an Infantry Officer*, p. 45.
37. *Parade's End* (New York, 1950), p. 543.
38. Clement Shorter, *Charlotte Brontë and Her Circle* (New York, 1896), p. 387.
39. *When the Going Was Good* (Harmondsworth, Eng., 1951), p. 70.
40. *The Long Week-End* (1940; 1971), p. 271.
41. *Surprised by Joy* (New York, 1955), p. 152.
42. *A Subaltern on the Somme*, p. 72.
43. *Spots of Time: A Retrospect of the Years 1897–1920* (1965), p. 114.
44. Donald Weeks, *Corvo: Saint or Madman?* (New York, 1971), p. 335.
45. *A Little Learning* (1954), p. 122.
46. Housman, ed., *War Letters of Fallen Englishmen*, p. 225.
47. Page 282.
48. Quoted by Alfred Perceval Graves, *To Return to All That* (1930), p. 323.
49. *Memoirs of a Fox-Hunting Man*, p. 303.
50. *Ibid.*, p. 320.
51. *Ibid.*, p. 365.
52. *Sherston's Progress* (1936; New York, 1937), p. 44.

53. *Ibid.*, p. 39.

54. Quoted by Jon Silkin, *Out of Battle: The Poetry of the Great War* (1972), p. 51.

55. *The Challenge of the Dead* (1921), p. 92.

56. *The Journal of Arnold Bennett* (New York, 1933), p. 537.

57. Arthur Innes Adam, in Housman, ed., *War Letters of Fallen Englishmen*, p. 22.

58. P. H. Pilditch, IWM.

59. Stuart Cloete, *How Young They Died* (1969), p. 13.

60. John Brophy and Eric Partridge, *The Long Trail: Soldiers' Songs and Slang, 1914–18* (1965; 1969), p. 22.

61. Adam, *Behind the Scenes*, p. 126.

62. Isherwood, *Kathleen and Frank*, p. 426.

63. *General Jack's Diary, 1914–1918*, ed. John Terraine (1964), p. 164.

64. *Memoirs of an Infantry Officer*, p. 52.

65. R. MacLeod, IWM.

66. IWM.

67. IWM.

68. *F. E.: The Life of F. E. Smith, by His Son* (1959), p. 265.

69. *A Brass Hat in No Man's Land* (1930), p. 154.

70. "A Doctor's War," in Panichas, ed., *Promise of Greatness*, p. 200.

71. *A Subaltern on the Somme*, p. 220.

72. *The City that Shone* (1969), p. 236.

73. Arthur Behrend, *As From Kemmel Hill* (1963), p. 32.

74. Owen and Bell, eds., *Wilfred Owen: Collected Letters*, p. 572.

75. *From Peace to War: A Study in Contrast, 1857–1918* (1968), p. 126.

76. *Call to Arms* (1968), p. 28.

77. *The Mind's Eye*, p. 36.

78. Carrington, *Life of Kipling*, p. 342.

79. New York, 1973, p. 86.

80. Clive Watts, IWM.

81. James Pope-Hennessy, *Queen Mary* (1959), p. 505.

82. *The Spanish Farm Trilogy* (1927), pp. 698–99.

83. IWM.

84. *Now It Can Be Told*, p. 408.

85. Quoted by Leon Wolff, *In Flanders Fields: The 1917 Campaign* (New York, 1960), p. 43.

86. IWM.

87. IWM.

88. "The Kaiser's War," in Panichas, ed., *Promise of Greatness*, p. 10.

89. *Undertones of War*, p. 16.

90. *Sherston's Progress*, p. 69.

91. *The Age of Death* (New York, 1970), pp. 8, 226.

92. *Bright Book of Life*, p. 81.

III. *Adversary Proceedings*

1. Robert Rhodes James, *Gallipoli* (New York, 1965), p. 86.

2. Page 196.

3. *Further Speculations*, ed. Sam Hynes (Minneapolis, 1955), p. 164.
4. IWM.
5. "Charles Edmonds," *A Subaltern's War* (1930), p. 25.
6. Terraine, ed., *General Jack's Diary*, p. 165.
7. Stanley Casson, *Steady Drummer*, p. 47.
8. *A Passionate Prodigality* (1933), p. 52.
9. "Broadchalk," in *Three Personal Records of the War* (1929), p. 266.
10. *In Parenthesis*, pp. 204–5.
11. *Storm of Steel* (New York, 1929), p. ix.
12. *Undertones of War*, p. 207.
13. *Memoirs of an Infantry Officer*, p. 80.
14. *Undertones of War*, p. 46.
15. *With a Machine Gun to Cambrai*, p. 89.
16. Herbert Fairlie Wood, *Vimy!* (1967), p. 32.
17. *Parade's End*, p. 578.
18. Blunden, *Undertones of War*, p. 164.
19. H. H. Cooper, IWM.
20. Page 343.
21. *The Contrary Experience*, p. 63.
22. *Soldier From the Wars Returning* (1965), p. 87.
23. *Old Men Forget* (1953), p. 82.
24. Introduction to Huntly Gordon, *The Unreturning Army* (1967), p. x.
25. Page 22.
26. Owen and Bell, eds., *Collected Letters*, pp. 421, 427.
27. *Sagittarius Rising* (New York, 1936; 1963), pp. 86–87.
28. Hynes, ed., *Further Speculations*, p. 157.
29. *First Day on the Somme*, p. 241.
30. Noted by Bernard Bergonzi, *Heroes' Twilight* (1965), p. 89.
31. *A Subaltern on the Somme*, p. 38.
32. IWM.
33. Gibbs, *Now It Can Be Told*, p. 245.
34. *The Battle for Europe, 1918* (1972), p. 19.
35. Wolff, *In Flanders Fields*, p. 228.
36. *Early Years*, p. 36.
37. *From Peace to War*, p. 135.
38. Terraine, ed., *General Jack's Diary*, p. 280.
39. Quoted by Silkin, *Out of Battle*, pp. 317–18.
40. Page 156.
41. Quoted by Silkin, *Out of Battle*, p. 52.
42. *Now It Can Be Told*, p. 143.
43. Quentin Bell, *Virginia Woolf* (2 vols., New York, 1972), II, 51–52.
44. Michael Macdonagh, *In London During the Great War: The Diary of a Journalist* (1935), p. 179.
45. John Terraine, *Impacts of War*, pp. 69–75.
46. *Lord Northcliffe's War Book*, p. 53.
47. *Call to Arms*, p. 26.
48. Ronald Blythe, *Akenfield* (New York, 1969), p. 42.
49. *Lord Northcliffe's War Book*, pp. 214–15.
50. IWM.
51. *The Long Weekend*, p. 10.

52. Francis Whiting Halsey, *The Literary Digest History of the Great War* (10 vols., New York, 1919), III, 287.

53. 1923; Harmondsworth, Eng., 1950, p. 241.

54. *Memoirs of an Infantry Officer*, p. 280.

55. *War Poets, 1914–1918* (1958), p. 25.

56. *Old Soldiers Never Die*, p. 271.

57. *Undertones of War*, p. 251.

58. Owen and Bell, eds., *Collected Letters*, pp. 485, 487.

59. *The Old Century and Seven More Years* (1938), p. 140.

60. *An Adequate Response: The War Poetry of Wilfred Owen and Siegfried Sassoon* (Detroit, 1972), pp. 110, 107.

61. Page references to the volume indicated of *The Memoirs of George Sherston* (New York, 1937).

62. Page 178.

63. *Good-bye to All That*, p. 275.

64. *The Works in Verse and Prose of William Shenstone, Esq.* (3 vols., 1791), II, 152.

65. Page 69.

66. *The City That Shone*, pp. 220, 226.

67. Page 56.

68. *An Adequate Response*, p. 92.

69. *A Chapter of Accidents* (1972), p. 95.

70. New York, 1958, p. 47.

71. New York, 1926, p. 95.

72. *Poetries and Sciences: A Reissue of Science and Poetry (1926, 1935) with Commentary* (New York, 1970), p. 77.

73. *Listener*, April 5, 1973, pp. 452–54.

74. "The Illogical Promise," in Panichas, ed., *Promise of Greatness*, p. 571.

75. *The Collected Essays, Journalism, and Letters of George Orwell*, ed. Sonia Orwell and Ian Angus (4 vols., New York, 1968), I, 517–18.

76. Page references to the New Directions edition (1947).

77. *Memories, Dreams, Reflections*, recorded and ed. by Aniela Jaffé, trans. Richard and Clara Winston (New York, 1963), p. 203.

IV. *Myth, Ritual, and Romance*

1. "Holy Ground," in Panichas, ed., *Promise of Greatness*, pp. 313–14.

2. *A Subaltern on the Somme*, p. 54.

3. Foreword to *The Wipers Times*, ed. Patrick Beaver (1973), p. x.

4. *Into Battle, 1914–1918* (1964), p. 20.

5. "Some Soldiers," in Panichas, ed., *Promise of Greatness*, p. 157.

6. *Undertones of War*, p. 172.

7. *Memoirs of an Infantry Officer*, p. 128.

8. *The Historian's Craft*, trans. Peter Putnam (New York, 1953), pp. 107, 109.

9. IWM.

10. See, for example, Harvey Swados, Preface to Erich Maria Remarque, *All Quiet on the Western Front* (New York, 1929; 1967), p. xii.

11. *The First Hundred Thousand*, p. 199.

12. *Johnny Got His Gun* (New York, 1939; 1970), p. 23.

13. *The Challenge of the Dead*, p. 33.

14. Keynes, ed., *Letters*, p. 637.

15. *A Subaltern on the Somme*, p. 29.

16. *5 Pens in Hand* (New York, 1958), p. 123.

17. Owen and Bell, eds., *Letters*, p. 562.

18. Timothy d'Arch Smith, *Love in Earnest: Some Notes on the Lives and Writings of English "Uranian" Poets from 1889 to 1930* (1970), pp. 140–41.

19. IWM.

20. *A Victorian Son* (1972), p. 217.

21. *The Mind's Eye*, p. 59.

22. *S.O.S. Stand To!*, p. 141.

23. *Ibid.*, p. 47.

24. *Steady Drummer*, p. 80.

25. Page 115.

26. *With a Machine Gun to Cambrai*, p. 69.

27. *The Wet Flanders Plain*, p. 58.

28. *Laughter in the Next Room* (New York, 1948), pp. 8–9.

29. Quoted by H. M. Tomlinson, *Waiting for Daylight* (New York, 1922), pp. 180–81.

30. Page 614.

31. "First Amorous Adventure," *Playboy* (January, 1972), p. 247.

32. *Now It Can Be Told*, pp. 398–99.

33. *A Subaltern on the Somme*, p. 87.

34. *A Victorian Son*, p. 244.

35. *How Young They Died*, p. 135.

36. Edward G. D. Liveing, *Attack* (1918), p. 66.

37. *The Letters of Charles Sorley* (Cambridge, Eng., 1919), p. 305.

38. *Old Soldiers Never Die*, p. 58.

39. *The Forgotten Soldier*, trans. Lily Emmet (New York, 1971), p. 236.

40. J. Brough, "The Tripartite Ideology of the Indo-Europeans: An Experiment in Method," *Bulletin of the School of Oriental and African Studies*, XXII (1959), 69–85.

41. *Anatomy of Criticism* (Princeton, N.J., 1957), pp. 187, 194.

42. *How to Read Shakespeare* (New York, 1971), p. 105.

43. "Rumors at the Front," in *A Martial Medley* (1931), p. 114.

44. *Anatomy*, p. 187.

45. R. E. Vernède, *Letters to His Wife*, p. 4.

46. Middlebrook, *First Day on the Somme*, p. 314.

47. *Into Battle*, p. 56.

48. *Heroes' Twilight*, p. 198.

49. "Rumors at the Front," p. 114.

50. *Old Soldiers Never Die*, pp. 279–80.

51. *A Subaltern's War*, p. 40.

52. Quoted by Farrar-Hockley, *The Somme*, p. 144.

53. *Now It Can Be Told*, p. 287.

54. IWM.

55. *A Subaltern on the Somme*, pp. 38–39.

56. Aitken, *Gallipoli to the Somme*, p. 126; Masefield, *The Old Front Line*, p. 15.

57. *The Challenge of the Dead*, p. 92.
58. *Laughter in the Next Room*, pp. 3–5.
59. *Mimesis: The Representation of Reality in Western Literature*, trans. Willard Trask (Princeton, N.J., 1953; New York, 1957), pp. 107–24.
60. *Letters from the Front*, ed. John Laffin (1973), p. 12.
61. *Surprised by Joy*, p. 164.
62. *Anatomy*, p. 149.
63. *Memoirs of a Fox-Hunting Man*, pp. 372–73.
64. *Memoirs of an Infantry Officer*, p. 21.
65. *Sherston's Progress*, p. 41.
66. *Undertones of War*, p. 162.
67. *A Subaltern's War*, p. 117.
68. *A Passionate Prodigality*, p. 43.
69. *Passchendaele and the Somme*, p. 159.
70. *Ibid.*, pp. 128–29.
71. *Ibid.*, p. 77.
72. *Gone for a Soldier* (Kineton, Eng., 1972), p. 79.
73. *The Attack and Other Papers* (1953), pp. 13–14.
74. *People at War, 1914–1918*, ed. Michael Moynihan (1973), p. 148.
75. *Heroes' Twilight*, p. 41.
76. *Validity in Interpretation* (New Haven, Conn., 1967), p. 105.
77. IWM.
78. Laffin, ed., *Letters from the Front*, p. 15.
79. Housman, ed., *War Letters of Fallen Englishmen*, p. 69.
80. *The Wet Flanders Plain*, p. 46.
81. *A Subaltern's War*, pp. 135–36.
82. *Gone for a Soldier*, p. ix.
83. IWM.
84. New York, 1969; 1971, p. 148.
85. Preface to *The Anathémata* (New York, 1963), p. 41.
86. Preface to *In Parenthesis*, p. xv.
87. Pages x–xi.
88. David Blamires, *David Jones: Artist and Writer* (Toronto, 1972), p. 59.
89. *Ibid.*, p. 153.
90. Preface to *The Anathémata*, p. 24.
91. *Soldier from the Wars Returning*, p. 87.
92. Blamires, *David Jones*, p. 3.
93. Page xiv.
94. Page references to *In Parenthesis* (New York, 1963).

V. *Oh What a Literary War*

1. *From Peace to War*, p. 164.
2. *A Private in the Guards* (New York, 1919), p. 135.
3. Alan Moorehead, *Gallipoli* (New York, 1956), p. 301.
4. Keynes, ed., *Letters*, p. 670.
5. *War in the Modern World* (Durham, N.C., 1959), p. 162.
6. *In Parenthesis*, p. 95.

7. *Undertones of War*, pp. 89–90.
8. Preface to the *Dictionary* (1755).
9. *In Parenthesis*, pp. 139–40.
10. *Disenchantment*, p. 100.
11. *Into Battle*, p. 18.
12. New York, 1922; 1934, pp. 53–54.
13. *Vain Glory*, ed. Guy Chapman (1968), p. 160.
14. *Johnson's Journey to the Western Islands of Scotland and Boswell's Journal of a Tour to the Hebrides*, ed. R. W. Chapman (Oxford, 1924), p. vii.
15. *Early Years*, p. 108.
16. *Undertones of War*, pp. 280, 222–23.
17. *The Contrary Experience*, p. 99.
18. *Ibid.*, p. 130.
19. *A Subaltern's War*, pp. 127–28.
20. IWM.
21. *Memoirs of an Infantry Officer*, pp. 107–8.
22. *A Subaltern on the Somme*, p. 161.
23. IWM.
24. J. S. Wane, IWM.
25. *Anglo-Welsh Review*, XX, No. 45 (Autumn, 1971), 23.
26. *A Passionate Prodigality*, p. 276.
27. Housman, ed., *War Letters of Fallen Englishmen*, p. 216.
28. *From Peace to War*, p. 154.
29. Housman, ed., *War Letters of Fallen Englishmen*, p. 81.
30. *A Subaltern's War*, p. 106.
31. *Undertones of War*, p. 173.
32. *Ibid.*, p. 158.
33. *Ibid.*, pp. 293, 235.
34. IWM.
35. "War Service in Perspective," Panichas, ed., *Promise of Greatness*, p. 373.
36. Laffin, ed., *Letters from the Front*, pp. 14–15.
37. *Memoirs of an Infantry Officer*, p. 31.
38. *Gone for a Soldier*, p. 70.
39. "The Best of My War," Panichas, ed., *Promise of Greatness*, p. 235.
40. *The Poetry of War, 1939–1945*, ed. Ian Hamilton (1965), p. 172.
41. D. S. R. Welland, *Wilfred Owen: A Critical Study* (1969), p. 30.
42. IWM.
43. *Listener*, July 15, 1971, p. 74.
44. Housman, ed., *War Letters of Fallen Englishmen*, p. 199.
45. *Gallipoli to the Somme*, p. 169.
46. Page 127.
47. 2 vols., 1923; I, 225.
48. *World Within World* (1951), pp. 273–74.
49. Page 23.
50. Page 86.
51. *Gallipoli to the Somme*, p. 171.
52. Wolff, *In Flanders Fields*, p. 244.
53. *Stilwell and the American Experience in China, 1911–1945* (New York, 1971; 1972), p. 557.

54. Graves and Hodge, *The Long Weekend*, p. 8.
55. *A Little Learning*, p. 88.
56. *A Sort of Life* (1971), p. 64.
57. *Gallipoli to the Somme*, pp. 82, 94.
58. A. G. Reeve, IWM.
59. F. T. H. Bennett, IWM.
60. *Ypres, 1917*, p. 67.
61. Terraine, ed., *General Jack's Diary*, pp. 187–88.
62. Quoted by Moore, *The Thin Yellow Line*, p. 143.
63. *Ibid.*, p. 144.
64. Quoted by Brian Gardner, *The Big Push* (1961), p. 82.
65. Page 386.
66. William Collins, IWM.
67. *Margin Released* (1962), p. 111.
68. *Gone for a Soldier*, p. 68.
69. Laffin, ed., *Letters from the Front*, p. 96.
70. Middlebrook, *First Day on the Somme*, p. 249.
71. Trans. Andrew Sinclair.
72. Bernard Bergonzi, "Kipling and The First World War," *The Age of Kipling*, ed. John Gross (New York, 1972), p. 137.
73. IWM.
74. IWM.
75. IWM.
76. IWM.
77. Terraine, ed., *General Jack's Diary*, p. 126.
78. Introduction, *Old Soldiers Never Die*, p. 5.
79. IWM.
80. IWM.
81. *How Young They Died*, p. 105.
82. Introduction to Richards, *Old Soldiers Never Die*, p. 2.
83. *The Contrary Experience*, p. 108.
84. *Letters to His Wife*, p. 162.
85. Page 189.
86. Page x.
87. *A Personal Anthology*, ed. Anthony Kerrigan (New York, 1967), p. 151.
88. *With a Machine Gun to Cambrai*, p. 36.
89. Owen and Bell, eds., *Collected Letters*, p. 421.
90. *Up to Mametz* (1931), p. 156; *How Young They Died*, p. 143; *Not for Glory* (1969), pp. 53–54.
91. Owen and Bell, eds., *Collected Letters*, p. 578.
92. Frances Donaldson, *Evelyn Waugh: Portrait of a Country Neighbor* (1967), p. 45.
93. Page 338.
94. Pages 521, 252.
95. *Free Fire Zone: Short Stories by Vietnam Veterans*, ed. Wayne Karlin *et al.* (New York, 1973), pp. 80–96.
96. 1918; 1948, p. 6.
97. Page 231.
98. Bergonzi, *Heroes' Twilight*, p. 127.

99. *World Within World*, p. 7.
100. *Boy in the Blitz*, p. 36.
101. *Ibid.*, pp. 172, 175.

VI. *Theater of War*

1. Alexander McKee, *Vimy Ridge* (1966), p. 184.
2. "Reflections upon War and Death," trans. E. Colburn Mayne, in *Sigmund Freud: Character and Culture*, ed. Philip Rieff (New York, 1963), p. 122.
3. IWM.
4. *A Subaltern's War*, pp. 133, 137, 160.
5. *How Young They Died*, p. 202.
6. *Memoirs of an Infantry Officer*, pp. 236–37.
7. *Sherston's Progress*, p. 77.
8. See Samuel Hynes, *The Edwardian Turn of Mind* (Princeton, N.J., 1968), pp. 46ff.
9. *Steady Drummer*, p. 38.
10. IWM.
11. "Forced to Think," in Panichas, ed., *Promise of Greatness*, p. 103.
12. IWM.
13. *Undertones of War*, pp. 9–10.
14. Quoted by Chapman, ed., *Vain Glory*, pp. 143–44.
15. Terraine, *Impacts of War*, pp. 58–59.
16. *The Battle for Europe, 1918*, p. 206.
17. *Mary McCarthy's Theatre Chronicles, 1937–1962* (New York, 1963), p. 194.
18. *Storm of Steel*, p. 229.
19. Orwell and Angus, eds., *Collected Essays*, II, 77.
20. Longworth, *The Unending Vigil*, p. 33.
21. *Kathleen and Frank*, pp. 306–7.
22. *New Statesman* (March 17, 1972), p. 344.
23. *In Parenthesis*, p. xi.
24. IWM.
25. Middlebrook, *First Day on the Somme*, p. 123.
26. *Memoirs of an Infantry Officer*, p. 159.
27. Quoted by Chapman, ed., *Vain Glory*, p. 357.
28. *My Father and Myself*, pp. 61, 63.
29. "Some Soldiers," in Panichas, ed., *Promise of Greatness*, p. 160.
30. *A Passionate Prodigality*, p. 37.
31. IWM.
32. *Undertones of War*, p. 41.
33. *Memoirs of an Infantry Officer*, pp. 27, 32.
34. *A Passionate Prodigality*, p. 183.
35. *Sagittarius Rising*, p. 93.
36. *Now It Can Be Told*, p. 76.
37. *Passchendaele and the Somme*, p. 177.
38. Quoted by Bergonzi, *Heroes' Twilight*, p. 62.
39. *Steady Drummer*, p. 76.
40. *Gallipoli to the Somme*, p. 42.

41. *In Parenthesis*, pp. 97–98.
42. *The Challenge of the Dead*, p. 97.
43. *Blasting and Bombardiering*, pp. 132–33.
44. IWM.
45. *Short Stories of the First World War*, ed. George Bruce (1971), p. 18.
46. *Ypres, 1917*, pp. 58–59.
47. *Problems of the Theatre*, trans. Gerhard Nellhaus (New York, 1964), p. 31.
48. *The Third Book of Criticism* (New York, 1969), p. 86.
49. *The Crowning Privilege* (1955), p. 13.
50. *Robert Graves* (New York, 1961), pp. 68–70, 75.
51. *But It Still Goes On* (New York, 1931), pp. 3–6, 12.
52.' 1973, p. 161.
53. Lecture, Princeton University, Dec. 2, 1971.
54. Part I, Chap. ii.
55. Paragraph 4.
56. *The Third Book of Criticism*, p. 78.
57. *Difficult Questions, Easy Answers* (1972), p. 13.
58. *Ibid.*, p. 195.
59. Page references to the 1957 (Anchor Books paperback) edition.
60. *With a Machine Gun to Cambrai*, p. 91.
61. *The Rights of Man* (1791), Part the First, paragraph 64.
62. *But It Still Goes On*, pp. 32–33.
63. *The Crowning Privilege*, p. 117.
64. *Good-bye to All That* (New York, 1930), p. 102.
65. *5 Pens in Hand*, p. 46.
66. Page 28.
67. *Robert Graves* (New York, 1967), pp. 10–11.
68. New York, 1930, pp. 21–24.
69. *But It Still Goes On*, p. 26.
70. First printing (New York, 1957), p. 264.
71. Joseph Cohen, "Owen Agonistes," *English Literature in Transition*, VIII, No. 5 (Dec., 1965), 254.
72. *Memoirs of an Infantry Officer*, p. 148.
73. *Good-bye to All That*, p. 158; *Officers and Gentlemen* (New York, 1955), p. 226.
74. *Officers and Gentlemen*, pp. 262–63.
75. *Undertones of War*, p. ix.
76. Page 36.
77. New York, 1921; 1932, p. 11.
78. *Ibid.*, pp. 34–35.
79. Page 577.
80. Pages 691, 760.
81. *The Fortress* (1956; 1972), p. 85.
82. Karlin, *et al.*, ed., *Free Fire Zone*, p. 17.
83. *The Ticket That Exploded* (Paris, 1962; New York, 1968), p. 50.
84. Page references to *The Wanting Seed* (New York, 1964).
85. Robert Jay Lifton, *History and Human Survival* (New York, 1970), p. 200.
86. Page 756.
87. IWM.

88. Parlophone Record No. R. 517, Sides E2935 and E2936. The kind gift of Mr. Leo Cooper, this record is in my collection.

89. New York, 1923; 1959, p. 62.

VII. *Arcadian Recourses*

1. *Mars His Idiot* (New York, 1935), p. 11.
2. Owen and Bell, eds., *Collected Letters*, p. 520.
3. *From Peace to War*, p. 164; *From Alamein to Zem Zem* (New York, 1966), p. 98.
4. *Akenfield*, p. 16.
5. *The Country and the City* (New York, 1973), pp. 281–82.
6. *The Diary of a Desert Rat* (1971), p. 132.
7. "Pastoralism as Culture and Counter-Culture in English Fiction, 1800–1928," *Novel*, VI, No. 1 (Fall, 1972), 20.
8. *Akenfield*, p. 116.
9. IWM.
10. *John Thomas and Lady Jane* (New York, 1972), p. 254.
11. *A Private in the Guards*, pp. 166–67, 162.
12. *Gone for a Soldier*, p. 50.
13. *Siegfried's Journey*, p. 79.
14. Slater, ed., *My Warrior Sons*, p. 72.
15. *Hudson Review*, XXVI, No. 2 (Summer, 1973), p. 425.
16. Bruce, ed., *Short Stories of the First World War*, pp. 121–27.
17. "Mr. Sassoon's War Verses," *The Evolution of an Intellectual* (1920), pp. 73–74.
18. Page 16.
19. *Memoirs of a Fox-Hunting Man*, p. 376.
20. "The Pastoral of the Self," *Daedalus*, 88 (Winter, 1959), 686–89.
21. *A Passionate Prodigality*, pp. 52–53.
22. *The Void of War: Letters from Three Fronts* (New York, 1918), p. 82.
23. "Conjuring with Nature: Some Twentieth-Century Readings of Pastoral," *Twentieth-Century Literature in Retrospect* (Cambridge, Mass., 1971), p. 249.
24. *A Scholar's Letters from the Front* (1918), p. 66.
25. *From Peace to War*, p. 166.
26. Richard Ellmann, *James Joyce* (New York, 1959), p. 438.
27. *Ypres, 1917*, p. 96.
28. Quoted by Chapman, ed., *Vain Glory*, p. 323.
29. Richard M. Watt, *Dare Call It Treason* (New York, 1963), p. 194.
30. Page 27.
31. Pages 18–19.
32. IWM.
33. Housman, ed., *War Letters of Fallen Englishmen*, p. 164.
34. Laffin, ed., *Letters from the Front*, p. 64.
35. *Old Soldiers Never Die*, p. 137.
36. *Recollections of Rifleman Bowlby, Italy, 1944* (1969), p. 50.
37. *A Victorian Son*, p. 182.
38. *Sherston's Progress*, pp. 169–70.

39. *Recollections*, pp. 160–61.

40. *The Sleeping Lord and Other Fragments* (1974), p. 57.

41. C. S. Peel, *How We Lived Then, 1914–1918* (1929), p. 16.

42. *Anatomy of Criticism*, p. 144.

43. *Now It Can Be Told*, p. 153.

44. *The Anathémata*, p. 22.

45. *Memoirs* (2 vols., 1965), I, 16.

46. Owen and Bell, eds., *Collected Letters*, p. 423.

47. *The Challenge of the Dead*, pp. 80–81.

48. "*Et in Arcadia Ego,*" *Philosophy and History: Essays Presented to Ernst Cassirer* (Oxford, 1936), pp. 295–320.

49. *Memoirs of an Infantry Officer*, p. 60.

50. *Disenchantment*, p. 162.

51. *How Young They Died*, p. 103.

52. Vonnegut, *Slaughterhouse-Five*, p. 4.

53. "Myth, Fiction, and Displacement," *Fables of Identity* (New York, 1963), p. 25.

54. *Memoirs of an Infantry Officer*, p. 61.

55. *The Collected Works of Isaac Rosenberg*, ed. Gordon Bottomley and Denys Harding (1937), p. 311.

56. *Ibid.*

57. *Out of Battle*, p. 280.

58. Page 113.

59. *Poetry of the First World War*, ed. Maurice Hussey (1967), p. 30.

60. Page references to the World's Classics edition of *Undertones of War* (Oxford, 1956).

61. *A Passionate Prodigality*, p. 207.

62. *Let the Poet Choose*, ed. James Gibson (1973), p. 37.

63. *Votive Tablets* (1931), p. 255.

64. *Times Literary Supplement*, Feb. 9, 1951, p. 84.

65. *Poems of Many Years* (1957), p. 192.

66. *Out of Soundings* (New York, 1931), p. 245.

VIII. *Soldier Boys*

1. New York, 1947, p. 111.

2. *A Chapter of Accidents*, p. 155.

3. *Kathleen and Frank*, p. 350.

4. Owen and Bell, eds., *Collected Letters*, p. 458.

5. New York, 1962, pp. 275–77.

6. *Eminent Victorians*, p. 234.

7. H. Montgomery Hyde, *The Other Love* (1970), p. 161.

8. Page 117.

9. Page 115.

10. D. R. Pye, "Memoir," in R. J. E. Tiddy, *The Mummers' Play* (Oxford, 1923), p. 30.

11. Pages 71–72.

12. *Margin Released*, p. 89.

13. *Disenchantment*, p. 30.
14. *Gone for a Soldier*, pp. 31, 32.
15. *Ibid.*, pp. 102–3.
16. *But It Still Goes On*, p. 244.
17. *Memoirs of a Fox-Hunting Man*, p. 343.
18. Jon Stallworthy, *Wilfred Owen* (1974), p. 142.
19. "Some Soldiers," Panichas, ed., *Promise of Greatness*, p. 164.
20. *Memoirs of an Infantry Officer*, p. 87.
21. *Letters of D. H. Lawrence*, ed. Aldous Huxley (1932), p. 226.
22. Quoted by Reginald Pound, *The Lost Generation of 1914* (New York, 1965), p. 50.
23. *Rupert Brooke* (1964), p. 390.
24. *Anthology of War Poetry, 1914–1918* (1943), p. 34.
25. Page 616.
26. Owen and Bell, eds., *Collected Letters*, p. 571.
27. *Anatomy of Criticism*, p. 200.
28. *Good-bye to All That*, p. 19.
29. *Poems and Prose of Gerard Manley Hopkins*, ed. W. H. Gardner (Baltimore, 1953), p. 242.
30. Brian Reade, *Sexual Heretics* (New York, 1971), p. 22.
31. *Ibid.*, p. 272.
32. *Ibid.*, p. 160.
33. *Difficult Questions, Easy Answers*, p. 183.
34. Quoted in *Times Literary Supplement* (January 14, 1972), p. 27.
35. Quoted by Smith, *Love in Earnest*, p. 193.
36. "Modern War Poetry," *Southwest Review*, XLVII, No. 4 (Autumn, 1962), 286.
37. Quoted by Smith, *Love in Earnest*, p. 22.
38. *Poems of Gerard Manley Hopkins*, ed. W. H. Gardner (3rd ed., 1948), p. 215.
39. *Anthology of War Poetry*, p. 29.
40. *New Statesman* (July 30, 1971), p. 152.
41. *Sexual Heretics*, pp. 48–49.
42. *Encounter* (May, 1973), p. 68.
43. Michael Holroyd, *Lytton Strachey: A Biography* (Baltimore, 1971), p. 575; Tiddy, *The Mummers' Play*, p. 55.
44. IWM.
45. *Now It Can Be Told*, pp. 288, 289.
46. *Love in Earnest*, pp. 221, 223.
47. *Ibid.*, p. 182.
48. Reade, ed., *Sexual Heretics*, p. 156.
49. Smith, *Love in Earnest*, p. 52.
50. Reade, ed., *Sexual Heretics*, p. 156.
51. *English Poetry of the First World War* (Princeton, N.J., 1964), p. 167.
52. Owen and Bell, eds., *Collected Letters*, p. 456.
53. *Good-bye to All That*, p. 264.
54. *Heroes' Twilight*, p. 128.
55. *Ibid.*, p. 131.
56. Owen and Bell, eds., *Collected Letters*, pp. 580, 581.
57. *Ibid.*, p. 511.

58. *Good-bye to All That*, pp. 295–96.
59. Smith, *Love in Earnest*, p. 143.
60. *Ibid.*, p. 147.
61. Although I have normally relied on C. Day Lewis's edition, in this poem I adopt two of Stallworthy's readings (p. 176).
62. *Anthology of War Poetry*, p. 85.
63. *Memoirs of an Infantry Officer*, p. 55.
64. *Undertones of War*, p. 93.
65. *A Brass Hat in No Man's Land*, pp. 92–93.
66. *The Void of War*, pp. 99–100.
67. Spender, *World Within World*, p. 156.
68. Keynes, ed., *Letters of Rupert Brooke*, p. 621.
69. John Peter, "A New Interpretation of *The Waste Land*," *Essays in Criticism*, XIX (1969), 140–75; and see Robert Sencourt, *T. S. Eliot: A Memoir* (New York, 1971), pp. 32, 243.
70. *The Collected Poems of Wilfred Owen*, ed. C. Day Lewis (New York, 1964), pp. 53–54.
71. *Memoirs of a Fox-Hunting Man*, pp. 369–70.
72. J. H. Morris, IWM.
73. *The Challenge of the Dead*, pp. 100, 101.
74. *The Wet Flanders Plain*, p. 81.
75. Kenneth Clark, *The Nude: A Study in Ideal Form* (Washington, D.C., 1956; New York, 1959), pp. 265ff.
76. *The Contrary Experience*, p. 177.
77. Reade, *Sexual Heretics*, p. 348.
78. *Ibid.*, pp. 225–26.
79. Day Lewis, ed., *Collected Poems of Wilfred Owen*, p. 140.
80. *Sexual Heretics*, pp. 13, 36.
81. *Ibid.*, p. 228.
82. Smith, *Love in Earnest*, p. 108.
83. *Ibid.*, p. 171.
84. *Edwardian Occasions* (New York, 1972), p. 116.
85. Pages 159–66.
86. Page 18.
87. Pages 330–32.

IX. *Persistence and Memory*

1. Ellmann, *James Joyce*, p. 427.
2. *An Unfinished Woman* (New York, 1969), p. 130.
3. "Mercury on a Fork," *Listener* (Feb. 18, 1971), p. 208.
4. Page 56.
5. *Anatomy of Criticism*, pp. 33–42.
6. *Ibid.*, pp. 147–50.
7. Dec. 16, 1971, p. 831.
8. *Mars His Idiot*, pp. 207–8.
9. *Memoirs*, I, 21.
10. *The Big Push*, p. 85.
11. *Akenfield*, p. 62.

12. John Brophy and Eric Partridge, *Songs and Slang of the British Soldier* (1931), p. 121.

13. *World Within World*, p. 16.

14. *Disenchantment*, p. 77.

15. *The Great War, 1914–1918*, trans. Nicole Stone (1973), p. 223.

16. *Letters of Aldous Huxley*, ed. Grover Smith (New York, 1969), p. 124.

17. Albert Speer, *Inside the Third Reich*, trans. Richard and Clara Winston (New York, 1970), pp. 232, 241, 366.

18. Quoted by Jacques Nobécourt, *Hitler's Last Gamble: The Battle of the Bulge*, trans. R. H. Barry (New York, 1967), pp. 43–45.

19. *The Gathering Storm* (Boston, 1948), p. iii; *Völkische Beobachter* (Sept. 1, 1942).

20. Boston, 1952, p. 15.

21. Pages 119, 121.

22. Crimp, *The Diary of a Desert Rat*, p. 140.

23. "Poets in This War," *Times Literary Supplement* (April 23, 1971), p. 478.

24. *Boy in the Blitz*, pp. 25, 127.

25. *Mailer* (1972), p. 28.

26. *Advertisements for Myself* (1968), p. 318.

27. *Mailer*, pp. 26, 28.

28. *Advertisements*, p. 262.

29. *Cannibals and Christians* (New York, 1966), p. 130.

30. Page 324.

31. *Bright Book of Life*, p. 74.

32. *Mars His Idiot*, p. 127.

33. *City of Words: American Fiction, 1950–1970* (New York, 1971), p. 344.

34. *Anatomy of Criticism*, p. 62.

35. J. Stuart Castle, *Listener* (Feb. 3, 1972), p. 152.

36. *The Tiger and the Rose: An Autobiography* (1971), p. 71.

37. *Air with Armed Men*, pp. 26, 43–44.

38. *Listener* (Sept. 21, 1972), p. 374.

39. *Steady Drummer*, pp. 269–70.

40. Trans. H. T. Lowe-Porter (New York, 1944), p. 709.

41. Quoted by Silkin, *Out of Battle*, pp. 347–48.

42. *A Passionate Prodigality*, p. 222.

43. *Memoirs of a Public Baby* (1958), p. 123.

44. *Autobiography* (3 vols., 1967–69), II, 18.

45. *The Pound Era* (Berkeley, Calif., 1971), p. 416.

46. *A Scholar's Letters from the Front*, p. 85.

47. *A Subaltern on the Somme*, p. 36.

48. *From Peace to War*, p. 152.

49. Owen and Bell, eds., *Collected Letters*, pp. 533–34.

50. *The Void of War*, p. 152.

51. *Mars His Idiot*, p. 167.

52. May 18, 1965.

53. Page references to *Gravity's Rainbow* (New York, 1973).

54. *Bright Book of Life*, p. 155.

55. *The Tiger and the Rose*, pp. 83–84.

56. *Anatomy of Criticism*, p. 346.

Index